Lecture Notes in Computer Science 5833

Commenced Publication in 1973
Founding and Former Series Editors:
Gerhard Goos, Juris Hartmanis, and Jan van Leeuwen

Carlos Alberto Heuser Günther Pernul (Eds.)

Advances in Conceptual Modeling - Challenging Perspectives

ER 2009 Workshops CoMoL, ETheCoM, FP-UML, MOST-ONISW, QoIS, RIGiM, SeCoGIS
Gramado, Brazil, November 9-12, 2009
Proceedings

Volume Editors

Carlos Alberto Heuser
Federal University of Rio Grande do Sul
Instituto de Informática
Porto Alegre, Brazil
E-mail: heuser@inf.ufrgs.br

Günther Pernul
University of Regensburg
Department of Management Information Systems
93053 Regensburg, Germany
E-mail: guenther.pernul@wiwi.uni-regensburg.de

Library of Congress Control Number: 2009935903

CR Subject Classification (1998): D.2, D.3, D.2.2, D.3.2, F.3.3, H.2.8, D.1.2

LNCS Sublibrary: SL 3 – Information Systems and Application, incl. Internet/Web and HCI

ISSN 0302-9743
ISBN-10 3-642-04946-X Springer Berlin Heidelberg New York
ISBN-13 978-3-642-04946-0 Springer Berlin Heidelberg New York

springer.com

Printed in Germany

Typesetting: Camera-ready by author, data conversion by Scientific Publishing Services, Chennai, India
Printed on acid-free paper SPIN: 12775286 06/3180 5 4 3 2 1 0

Preface

This book contains the papers accepted for presentation and publication in the workshop proceedings of the 28th edition of the International Conference on Conceptual Modeling (ER Conference), held during November 9–12, 2009, in Gramado, Brazil. The ER workshops complement the main ER conference and are intended to serve as an intensive collaborative forum for exchanging late-breaking ideas and theories in an evolutionary stage and related to conceptual modeling.

For the 2009 edition the workshop committee received 14 excellent proposals from which the following were selected:

- ACM-L: Active Conceptual Modeling of Learning
- CoMoL: Conceptual Modeling in the Large
- ETheCoM: Evolving Theories of Conceptual Modeling
- FP-UML: Workshop on Foundations and Practices of UML
- MOST-ONISW: Joint International Workshop on Metamodels, Ontologies, Semantic Technologies, and Information Systems for the Semantic Web
- QoIS: Quality of Information Systems
- RIGiM: Requirements, Intentions and Goals in Conceptual Modeling
- SeCoGIS: Semantic and Conceptual Issues in Geographic Information Systems

These workshops attracted 100 submissions from which the workshop program committees selected 33 papers, maintaining a highly competitive acceptance rate of 30%.

The workshop co-chairs are highly indebted to the workshop organizers and program committees for their work.

July 2009

Carlos A. Heuser
Günther Pernul

Organization

ER 2009 Workshops Chairs

Carlos A. Heuser	Universidade Federal do Rio Grande do Sul, Brazil
Günther Pernul	Universität Regensburg, Germany

CoMoL 2009 Program Chairs

Stefan Jablonski	University Bayreuth, Germany
Roland Kaschek	Kazakhstan Institute of Management, Economics and Strategic Research, Kazakhstan
Bernhard Thalheim	Christian-Albrechts-University Kiel, Germany

CoMoL 2009 Program Committee

Sabah S. Al-Fedaghi	Kuwait University, Kuwait
Karen C. Davis	University of Cincinnati, USA
Valeria De Antonellis	Brescia University, Italy
Lois Delcambre	Portland State University, USA
Ulrich Frank	University Duisburg-Essen, Germany
Nicola Guarino	LOA-CNR, Italy
Klaus P. Jantke	Fraunhofer IDMT, Germany
John Krogstie	Norwegian University of Science and Technology, Norway
Sebastian Link	Victoria University of Wellington, New Zealand
Heinrich C. Mayr	Alpen-Adria University Klagenfurt, Austria
Andreas Oberweis	University Karlsruhe, Germany
Antoni Olive	UPC Barcelona Tech, Spain
Andreas L. Opdahl	University of Bergen, Norway
Erich Ortner	Technical University of Darmstadt, Germany
Fabian Pascal	Database Debunkings, USA
Oscar Pastor	DSIC - Universidad Politécnica de Valencia, Spain
Klaus Pohl	University Duisburg-Essen, Germany
Klaus-Dieter Schewe	Information Science Research Centre, New Zealand
Michael Schrefl	Johannes Kepler University Linz, Austria
Markus Stumptner	University of South Australia, Australia
Yuzuru Tanaka	Hokkaido University, Japan
Susan Urban	Texas Tech University, USA
Mathias Weske	HPI University Potsdam, Germany
Roel Wieringa	University of Twente, The Netherlands

CoMoL 2009 External Reviewers

Tayyeb Amin
Devis Bianchini
Bernd Neumayr

ETheCoM 2009 Program Chairs

Markus Kirchberg	Institute for Infocomm Research, A*STAR, Singapore
Klaus-Dieter Schewe	Information Science Research Centre, New Zealand

ETheCoM 2009 Program Committee

Sabah S. Al-Fedaghi, Kuwait
Shawn Bowers, USA
Stefan Brass, Germany
Andrea Calì, UK
Gill Dobbie, New Zealand
David W. Embley, USA
Flavio A. Ferrarotti, Chile
Aditya K. Ghose, Australia
Guido Governatori, Australia
Sven Hartmann, Germany
Roland Hausser, Germany
Edward Hermann Haeusler, Brazil
Stephen Hegner, Sweden
Henning Koehler, Australia
Leandro Krug Wives, Brazil
Sebastian Link, New Zealand
Chengfei Liu, Australia
Hui Ma, New Zealand
Wilfred Ng, Hong Kong
Jaroslav Pokorny, Czech Republic
Letizia Tanca, Italy
James F. Terwilliger, USA
Bernhard Thalheim, Germany
Alex Thomo, Canada
Thu Trinh, Germany
Millist Vincent, Australia
Junhu Wang, Australia
Qing Wang, New Zealand
Jeffrey Xu Yu, Hong Kong

ETheCoM 2009 External Reviewers

Henrik Björklund
Fernando Bordignon
Thomas Packer

FP-UML 2009 Program Chairs

Juan Trujillo	University of Alicante, Spain
Dae-Kyoo Kim	Oakland University, USA

FP-UML 2009 Program Committee

Doo-Hwan Bae	KAIST, South Korea
Michael Blaha	OMT Associates Inc., USA
Cristina Cachero	Universidad de Alicante, Spain
Brian Dobing	University of Lethbridge, Canada
Joerg Evermann	Victoria University Wellington, New Zealand
Eduardo Fernández	Universidad de Castilla-La Mancha, Spain
Robert France	Colorado State University, USA
Irene Garrigos	Fernandez University of Alicante, Spain
Jens Lechtenböorger	Universität Münster, Germany
Pericles Loucopoulos	University of Manchester, UK
Hui Ma Massey	University, New Zealand
Oscar Pastor	Universidad Politècnica de València, Spain
Jose Norberto Mazon Lopez	University of Alicante, Spain
Heinrich C. Mayr	Universität Klagenfurt, Austria
Sooyong Park	Sogang University, Korea
Jeff Parsons	Memorial University of Newfoundland, Canada
Colette Rolland	Université Paris 1-Panthéon Sorbonne, France
Matti Rossi	Helsingin kauppakorkeakoulu, Finland
Manuel Serrano	Universidad de Castilla-La Mancha, Spain
Keng Siau	University of Nebraska-Lincoln, USA
Il-Yeol Song	Drexel University, USA
Ling Tok Wang	National University of Singapore, Singapore
Ambrosio Toval	Universidad de Murcia, Spain
Panos Vassiliadis	University of Ioannina, Greece

FP-UML 2009 External Reviewers

M. Kirchberg
A. Tretiakov
O. Thonggoom G. Abraham

MOST-ONISW 2009 Program Chairs

Martin Doerr	Foundation for Research and Technology, Greece
Fred Freitas	Federal University of Pernambuco, Brazil
Giancarlo Guizzardi	Federal University of Espírito Santo, Brazil
Hyoil Han	Drexel University, USA

MOST-ONISW 2009 Program Committee

Mara Abel	Federal University of Rio Grande do Sul, Brazil
Jinli Cao	La Trobe University, Australia
Oscar Corcho	Universidad Politécnica de Madrid, Spain
Stefan Conrad	Heinrich-Heine-Universität Düsseldorf, Germany

Evandro Costa	UFAL, Brazil
Stefania Costache	L3S Research Center, Germany
Cléver Ricardo G. de Farias	University of São Paulo, Brazil
Yihong Ding	Brigham Young University, USA
Martin Doerr	Foundation for Research and Technology, Greece
Vadim Ermolayev	Zaporozhye National University, Ukraine
Fred Freitas	Federal University of Pernambuco, Brazil
Dragan Gasevic	Athabasca University, Canada
Giancarlo Guizzardi	Federal University of Espírito Santo, Brazil
Hele-Mai Haav	University of Technology, Estonia
Hyoil Han	Drexel University, USA
Siegfried Handschuh	National University of Ireland / DERI, Ireland
Kenji Hatano	Doshisha University, Japan
Ralf Heese	Freie Universität Berlin, Germany
Ramón Hermoso Traba	Rey Juan Carlos University, Spain
Wolfgang Hesse	University of Marburg, Germany
Haklae Kim	National University of Ireland, Ireland
Steffen Lamparter	Institute AIFB, University of Karlsruhe, Germany
Yugyung Lee	University of Missouri at Kansas City, USA
Abdul-Rahman Mawlood-Yunis	Carleton University, Canada
Jun Miyazaki	Nara Institute of Science and Technology, Japan
Feng Pan	Microsoft, USA
Jeffrey Parsons	Memorial University of Newfoundland, Canada
Oscar Pastor	Technical University of Valencia, Spain
Fabio Porto	Ecole Polytechnique Fédérale de Lausanne, Switzerland
Xiaojun Qi	Utah State University, USA
Tarmo Robal	Tallinn University of Technology, Estonia
Melike Sah	University of Southampton, UK
Daniel Schwabe	PUC-RJ, Brazil
Krishnaprasad Thirunarayan	Wright State University, USA
Renata Vieira	PUC-RS, Brazil
Renata Wassermann	University of São Paulo, Brazil
Xian Wu	IBM China Research Lab., China
Leyla Zhuhadar	University of Louisville, USA

QoIS 2009 Program Chairs

Isabelle Comyn-Wattiau	CEDRIC-CNAM and ESSEC Business School, France
Bernhard Thalheim	Christian Albrechts University Kiel, Germany

QoIS 2009 Program Committee

Jacky Akoka	Conservatoire National des Arts et Métiers and TMSP, France
Laure Berti	IRISA, France
Mokrane Bouzeghoub	University of Versailles, France
Andrew Burton-Jones	University of British Columbia, Canada
Cristina Cachero	University of Alicante, Spain
Tiziana Catarci	Università di Roma "La Sapienza", Italy
Corinne Cauvet	University of Aix-Marseille 3, France
Marcela Genero	Universidad de Castilla La Mancha, Spain
Virginie Goasdoue-Thion	University of Dauphine, Paris, France
Paul Johannesson	Stockholm University, Sweden
Zoubida Kedad	University of Versailles, France
Jim Nelson	Southern Illinois University, USA
Jeffrey Parsons	Memorial University of Newfoundland, Canada
Oscar Pastor	Valencia University of Technology, Spain
Geert Poels	University of Ghent, Belgium
Farida Semmak	University of Paris XII, IUT Sénart Fontainebleau, France
Houari Sahraoui	University of Montréal, Canada
Keng Siau	University of Nebraska, USA
Samira Si-Said	Conservatoire National des Arts et Métiers, France
Monique Snoeck Katholieke	Universiteit Leuven, Belgium
David Tegarden	Virginia Polytechnic Institute, USA
Dimitri Theodoratos	NJ Institute of Technology, USA
Juan Trujillo	University of Alicante, Spain
Guttorm Sindre	Norwegian University of Science and Technology, Norway

QoIS 2009 External Reviewers

Carola Aiello
Marouane Kessenteni
Jesus Pardillo
Manuel Serrano
Stephane Vaucher

RIGiM 2009 Program Chairs

Colette Rolland	Université Paris 1 Panthéon Sorbonne, France
Jaelson Castro	Universidade Federal de Pernambuco, Brazil
Camille Salinesi	Université Paris 1 Panthéon Sorbonne, France
Eric Yu	University of Toronto, Canada

RIGiM 2009 Program Committee

Fernanda Alencar	Universidade Federal de Pernambuco, Brazil
Ian Alexander	Scenario Plus, UK

Daniel Amyot	University of Ottawa, Canada
Raquel Anaya	Universidad EAFIT, Colombia
Mikio Aoyoma	Nanzan University, Japan
Aybuke Aurum	University of New South Wales, Australia
Franck Barbier	University of Pau, France
Sjaak Brinkkemper	Utrecht University, The Netherlands
Luca Cernuzzi	Universidad Católica "Nuestra Señora de la Asunción", Paraguay
Lawrence Chung	University of Texas at Dallas, USA
Luiz Cysneiros	York University, Canada
Eric Dubois	Centre de Recherche Public Henri Tudor, Luxembourg
Vincenzo Gervasi	University of Pisa, Italy
Aditya K. Ghose	University of Wollongong, Australia
Paolo Giorgini	University of Trento, Italy
Peter Haumer	IBM Rational, USA
Patrick Heymans	University of Namur, Belgium
Aneesh Krishna	University of Wollongong, Australia
John Krogstie	Norwegian University of Science and Technology, Norway
Lin Liu	Tsinghua University, China
Julio Leite	Pontifícia Universidade Católica do Rio de Janeiro, Brazil
Peri Loucopoulos	University of Manchester, UK
John Mylopoulos	University of Toronto, Canada
Selmin Nurcan	Université Paris 1 Panthéon - Sorbonne, France
Andreas Opdahl	University of Bergen , Norway
Yves Pigneur	University of Lausanne, Switzerland
Klaus Pohl	University of Duisburg-Essen, Germany
Jolita Ralyte	University of Geneva, Switzerland
Bjorn Regnell	Lund University, Sweden
Motoshi Saeki	Tokyo Institute of Technology, Japan
Pnina Soffer	University of Haifa, Israel
Carine Souveyet	Université Paris 1 Panthéon - Sorbonne, France
Yair Wand	University of British Columbia, Canada
Roel Wieringa	University of Twente, The Netherlands
Carson Woo	University of British Columbia, Canada

SeCoGIS 2009 Program Chairs

Claudia Bauzer Medeiros	University of Campinas, Brazil
Esteban Zimányi	Université Libre de Bruxelles, Belgium

SeCoGIS 2009 Program Committee

Alia I. Abdelmoty	Cardiff University, UK
Gennady Andrienko	Fraunhofer Institute AIS, Germany

Natalia Andrienko	Fraunhofer Institute AIS, Germany
Claudio Baptista	Universidade Federal de Campina Grande, Brazil
Spiridon Bakiras	City University of New York , USA
Michela Bertolotto	University College Dublin, Ireland
Roland Billen	Université de Liège, Belgium
Jean Brodeur	Natural Resources Canada, Canada
Bendicte Bucher	Institut Gographique National, France
James D. Carswell	Dublin Institute of Technology, Ireland
Marco Antonio Casanova	Pontifícia Universidade Católica do Rio de Janeiro, Brazil
Christophe Claramunt	Naval Academy Research Institute, France
Eliseo Clementini	University of L'Aquila, Italy
Maria Luisa Damiani	University of Milan, Italy
Clodoveu Davis	Universidade Federal de Minas Gerais, Brazil
Max Egenhofer	NCGIA, USA
Fernando Ferri	Istituto di Ricerche sulla Popolazione e le Politiche Sociali, Italy
Frederico Fonseca	Penn State University, USA
Anders FriisChristensen	European Commission Joint Research Centre, Italy
Antony Galton	University of Exeter, UK
Bo Huang	University of Calgary, Canada
Marc Joliveau	CIRRELT, Canada
KiJoune Li	Pusan National University, South Korea
Thérèse Libourel	Université de Montpellier II, France
Jugurta Lisboa Filho	Universidade Federal de Viçosa, Brazil
Miguel R. Luaces	Universidade da Coruña, Spain
José Macedo	Ecole Polytechnique Fédérale de Lausanne, Switzerland
Kyriakos Mouratidis	Singapore Management University, Singapore
Pedro Rafael Muro	Medrano Universidad de Zaragoza, Spain
Peter van Oosterom Delft	University of Technology, The Netherlands
Dimitris Papadias	Hong Kong University of Science and Technology, China
Guenther Pernul	Universität Regensburg, Germany
Dieter Pfoser	Institute for the Management of Information Systems, Greece
Ricardo R. Ciferri	Universidade Federal de São Carlos, Brazil
Andrea Rodriguez	University of Concepción, Chile
Sylvie Servigne	Martin INSA de Lyon, France
Emmanuel Stefanakis	Harokopio University of Athens, Greece
Kathleen Stewart	Hornsby University of Iowa, USA
Christelle Vangenot	Ecole Polytechnique Fédérale de Lausanne, Switzerland
Alejandro Vaisman	University of Hasselt, Belgium, and Universidad de Buenos Aires, Argentina
Lubia Vinhas	Instituto National de Pesquisas Espaciais, Brazil
Nancy Wiegand	University of Wisconsin-Madison, USA

Table of Contents

CoMoL 2009 – Conceptual Modeling in the Large

ETheCoM 2009 – First International Workshop on Evolving Theories of Conceptual Modelling

Invited Talks

Accepted Papers

FP-UML 2009 – Fifth International Workshop on Foundations and Practices of UML

Dependability and Agent Modeling

Semantics Representation and Tools

MOST-ONISW 2009 – The Joint International Workshop on Metamodels, Ontologies, Semantic Technologies, and Information Systems for the Semantic Web

Modelling

Elicitation Issues

SECOGIS 2009 – Third International Workshop on Semantic and Conceptual Issues in Geographic Information Systems

Foundational Aspects

Semantical Aspects

Preface to CoMoL 2009

Stefan Jablonski[1], Roland Kaschek[2], and Bernhard Thalheim[3]

[1] University Bayreuth, Germany
[2] Kazakhstan Institute of Management, Economics and Strategic Research, Kazakhstan
[3] Christian-Albrechts-University Kiel, Germany

Conceptual modelling has changed over years. Database applications form an integral part of most computational infrastructures. Applications are developed, changed and integrated by specialists in conceptual modelling, by computer engineers, or by people who do not have sufficient related background knowledge. Conceptual databases models are everywhere in applications and are likely to interfere with other models such as functionality models, distribution and collaboration models, and user-interface models. Models also depend on the cultural and educational background of their developers and users. Models typically follow applications, infrastructures, currently existing systems, theoretical and technological insight, and reference models provided by successful applications or literature. This basis of conceptual models is constantly changing and therefore models are constantly evolving or quickly become outdated. Applications are starting in a separated form and are later integrated into new applications. The coherence and consistency of the many coexisting models at best is partially addressed. Furthermore, models not necessarily share their targeted level of abstraction. Recently modelling is challenged by liberation of data from structure and the integration of derived or aggregated data, e.g. in streaming databases, data warehouses and scientific applications. Typically models for applications start at an intermediate level and size. Later they evolve, grow, and tend to become incomprehensible. Nowadays conceptual modelling in the small has become state of the art for specialists and educated application engineers. Conceptual modelling in the large has been mainly developed within companies that handle large and complex applications. It covers a large variety of aspects such as models of structures, of business processes, of interaction among applications and with users, of components of systems, and of abstractions or of derived models such as data warehouses and OLAP applications.

Conceptual modelling in the large is typically performed by many modelers and teams. It also includes architectural aspects within applications. At the same time quality, configuration and versioning of models developed so far become an issue.

We selected for the workshop three papers from nine submitted. We are sure that these papers reflect in a very good form the current state of the art. We thank the program committee members and additional reviewers for their support in evaluating the papers submitted to CoMoL'09. We are very thankful to the ER'09 organisation team for taking care of workshop proceedings.

C.A. Heuser and G. Pernul (Eds.): ER 2009 Workshops, LNCS 5833, p. 1, 2009.

Semantic Service Design for Collaborative Business Processes in Internetworked Enterprises

Devis Bianchini[1], Cinzia Cappiello[2], Valeria De Antonellis[1], and Barbara Pernici[2]

[1] University of Brescia, Dipartimento di Elettronica per l'Informazione, via Branze, 38, 25123 - Brescia, Italy
{bianchin,deantone}@ing.unibs.it

[2] Politecnico of Milan, Dipartimento di Elettronica e Informazione, via Ponzio, 34/5, 20133 - Milano, Italy
{cappiell,pernici}@elet.polimi.it

Abstract. Modern collaborating enterprises can be seen as borderless organizations whose processes are dynamically transformed and integrated with the ones of their partners (Internetworked Enterprises, IE), thus enabling the design of collaborative business processes. The adoption of Semantic Web and service-oriented technologies for implementing collaboration in such distributed and heterogeneous environments promises significant benefits. IE can model their own processes independently by using the Software as a Service paradigm (SaaS). Each enterprise maintains a catalog of available services and these can be shared across IE and reused to build up complex collaborative processes. Moreover, each enterprise can adopt its own terminology and concepts to describe business processes and component services. This brings requirements to manage semantic heterogeneity in process descriptions which are distributed across different enterprise systems. To enable effective service-based collaboration, IEs have to standardize their process descriptions and model them through component services using the same approach and principles. For enabling collaborative business processes across IE, services should be designed following an homogeneous approach, possibly maintaining a uniform level of granularity. In the paper we propose an ontology-based semantic modeling approach apt to enrich and reconcile semantics of process descriptions to facilitate process knowledge management and to enable semantic service design (by discovery, reuse and integration of process elements/constructs). The approach brings together Semantic Web technologies, techniques in process modeling, ontology building and semantic matching in order to provide a comprehensive semantic modeling framework.

1 Introduction

Modern collaborating enterprises can be seen as large and complex borderless organizations whose processes are transformed and integrated with the ones of

C.A. Heuser and G. Pernul (Eds.): ER 2009 Workshops, LNCS 5833, pp. 2–11, 2009.

their partners (Internetworked Enterprises, IE), thus enabling the design of collaborative business processes [11]. This enables the creation of a "virtual corporation" that operates through an integrated network that connects the company's employees, suppliers, distributors, retailers and customers. Prior to the advent of the Internet, a number of companies developed their own intranets by using electronic data interchange and client/server computing technologies in order to simplify communications and documents exchange. Heterogeneity of such collaborative environments implies the adoption of standards and infrastructures to communicate. With the advent of Service Oriented Architecture (SOA), organizations have experienced services as a platform-independent technology to develop and use simple internal applications or outsource activities by searching for external services, thus enabling more efficient and easier inter-organizational interactions. IE can model their own processes independently by using the Software as a Service paradigm (SaaS). Each enterprise maintains a catalog of available services that can be shared across IE and reused to build up collaborative processes. Each enterprise describes business processes and component services with its own terminology and concepts. This brings requirements to manage semantic heterogeneity in process descriptions which are distributed across different enterprise systems. To enable effective service-based collaboration, IEs have to standardize their process descriptions and model them through component services using the same approach and principles. In particular, services should be designed following an homogeneous approach, possibly maintaining a uniform level of granularity.

In the literature, the vision of Semantic Business Process Management (SBPM) has been proposed to bridge the gap between the business and IT views of the business processes [6]. In SBPM more automation is achieved in process modeling using ontologies and Semantic Web service technologies. A business process semantic framework has been also proposed in [9], where the notion of goal ontology has been introduced to conceptualize the goals to be achieved through business process execution, while in [8] a survey of Semantic Web Service and BPM technologies and existing efforts on their joint use is given. In this paper we propose an ontology-based semantic modeling approach apt to enrich and reconcile semantics of process descriptions to facilitate process knowledge management through the semi-automatic identification of component services, thus providing enabling techniques to solve the business-IT gap. In particular, we work in the field of SBPM focusing on the process implementation phase. Other approaches [10,12,13] provide guidelines for the service identification without giving an operational support based on service semantics, to enable the development of SBPM systems. Our goal is to enable computer-aided semantic process management and service design (by discovery, reuse and integration of process elements/constructs). The approach brings together Semantic Web technologies, techniques in process modeling, ontology building and semantic matching in order to provide a comprehensive semantic methodological framework. The paper is organized as follows: Section 2 introduces the methodological framework and a running example; Sections 3-5 illustrate in details the application of the proposed methodology; finally, Section 6 closes the paper.

2 The Methodological Framework

Processes are usually represented using a workflow-based notation (e.g., BPMN[1]), independently from implementation technologies and platforms. A business process can be defined as a set of simple tasks, combined through control structures (e.g., sequence, choice, cycle or parallel) to form composite tasks, also denoted as sub-processes. Each simple task is described through the performed operation and I/O data. We define an I/O parameter as a pair $\langle n, \mathcal{P} \rangle$, where n is the I/O name and $\mathcal{P} = \{p_i\}$ a set of I/O properties. Simple task constitutes the minimal unit of work and it can have transactional properties. Data exchanged between tasks and control flows connecting them are modelled as data dependencies and control flow dependencies, respectively. Collaborative business processes are designed as processes spanning over different actors. Actors are represented as abstract entities that interact each other as responsible of one or more simple tasks. According to a service-oriented perspective, processes can be seen as a set of component services that must be properly composed and executed. Services constitute a particular kind of sub-process, reflecting some additional constraints: (i) services are self-contained and interact each other using decoupled message exchanges; (ii) each service is the minimal set of tasks that performed together create an output that is a tangible value for the service requester. Tangible values are data associated to interactions between the process actors. We propose a methodological framework for component service design starting from process descriptions. The phases of the methodology are the following:

Semantic Process Annotation - In a distributed heterogeneous environment, where different IEs provide independently developed process representations, business process elements (inputs and outputs, task names) must be semantically annotated with concepts extracted from shared ontologies;

Identification of Candidate Services - Candidate component services must be identified ensuring the same decomposition granularity, thus enabling better service comparison for sharing and reuse purposes;

Reconciliation of Similar Services - Component services must be clustered on the basis of the similarity of their tasks and I/O data, in order to identify similar services on different processes and enable the design of reusable component services.

The methodological phases are described in the following sections, with reference to the example of collaborative business process shown in Figure 1. We distinguish between the *process level*, where collaborative business processes are represented, and the *semantic service level*, where component services are identified as semantic-enriched service descriptions. In the example shown in figure, we present a cooperative scenario in which a sofa manufacturer produces all the textiles sofa components and purchases backbones from trusted suppliers.

[1] http://www.bpmn.org/

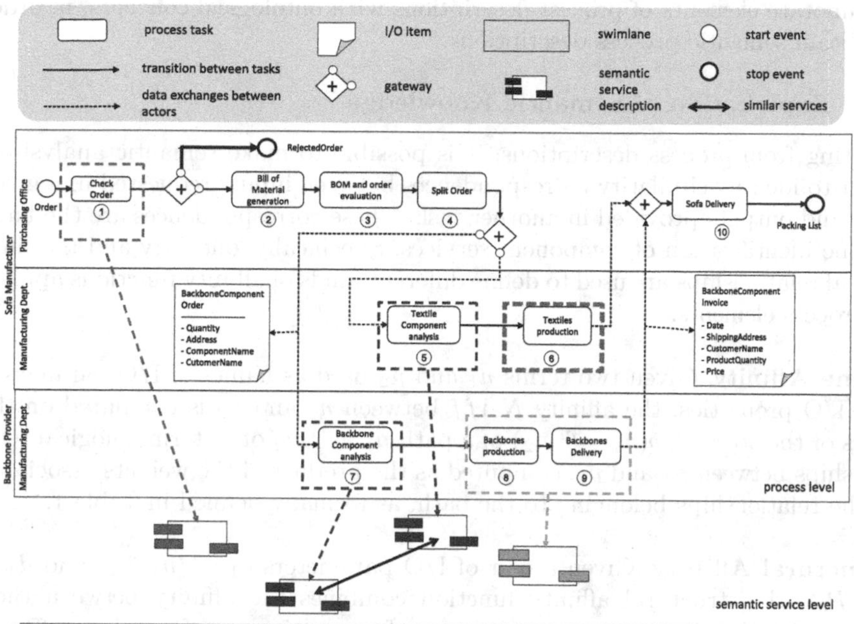

Fig. 1. Reference scenario

3 Semantic Process Annotation

We assume that IEs participating in collaborative business processes agree on a common reference ontology, that provides atomic concept definitions and equivalence/subsumption relationships between them. However, in a distributed and heterogeneous environment, local terms used by different IEs for business process elements do not necessarily coincide with atomic concepts in the reference ontology or they may suffer from terminological discrepancies (e.g., synonymies or homonymies). The first phase of the methodology aims at solving these heterogeneities by extending the reference ontology with a common Thesaurus, extracted from a lexical system (e.g., WordNet [5]), where terms are related each other and with the names of ontological concepts by means of terminological relationships. In [1] the combined use of the Thesaurus and the reference ontology is detailed. A weight $\sigma_{rel} \in (0, 1]$ is associated with each kind of relationship. The following terminological relationships are considered: (i) *synonymy* (SYN), with $\sigma_{SYN} = 1.0$, established between two terms that can be used interchangeably in the process (e.g., `ShippingAddress` SYN `Address`); (ii) *narrower/broader term* (BT/NT), with $\sigma_{BT/NT} = 0.8$, established between a term n_1 and another term n_2 that has a more generic (resp., specific) meaning (e.g., `InvoicedQuantity` NT `Quantity`); (iii) *related term* RT, with $\sigma_{RT} = 0.5$, established between two terms whose meaning is related in the considered application scenario (e.g., `Order` RT `Invoice`). In [4] techniques apt to guide the process designer in the construction of the Thesaurus are explained. The Thesaurus constitutes a fundamental asset to annotate elements of process descriptions with ontological concepts, in order to obtain semantic process descriptions.

3.1 Exploitation of Semantic Knowledge

Starting from process descriptions, it is possible to make semantic analysis in order to identify similarity correspondences between inputs requested in a given task and outputs provided in another task. These correspondences are the basis for the identification of component services. Specifically, ontology and terminological relationships are used to define different kinds of affinity functions applied to process elements.

Name Affinity. Given two terms n_1 and n_2 used as names of I/O parameters and I/O properties, the affinity $NAff$ between n_1 and n_2 is computed on the basis of the *strength* $\rho(n_1 \rightarrow^m n_2)$ of a path $n_1 \rightarrow^m n_2$ of m terminological relationships between n_1 and n_2, computed as the product of the weights associated to the relationships belonging to the path, as formally defined in Table 1.

Structural Affinity. Given a pair of I/O parameters $d_1 = \langle n_1, \mathcal{P}_1 \rangle$ and $d_2 = \langle n_2, \mathcal{P}_2 \rangle$, the structural affinity function combines the affinity between their names with the affinity between each pair of properties $p_1 \in \mathcal{P}_1$ and $p_2 \in \mathcal{P}_2$, as shown in Table 1.

The total structural affinity $Aff_{TOT}(D_1, D_2)$ between two sets of I/O data is defined as the sum of structural affinity for each pair of items $d_1 \in D_1$ and $d_2 \in D_2$, normalized with respect to the cardinality of D_1 and D_2.

The process designer is supported for the identification of I/O similarities and their reconciliation following data integration techniques [4]. In the running example, the following affinity values are evaluated for `BackboneComponentOrder` (`BCO`) and `BackboneComponent` (`BC`) items:

$$NAff(\texttt{BackboneComponentOrder,BackboneComponent}) = 0.5\ (\textsc{rt})$$
$$NAff(\texttt{InvoicedQuantity,Quantity}) = 0.8\ (\textsc{nt})$$
$$NAff(\texttt{Address,ShippingAddress}) = 1.0\ (\textsc{syn})$$
$$SAff(\texttt{BCO,BC}) = \frac{1}{2}\left[0.5 + \frac{2*(1.0+0.8+1.0+1.0)}{10}\right] = 0.63$$

A threshold-based mechanism is applied to $SAff$ values to state that there is a relevant similarity between task inputs and outputs, as shown in the following section. The threshold allows for the identification of affinity also between I/O parameters that present slight terminological discrepancies.

4 Identification of Component Services

4.1 Value-Based Service Identification

A major goal of our approach is to enable homogeneous identification of services by analyzing process description. For the identification of candidate component

Table 1. Name and structural affinity coefficients

NAME AFFINITY FUNCTION
$NAff(n_1, n_2) = \begin{cases} 1 & if\ n_1 = n_2 \\ max_m(\rho(n_1 \rightarrow^m n_2)) & if\ n_1 \neq n_2\ \wedge max_m(\rho(n_1 \rightarrow^m n_2)) \geq \alpha \\ 0 & otherwise \end{cases}$
where α is an affinity threshold, that is used to filter out names with high affinity values
STRUCTURAL AFFINITY FUNCTION
$SAff(d_1, d_2) = \frac{1}{2} \cdot \left[NAff(n_1, n_2) + \frac{2 \cdot \sum_{p_1, p_2} NAff(p_1, p_2)}{\lvert \mathcal{P}_1 \rvert + \lvert \mathcal{P}_2 \rvert} \right] \in [0, 1]$
where $d_1 = \langle n_1, \mathcal{P}_1 \rangle$ and $d_2 = \langle n_2, \mathcal{P}_2 \rangle$ (either input or output of simple tasks)

services, we propose some heuristics that are derived from empirical observations about the features and the definition of service found in literature. According to definitions given in [7], services constitute units of work that are invoked by one of the actors involved in the process to obtain a tangible value. For each actor A, the outgoing and incoming data flows (that is, data transfers towards and from other actors, respectively) are considered as service requests/invocations and responses/values, respectively. For example, in the collaborative process shown in Figure 1, candidate services can be recognized in the two sets of tasks: (i) $\{t_5, t_6\}$, characterized by the data transfers (`BackboneComponentOrder` and `BackboneComponent`) between the actors Purchasing Office and the Manufacturing Department of the Sofa Manufacturer; (ii) $\{t_7, t_8, t_9\}$, characterized by the data transfers between the Purchasing Office of the Sofa Manufacturer and the Manufacturing Office of the Backbone Provider. Another service that can be identified in this phase is the one containing only the t_1 task, delimited by the `Order` and the `RejectedOrder` data exchange.

This phase is supported by an automatic tool module that evaluates a structural affinity between data tansfers by applying functions shown in Table 1. Given a process actor A, the Structural Affinity is evaluated between each request $\mathcal{D}req$ associated to an outgoing data flow and each response $\mathcal{D}res$ associated to an incoming data flow. In other words, given a request, the tool has to determine the corresponding response, on the basis of their similarity correspondence. The process designer is in charge of explicitly validating, modifying or excluding results of this evaluation, according to his/her own domain or process knowledge.

For each validated pair $\langle \mathcal{D}req, \mathcal{D}res \rangle$, a candidate service $\mathcal{S}_z$ is recognized whose first task t_i is such that $\mathcal{D}req \in IN(t_i)$ and whose last task t_j is such that $\mathcal{D}res \in OUT(t_j)$. In particular, the service $\mathcal{S}_z$ invoked from a service (if exists) containing the task t_k such that $\mathcal{D}req \in OUT(t_k)$. The list of candidate services is proposed to the process designer, that can validate or reject them. In the running example, $\langle \mathcal{D}req, \mathcal{D}res \rangle \equiv \langle$`BackboneComponentOrder`,`BackboneComponent`$\rangle$ since $SAff() = 0.63$ and the candidate component service $\{t_7, t_8, t_9\}$ is identified. Similarly, the other candidate services identified in this phase are shown in the following table:

$\langle \mathcal{D}req, \mathcal{D}res \rangle$	$SAff()$ value	Service's tasks
⟨Order,RejectedOrder⟩	$SAff() = 0.6272$	$\mathcal{S}_1 = \{t_1\}$
⟨TextileComponentOrder, TextileComponent⟩	$SAff() = 0.53572$	$\mathcal{S}_2 = \{t_5, t_6\}$
⟨BackboneComponentOrder, BackboneComponent⟩	$SAff() = 0.63$	$\mathcal{S}_3 = \{t_7, t_8, t_9\}$

All the pairs $\langle \mathcal{D}req, \mathcal{D}res \rangle$ such that $SAff(\mathcal{D}req, \mathcal{D}res) > \mu$ are identified and listed to the process designer, that can confirm or reject them. Note that the tasks $\{t_2, t_3, t_4, t_{10}\}$ are not included in any identified service. These tasks do not provide evident values to the process actors, but they will be further analyzed in the subsequent phases in order to evaluate their internal cohesion or mutual coupling to be elected as services.

4.2 Evaluation of Service Cohesion/Coupling

Once the candidate services have been identified, services are further analyzed in terms of defined cohesion/coupling criteria in order to better define service structure and granularity. Homogeneous granularity is a strong requirement for effective collaboration. Specifically, identified services must ensure high internal cohesion and low coupling. The adopted cohesion/coupling metrics have been inspired by their well-known application in software engineering field [14] and are used to evaluate the degree of similarity correspondence between I/O flows among tasks. They have been detailed in [3] and are summarized in Table 2. The cohesion coefficient evaluates how much tasks within a single service contribute to obtain a service output. The coupling coefficient evaluates how much tasks belonging to different services need to interact. The ratio Γ between coupling and cohesion coefficients must be minimized. It is used to guide iterative service decomposition until we obtain maximum intra-service cohesion and minimal inter-service coupling. The building block is the task coupling coefficient $\tau(t_i, t_j)$, with $t_i \neq t_j$, defined as follows:

$$\tau(t_i, t_j) = \begin{cases} Aff_{TOT}(OUT(t_j), IN(t_i))) & if\ t_j \mapsto t_i \\ Aff_{TOT}(IN(t_j), OUT(t_i))) & if\ t_i \mapsto t_j \\ \frac{Aff_{TOT}(IN(t_i), IN(t_j))) + Aff_{TOT}(OUT(t_i), OUT(t_j)))}{2} & if\ t_i \| t_j \\ 0 & otherwise \end{cases} \quad (1)$$

where $t_i \mapsto t_j$ means that there is a data dependency from t_i to t_j (see [3] for a formal definition of data dependency between tasks) and $t_i \| t_j$ means that t_i and t_j are executed in two parallel branches of the business process. In Figure 2 the matrix of task couplings for the running example is shown. By evaluating the coefficients given in Table 2, we obtain $\Gamma = 0.0715$. The process coupling/cohesion ratio Γ must be minimized by increasing the internal service cohesion. To this aim, an iterative procedure is applied. At each iteration, the pair of tasks $\langle t_i, t_j \rangle$, such that t_i and t_j belongs to the same service $\mathcal{S}$ and they present the minimum coupling value, are selected and the candidate service $\mathcal{S}$ is split into two services $\mathcal{S}_i$ and $\mathcal{S}_j$, with $t_i \in \mathcal{S}_i$ and $t_j \in \mathcal{S}_j$. The new value Γ' of the coupling/cohesion ratio is calculated and, if $\Gamma < \Gamma'$, the split is maintained. The procedure is repeated until the minimal ratio is reached or only services $\mathcal{S}_i$ such that $|\mathcal{S}_i| = 1$ are

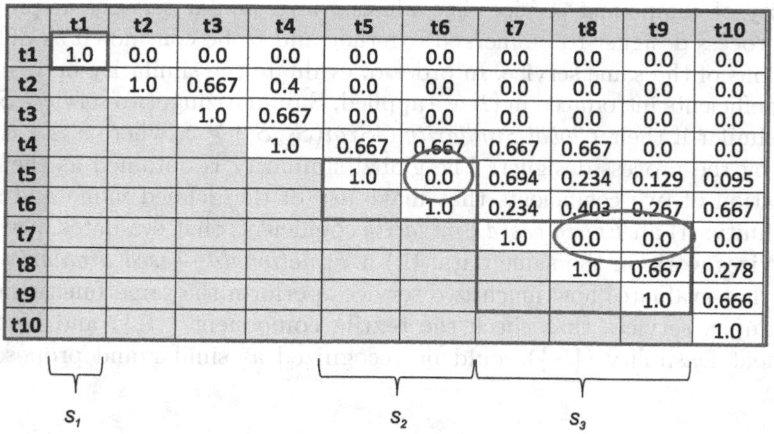

	t1	t2	t3	t4	t5	t6	t7	t8	t9	t10
t1	1.0	0.0	0.0	0.0	0.0	0.0	0.0	0.0	0.0	0.0
t2		1.0	0.667	0.4	0.0	0.0	0.0	0.0	0.0	0.0
t3			1.0	0.667	0.0	0.0	0.0	0.0	0.0	0.0
t4				1.0	0.667	0.667	0.667	0.667	0.0	0.0
t5					1.0	0.0	0.694	0.234	0.129	0.095
t6						1.0	0.234	0.403	0.267	0.667
t7							1.0	0.0	0.0	0.0
t8								1.0	0.667	0.278
t9									1.0	0.666
t10										1.0

S_1 S_2 S_3

Fig. 2. The matrix of task couplings for the considered running example

Table 2. Process coupling/cohesion coefficients

SERVICE COHESION COEFFICIENT

$$coh(\mathcal{S}) = \begin{cases} \frac{\sum_{i,j} \tau(t_i,t_j)}{\frac{|\mathcal{S}|\cdot(|\mathcal{S}|-1)}{2}+|\mathcal{S}|} & \forall\, t_i, t_j \in \mathcal{S}\ |\mathcal{S}| > 1,\ j > i \\ 1 & |\mathcal{S}| = 1 \end{cases}$$

where $|\mathcal{S}|$ is the number of tasks in $\mathcal{S}$

SERVICE COUPLING COEFFICIENT

$$coup(\mathcal{S}_1, \mathcal{S}_2) = \frac{\sum_{i,j} \tau(t_i,t_j)}{|\mathcal{S}_1|\cdot|\mathcal{S}_2|}\ \forall\, t_i \in \mathcal{S}_1\ \wedge\ t_j \in \mathcal{S}_2,\ \mathcal{S}_1 \neq \mathcal{S}_2$$

PROCESS COUPLING/COHESION RATIO

$$\Gamma = \frac{pcoup(\mathcal{BP})}{pcoh(\mathcal{BP})};\ pcoup(\mathcal{BP}) = \begin{cases} \frac{\sum_{i,j} coup(\mathcal{S}_i,\mathcal{S}_j)}{\frac{|\mathcal{BP}|\cdot(|\mathcal{BP}|-1)}{2}} & |\mathcal{BP}| > 1,\ j > i \\ 1 & |\mathcal{BP}| = 1 \end{cases} \quad pcoh(\mathcal{BP}) = \frac{\sum_i coh(\mathcal{S}_i)}{|\mathcal{BP}|}$$

where $|\mathcal{BP}|$ represents the actual number of candidate component services in the process

obtained. Finally, the system proposes to the designer the new candidate services for validation. For example, in Figure 2, task couplings highlighted with oval-shaped elements are considered for applying the split procedure. The candidate service $\{t_7, t_8, t_9\}$ is split into two services $\{t_7\}$ and $\{t_8, t_9\}$ and the candidate service $\{t_5, t_6\}$ is split into two services $\{t_5\}$ and $\{t_6\}$. The new value for the coupling/cohesion ratio is $\Gamma' = 0.0381 < \Gamma$.

5 Reconciliation of Similar Services

The set of services resulting from the previous phases includes all the component services of a considered business process. It is possible that some services in this set denote the invocation of the same service in different points of the process.

Similarity of component services is evaluated and similar services ar proposed to the process designer to be merged together and to be considered as different invocations of the same service. In order to evaluate the similarity between services, coefficients introduced in [2] are applied. Summarizing, two services $\mathcal{S}_1$ and $\mathcal{S}_2$ are similar if their *Global Similarity* $GSim(\mathcal{S}_1, \mathcal{S}_2) \geq \delta$, where δ is a threshold set by the process designer. The global similarity is obtained as the linear combination of two coefficients that make use of the defined name and structural affinity: (i) an *Entity-based Similarity* coefficient, that evaluates how much two services work on the same data; (ii) a *Functionality-based Similarity* coefficient, that evaluates how much two services perform the same functionalities. For example, services that check the textile component ($\{t_5\}$) and backbone component feasibility ($\{t_7\}$) could be recognized as similar and proposed for merging.

6 Conclusions

The methodology framework presented in this paper provides a semi-automatic support for the identification of the subset of functionalities that can be exported as component services to implement collaborative business processes in large and complex Internetworked Enterprises. The methodology starts from a process represented by means of a workflow-based language (e.g., BPMN) and supports the designer for the identification of component services. Semantic process description is supported by a given reference ontology and a Thesaurus to solve terminological discrepancies. A prototype tool environment has been developed to support the methodological framework. The tool works on a XPDL serialization of the BPMN process and provides functionalities for the affinity-based comparison of business process elements, cohesion/coupling evaluation and service similarity analysis. The process designer is supported by a Graphical User Interface during the validation of affinities between business process elements, the identification of candidate services and the validation of service similarities during the last phase of the methodology.

Experimentation and validation on real case scenarios will be performed and collaboration conditions that are suitable for the application of the methodology will be investigated. Finally, the proposed approach will be refined to consider issues of service orchestration and composition and will be extended to other models and languages for collaborative process representation.

Acknowledgements

This work has been partially supported by the TEKNE (Towards Evolving Knowledge-based internetworked Enterprise) FIRB Project (http://www.tekne-project.it/), founded by the Italian Ministry of Education, University and Research.

References

1. Bianchini, D., De Antonellis, V., Melchiori, M.: Flexible Semantic-based Service Matchmaking and Discovery. World Wide Web Journal 11(2), 227–251 (2008)
2. Bianchini, D., De Antonellis, V., Pernici, B., Plebani, P.: Ontology-based methodology for e-service discovery. Journal of Information Systems, Special Issue on Semantic Web and Web Services 31(4-5), 361–380 (2006)
3. Bianchini, D., Cappiello, C., De Antonellis, V., Pernici, B.: P2S: a methodology to enable inter-organizational process design through Web Services. In: van Eck, P., Gordijn, J., Wieringa, R. (eds.) CAiSE 2009. LNCS, vol. 5565, pp. 334–348. Springer, Heidelberg (2009)
4. Castano, S., De Antonellis, V., De Capitani di Vimercati, S.: Global Viewing of Heterogeneous Data Sources. IEEE Transactions on Knowledge and Data Engineering 13(2), 277–297 (2001)
5. Fellbaum, C.: Wordnet: An Electronic Lexical Database. MIT Press, Cambridge (1998)
6. Hepp, M., Leymann, F., Domingue, J., Wahler, A., Fensel, D.: Semantic Business Process Management: A Vision Towards Using Semantic Web Services for Business Process Management. In: Proc. of the IEEE Int. Conference on e-Business Engineering (ICEBE 2005), Beijing, China, pp. 535–540 (2005)
7. Dietz, J.L.G.: The Atoms, Molecules and Fibers of Organizations. Data & Knowledge Engineering 47(3), 301–325 (2003)
8. Krogstie, J., Veres, C., Sindre, G.: Integrating Semantic Web Technology, Web Services, and Workflow Modeling: Achieving System and Business Interoperability. Int. Journal of Enterprise Information Systems 3(1), 22–41 (2007)
9. Lin, Y., Sølvberg, A.: Goal Annotation of Process Models for Semantic Enrichment of Process Knowledge. In: Krogstie, J., Opdahl, A.L., Sindre, G. (eds.) CAiSE 2007 and WES 2007. LNCS, vol. 4495, pp. 355–369. Springer, Heidelberg (2007)
10. Mulye, R., Miller, J., Verma, K., Gomadam, K., Sheth, A.: A semantic template based designer for Web processes. In: Proc. of 2005 IEEE Int. Conf. on Web Services (ICWS 2005), Orlando, Florida, USA, pp. 461–469 (2005)
11. O'Brien, J.A.: Introduction to Information Systems: Essentials for the Internetworked Enterprise. McGraw-Hill Education, New York (2000)
12. Papazoglou, M.P., van den Heuvel, W.-J.: Business process development life cycle methodology. Communications of ACM 50(10), 79–85 (2007)
13. Sheng, Q.Z., Benatallah, B., Maamar, Z., Dumas, M., Ngu, A.H.H.: Enabling Personalized Composition and Adaptive Provisioning of Web Services. In: Persson, A., Stirna, J. (eds.) CAiSE 2004. LNCS, vol. 3084, pp. 322–337. Springer, Heidelberg (2004)
14. Vanderfeesten, I., Reijers, H.A., van der Aalst, W.M.P.: Evaluating workflow process designs using cohesion and coupling metrics. Computer in Industry 59(5), 420–437 (2008)

Algebraic Meta-structure Handling of Huge Database Schemata

Hui Ma[1], René Noack[2], and Klaus-Dieter Schewe[3]

[1] Victoria University of Wellington, School of Engineering and Computer Science, Wellington, New Zealand
hui.ma@ecs.vuw.ac.nz

[2] Christian-Albrechts-University Kiel, Department of Computer Science, Kiel, Germany
noack@is.informatik.uni-kiel.de

[3] Information Science Research Centre, Palmerston North, New Zealand
kdschewe@acm.org

Abstract. Practical experience shows that the maintenance of databases with a very large schema causes severe problems, and no systematic support is provided. In a recent study based on the analysis of a large number of very large database schemata twelve frequently recurring meta-structures were identified and classified into three categories associated with schema construction, lifespan and context. The systematic use of these meta-structures supports the modelling and maintenance of very large database schemata. In this paper we complement this study by a schema algebra that permits building larger schemata by composing smaller ones. Similar to abstraction mechanisms found in semantic data models the constructors can be classified into three groups for building associations and collections of subschemata, and for folding subschemata.

Keywords: meta-structure, schema algebra, schema modelling.

1 Introduction

Database schemata that are developed in practical projects tend to become very large and consequently hard to read and comprehend for many developers [5]. Examples are the relational SAP/R3 schema with more than 21,000 tables or the Lufthansa Cargo database schema, which repeats similar subschemata with respect to various transport data. Therefore, the common observation that very large database schemata are error-prone, hard to read and consequently difficult to maintain is not surprising at all.

Some remedies to the problem have already been discussed in previous work and applied in some database development projects. For instance, modular techniques such as *design by units* [11] allow schemata to be drastically simplified by exploiting principles of hiding and encapsulation that are known from Software Engineering. *Component engineering* [7,12] extends this approach by means of view-centered components with well-defined composition operators, and *hierarchy abstraction* [12] permits to model objects on various levels of detail.

C.A. Heuser and G. Pernul (Eds.): ER 2009 Workshops, LNCS 5833, pp. 12–21, 2009.

In order to contribute to a systematic development of very large schemata the *co-design* approach, which integrates structure, functionality and interactivity modelling, emphasises the initial modelling of skeletons of components, which is then subject to further refinement [13]. Thus, components representing subschemata form the building blocks, and they are integrated in skeleton schemata by means of connector types, which commonly are modelled by relationship types.

In [4] the structure of skeleton schemata and component subschemata was analysed more deeply. Based on a large number of very large schemata twelve frequently recurring meta-structures were identified and classified into three categories associated with schema construction, lifespan and context. It was further demonstrated how to apply these meta-structures in data modelling referring among others to *design-by-units* [11], *string-bag modelling* [14], and *incremental structuring* [6].

In this paper we continue this line of work, and investigate constructors for handling meta-structures in a systematic way, which leads to the definition of a schema algebra. Similar to abstraction mechanisms found in semantic data models [10] only three main groups of constructors are needed: *association* constructors that are used to combine schemata in a way that allows the original schemata to be regained, *folding* constructors that integrate schemata into a compact form, and *collection* constructors that deal with recurring similar subschemata. We formally describe these constructors and demonstrate their usage by means of examples.

In this paper we further develop the method for systematic schema development focussing on very large schemata. In Section 2 we briefly review frequently recurring meta-structures as determined and analysed in [4]. In Section 3 we elaborate on the schema algebra, i.e. the constructors for handling meta-schemata.

2 Meta-structures in Database Schema Modelling

In [4] based on an extensive study of a large number of conceptual database schemata frequently occurring meta-structures were identified and classified into three categories according to construction, lifespan and context. We briefly review these meta-structures here leaving details to previous work.

Construction meta-structures capture (almost) hierarchical schema structures, i.e. star- and snowflake-schemata as identified in [7,8], bulk meta-structures, and architecture and constructor-based meta-structures such as the waffle architecture in SAP R/3.

Star typing has been used already for a long time outside the database community. The star constructor permits to construct associations within systems that are characterized by complex branching, diversification and distribution alternatives. Such structures appear in a number of situations such as composition and consolidation, complex branching analysis and decision support systems.

A *star meta-structure* is characterized by a core entity (or relationship) type used for storing basic data, and a number of subtypes of the entity type that

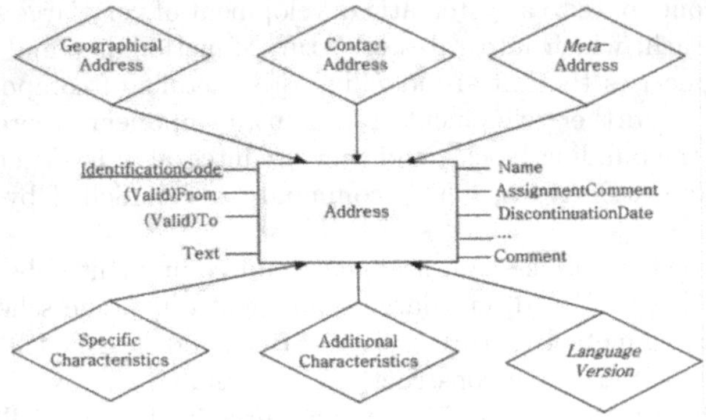

Fig. 1. The General Structure of Addresses

are used to capture additional properties [12]. A typical star structure is shown in Figure 1 with entity type "Address" as its core.

In the same fashion a *snowflake schema* represents the information structure of documented contributions of members of working groups during certain time periods.

A *bulk meta-structure* arises, when types that are used in a schema in a very similar way are clustered together on the basis of a classification. We can combine relationship types with the same components into a single type using the original relationships as values of a distinguishing attribute.

The evolution of an application over its lifetime is orthogonal to the construction. This leads to a number of *lifespan meta-structures. Evolution meta-structures* record life stages similar to workflows, *circulation meta-structures* display the phases in the lifespan of objects, *incremental meta-structures* permit the recording of the development, enhancement and ageing of objects, *loop meta-structures* support chaining and scaling to different perspectives of objects, and *network meta-structures* permit the flexible treatment of objects during their evolution by supporting to pass objects in a variety of evolution paths and enable multi-object collaboration.

According to [15] we distinguish between the *intext* and the *context* of things that are represented as objects. Intext reflects the internal structuring, associations among types and subschemata, the storage structuring, and the representation options. Context reflects general characterisations, categorisation, utilisation, and general descriptions such as quality. Therefore, we distinguish between *meta-characterisation meta-structures* that are usually orthogonal to the intext structuring and can be added to each of the intext types, *utilisation-recording meta-structures* that are used to trace the running, resetting and reasoning of the database engine, and *quality meta-structures* that permit to reason on the quality of the data provided and to apply summarisation and aggregation functions in a form that is consistent with the quality of the data. The dimensionality of a schema permits the extraction of other *context meta-structures* [3].

3 A Schema Algebra

In the following we present three groups of schema constructors dealing with associations, folding, and collections of schemata. This defines a (partial) schema algebra, as constructors are only applicable, if certain preconditions are satisfied. The composition operators presented in this section will permit the construction of any schema of interest, as they mimic all set operations similar to the structural approach in [1].

If $\mathcal{S}_1, \dots, \mathcal{S}_n$ are schemata and $\mathcal{O}$ is an n-ary operator applicable to them, the resulting schema $\mathcal{S}$ defines the equation $\mathcal{S} = \mathcal{O}(\mathcal{S}_1, \dots, \mathcal{S}_n)$. Using these equations in a directed way defines a graph-rewriting system [2,9] that can be used as foundation for schema development. The graph production rules are rather simple, and more complex rules can be derived by rule composition.

3.1 Renaming

As the names of types and clusters in ER-schemata must be unique, we must avoid name clashes when applying the schema constructors. Therefore, we have to provide a renaming constructor. For this, if $R_1, \dots, R_k$ and $R'_1, \dots, R'_k$ are pairwise distinct sequences of names, a *renaming* is a mapping $\{R_1 \mapsto R'_1, \dots$ $\dots, R_k \mapsto R'_k\}$. If $\mathcal{S}$ is a database schema, then replacing each occurrence of R_i in $\mathcal{S}$ by R'_i results in the schema

$$\varrho_{R_1 \mapsto R'_1, \dots, R_k \mapsto R'_k}(\mathcal{S}).$$

3.2 Association Constructors

The simplest form of a composition through association is by means of a direct sum, i.e. disjoint union constructor. More generally, we consider joins of two schema along input- and output-views [7]. For this let $\mathcal{S}_i$ be a schema with two views $\mathcal{I}_i$ called *input view*, and $\mathcal{O}_i$ called *output-view* ($i = 1, 2$). We request that $\mathcal{I}_i$ and $\mathcal{O}_j$ for $i = 1, j = 2$ or $i = 2, j = 1$ are isomorphic.

The *join schema*

$$\mathcal{S} = \mathcal{S}_1 \bowtie_{\mathcal{I}_1 := \mathcal{O}_2 \| \mathcal{I}_2 := \mathcal{O}_1} \mathcal{S}_2$$

results from the two given schemata by identifying in $\mathcal{S}_1 \cup \mathcal{S}_2$ the input-view of first schema with the output-view of the second one and vice versa.

The join of the schemata $\mathcal{S}_1$ and $\mathcal{S}_2$ along the input- and output-views shown in Figure 2 is the schema $\mathcal{S}$ shown in the same figure. We omitted all attributes, as these are preserved by the join.

A variant of the join operator is provided by means of a *reference-join.* The prerequisites are the same as for the join operator. In this case, however, the output-views $\mathcal{O}_i$ ($i = 1, 2$) of the original schemata are preserved within the resulting schema

$$\mathcal{S} = \mathcal{S}_1 \bowtie_{\mathcal{I}_1 \to \mathcal{O}_2 \| \mathcal{I}_2 \to \mathcal{O}_1} \mathcal{S}_2$$

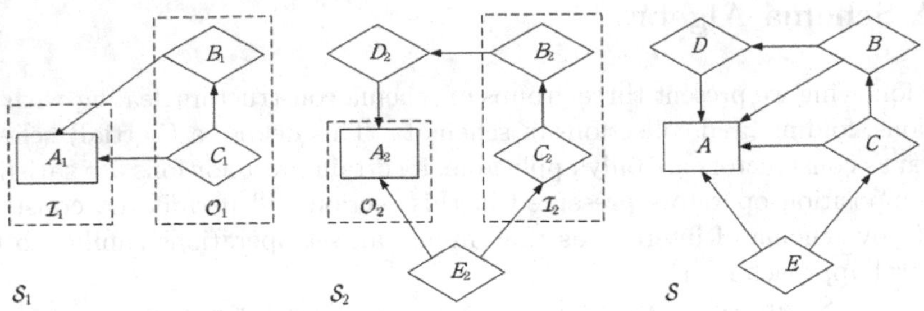

Fig. 2. The join operator on two schemat

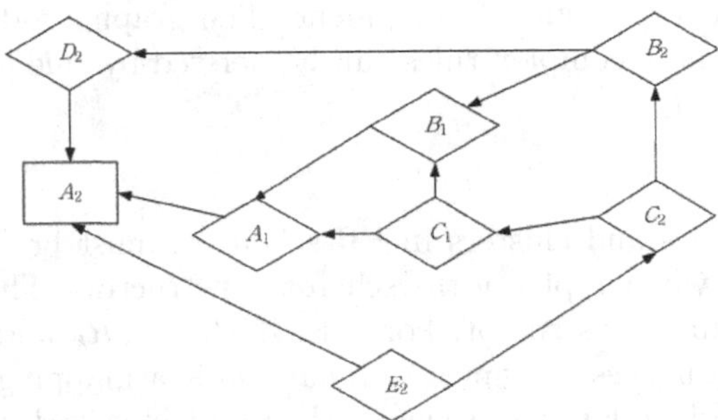

Fig. 3. The reference-join on two schemata

and references from the types in $\mathcal{I}_i$ to those in $\mathcal{O}_j$ for $(i,j) = (1,2)$ or $(2,1)$ are added. This requires that entity-types in the input-views be turned into relationship types. The schema shown in Figure 3 shows the result of the reference-join of the schemata $\mathcal{S}_1$ and $\mathcal{S}_2$ from Figure 2.

Another variant can be obtained, when cooperating views [11] are employed instead of merging input- and output-views or letting the former ones reference the latter ones. In this case the data exchange has to be specified explicitly. We omit the details.

Dual to the sum constructor we can define a product constructor. In this case let $(\mathcal{S}_i, \Sigma_i)$ be schemata with disjoint name sets $(i = 1,2)$. For types $R_i \in \mathcal{S}_i$ defined as $(comp(R_i), attr(R_i), k(R_i))$ $(i = 1,2)$ define their product $R_1 \times R_2$ by the type

$$R_{1,2} = (comp(R_1) \times \{R_2\} \cup \{R_1\} \times comp(R_2), attr(R_1) \cup attr(R_2), k(R_{12})),$$

i.e. if $comp(R_i) = \{r_{i1} : R_{i1}, \dots, r_{ik_i} : R_{ik_i}\}$, we obtain $comp(R_{12}) = \{r_{11} : R_{11,2}, \dots, r_{1k_1} : R_{1k_1,2}, r_{21} : R_{1,21}, \dots, r_{2k_2} : R_{1,2k_2}\}$, and the key $k(R_{12})$ is defined as $\{r_{1j} : R_{1j,2} \mid r_{1j} : R_{1j} \in k(R_1)\} \cup \{r_{2j} : R_{1,2j} \mid r_{2j} : R_{2j} \in k(R_2)\} \cup \{A \mid A \in attr(R_1) \cap k(R_1)\} \cup \{A \mid A \in attr(R_2) \cap k(R_2)\}$.

If R_1 is a cluster, say $R_1 = \{\ell_1 : R_{11}, \dots, \ell_{k_1} : R_{1k_1}\}$, and R_2 is a type as before, then their product is the cluster

$$R_1 \times R_2 = \{\ell_1 : R_{11,2}, \dots, \ell_{k_1} : R_{1k_1,2}\}.$$

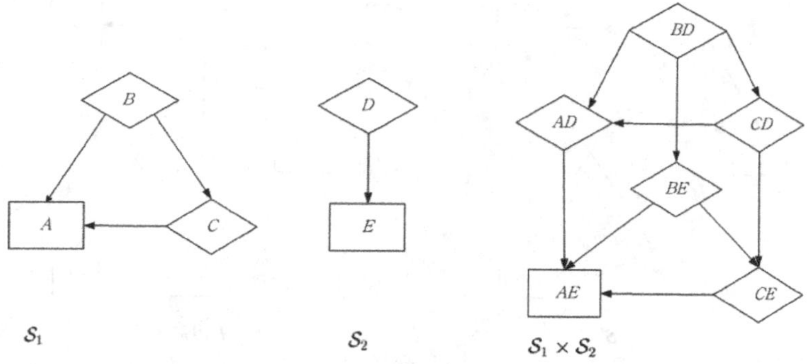

Fig. 4. The product operator on two schemata

The product of a type R_1 and a cluster R_2 is defined analogously. Finally, if both R_1 and R_2 are clusters, say $R_i = \{\ell_{i1} : R_{i1}, \dots, \ell_{ik_i} : R_{ik_i}\}$ for $i = 1, 2$, then their product is the cluster

$$R_1 \times R_2 = \{\ell_{1j_1,2j2} : R_{1j_1,2j_2} \mid 1 \leq j_1 \leq k_1,\ 1 \leq j_2 \leq k_2\}.$$

The *product schema* is defined as

$$\mathcal{S} = \mathcal{S}_1 \times \mathcal{S}_2 = \{R_1 \times R_2 \mid R_1 \in \mathcal{S}_1,\ R_2 \in \mathcal{S}_2\}.$$

Of course, in all cases we have to create new names for the new types (or clusters) $R_{i,j} = R_i \times R_j$, and also new names for labels in the clusters and roles in the components. Figure 4 shows the product $\mathcal{S} = \mathcal{S}_1 \times \mathcal{S}_2$ of the schemata in the same figure. We omitted all attributes.

While the dual of the direct sum constructor $\oplus$ is the product constructor $\times$, we can also define a dual *meet constructor* $\bullet_\varphi$ for the join constructor. In this case we need an additional *matching condition* φ, and we define the meet schema as

$$\mathcal{S} = \mathcal{S}_1 \bullet_\varphi \mathcal{S}_2 = \{R_1 \times R_2 \mid R_1 \in \mathcal{S}_1,\ R_2 \in \mathcal{S}_2 \text{ with } \varphi(R_1, R_2)\}.$$

Matching conditions can express requirements such as common attributes or inclusion constraints.

3.3 Folding and Unfolding of Schemata

As observed in [4] similar subschemata can be integrated by replacing a number of relationship types by a new relationship type plus an additional entity type. For this assume we have a schema $\mathcal{S}'$ with a central entity (or relationship) type C, and n relationship types $R_1, \dots, R_n$ that all relate C to a number $C_1, \dots, C_k$ of entity or relationship types as shown in the left hand part of Figure 5.

Then we can replace $R_1, \dots, R_n$ by a new relationship type R with a new additional component CA. This type must have an attribute "ContractionType" with domain $\{1, \dots, n\}$ that will be used to identify the original relation. It may further require an identifying attribute "Ident".

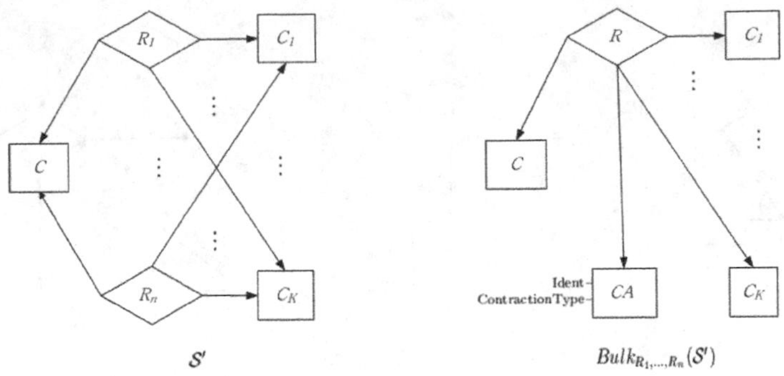

Fig. 5. The bulk operator on a database schema

The schema $\mathcal{S} = Bulk_{R_1,\dots,R_n}(\mathcal{S}')$ resulting from applying this *bulk* operator is illustrated in the right hand part of Figure 5. With respect to integrity constraints in Σ each occurrence of a type R_i has to be replaced by the projection $R[C, C_1, \dots, C_k]$ and the condition $R.CA.$ContractionType $= R_i$ has to be added.

The bulk constructor $Bulk_{R_1,\dots,R_n}$ can be refined to better handle attributes that are not common to all types $R_1, \dots, R_n$. Such an attribute A becomes an "optional" attribute of the type CA, i.e. its domain will be defined as $dom_{\mathcal{S}}(A) = dom_{\mathcal{S}'}(A) \cup \{undef\}$. If A is not an attribute of the type R_i, the constraint

$$CA.\text{ContractionType} = R_i \Rightarrow CA.A = undef$$

has to be added.

The *expansion* constructor $Expand_{E:A}$ is inverse to the bulk constructor. In this case we need an entity type E with $k(E) = \{ident\}$, and an attribute $A \in attr(E) - k(E)$ with a finite enumeration domain $dom(A) = \{v_1, \dots, v_n\}$. Furthermore, there must be a unique relationship type $R \in \mathcal{S}$ with a component E occurring once, i.e. $r : E \in comp(R)$, and for all r' and all R' with $r' : E \in comp(R')$ we must have $R' = R$ and $r' = r$.

In the resulting schema $Expand_{E:A}(\mathcal{S})$ the type R will be replaced by n types $R_1, \dots, R_n$ corresponding to the values $v_1, \dots, v_n$ of the attribute A. For each of these types we have $comp(R_i) = comp(R) - \{r : E\}$. Each attribute of R becomes an attribute of R_i, and each attribute $B \in attr(E) - \{ident, A\}$ is added as an attribute of R_i, unless Σ contains a constraint of the form above.

Component nesting can be applied to a schema $\mathcal{S}_1$ to replace a component C of a type R by a complete subschema $\mathcal{S}_2$ that is rooted at a type T. Identifying components $I_1, \dots, I_k$ and other components $C_1, \dots, C_\ell$ of C will become components of the root type T of $\mathcal{S}_2$ within the new schema. We denote the schema $\mathcal{S}$ resulting from the application of the nesting operator by $nest_{C:\mathcal{S}_2(T)}(\mathcal{S}_1)$. Figure 6 illustrates the application of the nesting operator.

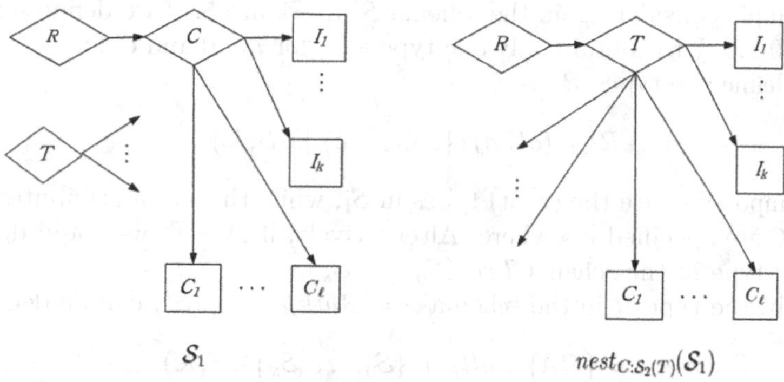

Fig. 6. The nesting operator on a database schema

Component nesting generalises entity model clustering, entity clustering, entity and relationship clustering, entity tree clustering in the design-by-units method [11].

3.4 Collection Constructors for Schemata

While all operators discussed so far have arity 1 or 2, the collection constructions apply to any number k of schemata. If $\mathcal{S}_1, \dots, \mathcal{S}_k$ are schemata without name clashes, we can build the *set schema* $\{\mathcal{S}_1, \dots, \mathcal{S}_k\}$, provided the element schemata are pairwise distinct, the *multiset schema* $\langle \mathcal{S}_1, \dots, \mathcal{S}_k \rangle$, the *list schema* $[\mathcal{S}_1, \dots, \mathcal{S}_k]$, and the *tree schema* $\langle\!\langle \mathcal{S}_1, \dots, \mathcal{S}_k \rangle\!\rangle$.

As schemata, the result of the first three constructions can be identified with the product, i.e. the join with empty views, of the element schemata, while a tree schema contains an additional relationship type with k components that are root types of the element schemata. As such, the collection constructions are only a mild extension.

However, they unfold their power by means of collection operators that can be applied to a set, multiset, list or tree schema $\mathcal{S}'$:

- $all\_of(\mathcal{S}')$ denotes the schemata that contains all schemata in the collection as subschemata. The construction can be used to specify that all $\mathcal{S}_1, \dots, \mathcal{S}_k$ (or their root types, respectively) must appear as components in some other construction, e.g. in the bulk or nesting construction we discussed above.
- Similarly, $any\_of(\mathcal{S}')$ denotes one arbitrary element schema, and $n\_of(\mathcal{S}')$ denotes an arbitrary selection of n of the element schemata.
- The selection of subschemata in the collection using any of the constructors all_of, any_of or n_of can be refined by adding selection criteria in form of a `where`-clause. For instance, $n\_of(\{\mathcal{S}_1, \dots, \mathcal{S}_k\})$ `where` φ would select n of the element schemata among those satisfying the condition φ.
- $n\_th(\mathcal{S}')$ for a list or tree schema denotes the n'th element schema, provided $1 \leq n \leq k$ is satisfied.

As an example consider again the schema $\mathcal{S}'$ in Figure 5. If we define schemata $\mathcal{S}_i$ for $i = 0, \dots, k$ to contain only one type – C_i for $i \neq 0$ and C for $i = 0$ – then we could define the types R_i as

$$R_i = (all\_of(\{\mathcal{S}_0, \dots, \mathcal{S}_k\}), \mathcal{S}, \mathcal{K}),$$

i.e. the components are the (root) types in $\mathcal{S}_i$, while the set of attributes $\mathcal{A}$ and the keys $\mathcal{K}$ are specified elsewhere. Alternatively, if $\mathcal{A} = \emptyset$, we could define R_i as the root type in the schema $Tree(\mathcal{S}_0, \dots, \mathcal{S}_k)$.

Similarly, the type R in the schema $\mathcal{S} = Bulk_{R_1,\dots,R_n}(\mathcal{S}')$ can be defined as

$$R = \{CA\} \cup all\_of(\{\mathcal{S}_0, \dots, \mathcal{S}_k\}), \mathcal{S}, \mathcal{K})$$

with the entity type $CA = (\{\text{Ident, ContractionType}\}, \{\text{Ident}\})$.

4 Conclusion

Very large database schemata with hundreds or thousands of types are usually developed over years, and then require sophisticated skills to read and comprehend them. However, lots of similarities, repetitions, and similar structuring elements appear in such schemata. In this paper we presented an algebra for handling meta-structures that occur frequently in such schemata. The meta-structures as such have already been identified in our previous work, and classified according to structure, lifespan and context. The work presented now highlights the solid formal background of the meta-structuring approach in database schema modelling and maintenance.

In particular, the meta-structuring approach aims at supporting component-based schema development, in which schemata are developed step-by-step on the basis of the skeleton of the meta-structure. The schema algebra adds formal rigidity to the refinements in this process, and thus further contributes to the development of industrial-scale database applications. In this way, the approach can be exploited to modularise schemata, and ease querying, searching, reconfiguration, maintenance, integration, extension, reengineering, and reuse. This enables systematic schema development, extension and implementation, and thus contributes to overcome the maintenance problems arising in practice from very large schemata.

References

1. Brown, L.: Integration Models – Templates for Business Transformation. SAMS Publishing, USA (2000)
2. Ehrig, H., Engels, G., Kreowski, H.-J., Rozenberg, G. (eds.): Handbook of Graph Grammars and Computing by Graph Transformations, Applications, Languages and Tools, vol. 2. World Scientific, Singapore (1999)
3. Feyer, T., Thalheim, B.: Many-dimensional schema modeling. In: Manolopoulos, Y., Návrat, P. (eds.) ADBIS 2002. LNCS, vol. 2435, pp. 305–318. Springer, Heidelberg (2002)

4. Ma, H., Schewe, K.-D., Thalheim, B.: Modelling and maintenance of very large database schemata using meta-structures. In: Yang, J., et al. (eds.) UNISCON 2009. LNBIP, vol. 20, pp. 17–28. Springer, Heidelberg (2009)
5. Moody, D.: Dealing with Complexity: A Practical Method for Representing Large Entity-Relationship Models. PhD thesis, University of Melbourne (2001)
6. Raak, T.: Database systems architecture for facility management systems. Master's thesis, Fachhochschule Lausitz (2002)
7. Schewe, K.-D., Thalheim, B.: Component-driven engineering of database applications. In: Conceptual Modelling - Proc. APCCM 2006. CRPIT, vol. 53, pp. 105–114. Australian Computer Society (2006)
8. Shoval, P., Danoch, R., Balaban, M.: Hierarchical ER diagrams (HERD) - the method and experimental evaluation. In: Olivé, À., Yoshikawa, M., Yu, E.S.K. (eds.) ER 2002. LNCS, vol. 2784, pp. 264–274. Springer, Heidelberg (2002)
9. Sleep, M.R., Plasmeijer, M.J., van Eekelen, M.C.J.D. (eds.): Term Graph Rewriting – Theory and Practice. John Wiley and Sons, Chichester (1993)
10. Smith, J.M., Smith, D.C.P.: Database abstractions: Aggregation and generalization. ACM ToDS 2(2), 105–133 (1977)
11. Thalheim, B.: Entity Relationship Modeling – Foundations of Database Technology. Springer, Heidelberg (2000)
12. Thalheim, B.: Component development and construction for database design. Data and Knowledge Engineering 54, 77–95 (2005)
13. Thalheim, B.: Engineering database component ware. In: Draheim, D., Weber, G. (eds.) TEAA 2006. LNCS, vol. 4473, pp. 1–15. Springer, Heidelberg (2007)
14. Thalheim, B., Kobienia, T.: Generating database queries for web natural language requests using schema information and database content. In: Applications of Natural Language to Information Systems – NLDB 2001. LNI, vol. 3, pp. 205–209. GI (2001)
15. Wisse, P.: Metapattern – Context and Time in Information Models. Addison-Wesley, Reading (2001)

On Computing the Importance of Entity Types in Large Conceptual Schemas

Antonio Villegas and Antoni Olivé

Dept. de Llenguatges i Sistemes Informàtics, Universitat Politècnica de Catalunya
{avillegas,olive}@lsi.upc.edu

Abstract. The visualization and the understanding of large conceptual schemas require the use of specific methods. These methods generate clustered, summarized or focused schemas that are easier to visualize and to understand. All of these methods require computing the importance of each entity type in the schema. In principle, the totality of knowledge defined in the schema could be relevant for the computation of that importance but, up to now, only a small part of that knowledge has been taken into account. In this paper, we extend six existing methods for computing the importance of entity types by taking into account all the relevant knowledge defined in the structural and behavioural parts of the schema. We experimentally evaluate the original and the extended versions of those methods with two large real-world schemas. We present the two main conclusions we have drawn from the experiments.

1 Introduction

Real information systems often have extremely complex conceptual schemas. The visualization and understanding of these schemas require the use of specific methods, which are not needed in small schemas [1]. These methods generate indexed, clustered, summarized or focused schemas that are easier to visualize and to understand [2].

Many of the above methods require computing the importance (also called relevance or score) of each type in the schema. The computed importance induces an ordering of the entity types, which plays a key role in the steps and result (output) of the method. For example, Castano, de Antonellis, Fugini and Pernici [3] propose a three-steps indexing method, in which the first step computes the importance of each entity type, based on the number and kind of relationships it has in the schema. Moody [4] proposes a clustering method in which the most important entity types are hypothesized to be those that have the higher connectivity, defined as the number of relationships in which they participate. Tzitzikas and Hainaut [5,6] propose methods for scoring each entity type in a schema, aiming at facilitating its understanding. As a last example we may mention Yu and Jagadish [7], who propose a metric of the importance of each entity type, which is used in order to automatically produce a good summary of a schema.

C.A. Heuser and G. Pernul (Eds.): ER 2009 Workshops, LNCS 5833, pp. 22–32, 2009.

Intuitively, it seems that an objective metric of the importance of an entity type in a given schema should be related to the amount of knowledge that the schema defines about it. The more (less) knowledge a schema defines about an entity type, the more (less) important should be that entity type in the schema. Adding more knowledge about an entity type should increase (or at least not decrease) the relative importance of that entity type with respect to the others. Note that in this paper we focus on objective metrics, which are independent from subjective evaluations of users and modelers.

As far as we know, the existing metrics for entity type importance are mainly based on the amount of knowledge defined in the schema, but only take into account the number of attributes, associations and specialization/generalization relationships. Surprisingly, none of the methods we are aware of take into account additional knowledge about entity types defined in a schema that, according to the intuition, could have an effect on the importance. A complete schema [8] includes also cardinalities, taxonomic constraints, general constraints, derivation rules and the specification of events, all of which contribute to the knowledge about entity types.

The main objective of this paper is to analyze the influence of that additional knowledge on a representative set of existing metrics for measuring the importance of the entity types. To this end, we have selected six methods from [3,5,6] and we have developed extended versions of all of them. We have experimentally evaluated both versions of each method using the conceptual schema of the osCommerce [9] and the UML metaschema [10]. The osCommerce is a popular industrial e-commerce system whose conceptual schema consists of 346 entity types (of which 261 are event types). The official 2.0 UML metaschema we have used consists of 293 entity types. The original and the extended versions give exactly the same results from the same input, but the extended versions can process the additional knowledge defined in the schema and then, of course, they give different results. We analyze the differences, and make conclusions on the effect of the additional knowledge on the metrics.

The rest of the paper is organized as follows. Section 2 introduces the concepts and notations. Section 3 briefly describes the seleted methods and explains the extensions we have done to them. Section 4 describes the experimentation with the methods, the results obtained and the conclusions we have drawn. Finally, Section 5 summarizes the paper and points out future work.

2 Basic Concepts and Notations

In this section we review the main concepts and the notation we have used to define the knowlege of conceptual schemas. In this paper, we deal with schemas written in the UML[10]/OCL[11]. Table 1 summarizes the notation (inspired by [6,12]) used in the rest of the paper.

A conceptual schema consists of a structural (sub)schema and a behavioral (sub)schema. The structural schema consists of a taxonomy of entity types (a set of entity types with their generalization/specialization relationships and the

Table 1. Schema Notations

Notation	Definition
$par(e)$	$= \{e' \in \mathcal{E} \mid e \; IsA \; e'\}$
$chi(e)$	$= \{e' \in \mathcal{E} \mid e' \; IsA \; e\}$
$gen(e)$	$= par(e) \cup chi(e)$
$attr(e)$	$= \{a \in \mathcal{A} \mid entity(a) = e\}$
$members(r)$	$= \{e \in \mathcal{E} \mid e \text{ is a participant of } r\}$
$assoc(e)$	$= \{r \in \mathcal{R} \mid e \in members(r)\}$
$conn(e)$	$= \biguplus_{r \in assoc(e)} \{members(r) \backslash \{e\}\}$[1]
$context(\alpha)$	$= e \in \mathcal{E} \mid \alpha \in \mathcal{SR} \wedge \alpha \; DefinedIn \; e$
$members(exp)$	$= \{e \in \mathcal{E} \mid e \text{ is a participant of } exp\}$
$expr(\alpha)$	$= \{expr \mid expr \text{ is contained in } \alpha\}$
$ref(\alpha)$	$= \bigcup_{exp \in expr(\alpha)} \{members(exp)\}$
$expr_{nav}(\alpha)$	$= \{expr \in expr(\alpha) \mid expr \text{ is a navigation expression}\}$
$nav_{expr}(\alpha)$	$= \bigcup_{exp \in exprnav(\alpha)} \{\{e, e'\} \subset \mathcal{E} \mid \{e, e'\} = members(exp)\}\})$
$nav_{context}(\alpha)$	$= \{\{e, e'\} \subset \mathcal{E} \mid e = context(\alpha) \wedge e' \in ref(\alpha)\}$
$nav(\alpha)$	$= nav_{context}(\alpha) \cup \; nav_{expr}(\alpha)$
$rconn(e)$	$= \biguplus_{\alpha \in \mathcal{SR}} \{e' \in \mathcal{E} \mid \{e, e'\} \subset nav(\alpha)\}$
$par_{inh}(e)$	$= par(e) \cup \{par_{inh}(e') \mid e' \in par(e)\}$
$chi_{inh}(e)$	$= chi(e) \cup \{chi_{inh}(e') \mid e' \in chi(e)\}$
$attr_{inh}(e)$	$= attr(e) \cup \{attr_{inh}(e') \mid e' \in par(e)\}$
$assoc_{inh}(e)$	$= assoc(e) \uplus \{assoc(e') \mid e' \in par_{inh}(e)\}$
$conn_{inh}(e)$	$= conn(e) \uplus \{conn(e') \mid e' \in par_{inh}(e)\}$
$rconn_{inh}(e)$	$= rconn(e) \uplus \{rconn(e') \mid e' \in par_{inh}(e)\}$

taxonomic constraints), a set of relationship types (either attributes or associations), the cardinality constraints of the relationship types, and a set of other static constraints formally defined in OCL.

We denote by $\mathcal{E}$ the set of entity types defined in the schema. For a given $e \in \mathcal{E}$ we denote by $par(e)$ and $chi(e)$ the set of directly connected ascendants and descendants of e, respectively, and by $gen(e)$ the union of both sets. The set of attributes defined in the schema is denoted by $\mathcal{A}$. If $a \in \mathcal{A}$ then $entity(a)$ denotes the entity type where a is defined. The set of attributes of an entity type e is denoted by $attr(e)$.

The set of associations defined in the schema is denoted by $\mathcal{R}$. If $r \in \mathcal{R}$ then $members(r)$ denotes the set of entity types that participate in association r, and $assoc(e)$ the set of associations in which e participates. Note that an entity type e may participate more than once in the same association, and therefore $members(r)$ and $assoc(e)$ are multisets (may contain duplicate elements). Moreover, $conn(e)$ denotes the multiset of entity types connected to e through associations. For example, if r_1 is the association HasComponent(assembly:Part, component:Part), then

[1] Note that "\" denotes the difference operation of multisets as in $\{a, a, b\} \backslash \{a\} = \{a, b\}$ and "$\uplus$" denotes the multiset (or bag) union that produces a multiset as in $\{a, b\} \uplus \{a\} = \{a, a, b\}$.

members(r_1)={Part, Part}, *assoc*(Part)={HasComponent, HasComponent} and *conn*(Part)={Part}.

The behavioural schema consists of a set of event types. We adopt the view that events can be modeled as a special kind of entity type. Event types have characteristics, constraints and effects. The characteristics of an event are its attributes and the associations in which it participates. The constraints are the conditions that events must satisfy to occur. Each event type has an operation called *effect()* that gives the effect of an event occurence. The effect is declaratively defined by the postcondition of the operation, which is specified in OCL (see chp. 11 of [8]). Furthermore, entity and relationship types may be base or derived. If they are derived, there is a formal derivation rule in OCL that defines their population in terms of the population of other types.

We denote by $\mathcal{SR}$ the set of constraints, derivation rules and pre- and postconditions. Each schema rule α is defined in the context of an entity type, denoted by $context(\alpha)$. In OCL, each rule α consists of a set of OCL expressions (see OCL [11]) which we denote by $expr(\alpha)$. An expression exp may refer to several entity types which are denoted by $members(exp)$. The set of entity types that are referred to in one or more expressions of a rule α is denoted by $ref(\alpha)$.

We also include in $\mathcal{SR}$ the schema rules corresponding to the equivalent OCL invariants of the cardinality constraints. For example, in Fig. 1 the cardinality "1.." between Company and Employee is transformed into the invariant:

```
context Company inv: self.employee->size()>0
```

A special kind of OCL expression is the navigation expression that define a schema navigation from an entity type to another through an association (see *NavigationCallExp* of OCL in [11]). We use $expr_{nav}(\alpha)$ to indicate the navigation expressions inside a rule α. Such expressions only contain two entity types as its participants, i.e. the *source* entity type and the *target* one (see the example in Fig. 1).

We denote by $nav_{expr}(\alpha)$ the set of pairs that participate in the navigation expressions of α. We also denote by $nav_{context}(\alpha)$ the sets of pairs of entity types composed by the context of the rule α and every one of the participant entity types of such rule ($e \in ref(\alpha)$). Finally, we define $nav(\alpha)$ as the union of $nav_{context}(\alpha)$ with $nav_{expr}(\alpha)$ and, $rconn(e)$ as the multiset of entity types

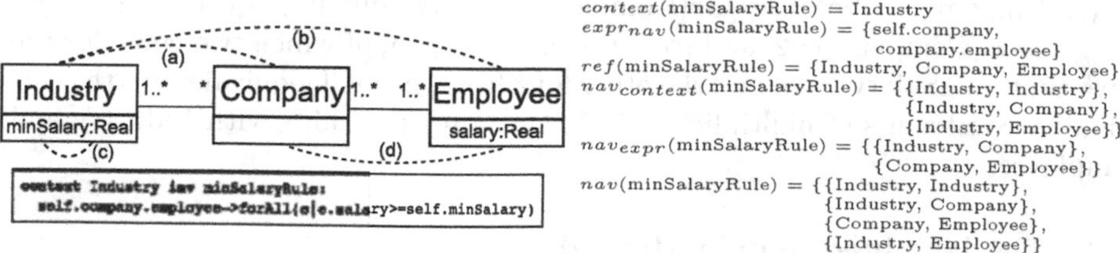
(b)
(a)
Industry
1..*
*
Company
1..*
1..*
Employee
minSalary:Real
salary:Real
(c)
(d)
context Industry inv minSalaryRule:
self.company.employee->forAll(e|e.salary>=self.minSalary)
context(minSalaryRule) = Industry
exprnav(minSalaryRule) = {self.company, company.employee}
ref(minSalaryRule) = {Industry, Company, Employee}
navcontext(minSalaryRule) = {{Industry, Industry}, {Industry, Company}, {Industry, Employee}}
navexpr(minSalaryRule) = {{Industry, Company}, {Company, Employee}}
nav(minSalaryRule) = {{Industry, Industry}, {Industry, Company}, {Company, Employee}, {Industry, Employee}}

Fig. 1. Example of navigations of minSalaryRule. Dashed lines (a), (b) and (c) represent the elements in $nav_{context}$(minSalaryRule) while (d) and (a) are the connections through navigation expressions (see nav_{expr}(minSalaryRule)).

that compose a pair with e in $nav(\alpha)$. Note that since we use $\uplus$, $rconn(e)$ may contain duplicates because it takes into account each rule α and an entity type e can be related to another one e' in two or more different rules. Intuitively, $rconn(e)$ is the multiset of entity types to which an entity type e is connected through schema rules.

The last row section in Table 1 defines the notation we use to take into account the inherited properties from the ancestors of entity types. As a special case, $chi_{inh}(e)$ is the set of descendants of e.

3 Methods and Their Extensions

In this section we briefly review the definition of six existing methods for computing the importance of entity types in a schema. Each method is followed by a brief description and formal definition of our extension to it.

The original version of the methods only takes into account the indicated elements of the structural schema while in the extended version we also take into account the rules and the complete behavioural schema.

3.1 The Simple Method

This method was introduced in [6] and takes into account only the number of directly connected elements. Formally, the importance $I_{SM}(e)$ of an entity type e is defined as:

$$I_{SM}(e) = |par(e)| + |chi(e)| + |attr(e)| + |assoc(e)|$$

Our extension to this method follows the same idea but also including the number of participations of each entity type in the navigation relationships represented in the schema rules specification, i. e., derivation rules, invariants and pre- and postconditions (and cardinality constraints). On the other hand, we now take into account (in $|assoc(e)|$) the associations of each entity type e with the event types of the behavioural schema. Formally:

$$I^{+}_{SM}(e) = |par(e)| + |chi(e)| + |attr(e)| + |assoc(e)| + |rconn(e)|$$

For example, in the schema shown in Fig.1 we would have I_{SM}(Company)=2 and I^{+}_{SM}(Company)=8, because $|par$(Company)$|$=$|chi$(Company)$|$=$|attr$(Company)$|$ =0, $|assoc$ (Company)$|$=2, and $|rconn$(Company)$|$=6, of which two come for the invariant (minSalaryRule) and the other four from the OCL equivalent to the cardinality constraints of multiplicity "1..*" in its relationships with Industry and Employee.

3.2 The Weighted Simple Method

This is a variation to the simple method that assigns a strength to each kind of component of knowledge in the equation, such that the higher the strength, the greater the importance of such component [3]. The definition of importance here is:

$$I_{WSM}(e) = q_{inh}(|par(e)| + |chi(e)|) + q_{attr}|attr(e)| + q_{assoc}|assoc(e)|$$

where q_{attr} is the strength for attributes, q_{inh} is the strength for generalization/specialization relationships, and q_{assoc} is the strength for associations. Each of them with values in the interval *[0,1]*.

Our extension to this method consists on adding the *schema rules navigation* component to the importance computation. In the same way as the other components, we selected a strength (q_{rule}) to specify the weight of navigation relationships in the schema rules. The definition is now:

$$I^{+}_{WSM}(e) = q_{inh}(|par(e)| + |chi(e)|) + q_{attr}|attr(e)| + q_{assoc}|assoc(e)| + q_{rule}|rconn(e)|$$

3.3 The Transitive Inheritance Method

This is a variation of the simple method taking into account both directly defined features and inherited ones [6]. For each entity type the method computes the number of ascendants and descendants and all specified attributes and accessible associations from it or any of its ascendants. Formally:

$$I_{TIM}(e) = |par_{inh}(e)| + |chi_{inh}(e)| + |attr_{inh}(e)| + |assoc_{inh}(e)|$$

In the same way as before, we extend it with the *schema rules navigation* component. This time the computation of such component also takes into account the *rconn* of the ancestors:

$$I^{+}_{TIM}(e) = |par_{inh}(e)| + |chi_{inh}(e)| + |attr_{inh}(e)| + |assoc_{inh}(e)| + |rconn_{inh}(e)|$$

3.4 EntityRank

The EntityRank method [5,6] is based on link analysis following the same approach than Google's PageRank [13]. Roughly, each entity type is viewed as a state and each association between entity types as a bidirectional transition between them.

The importance of an entity type is the probability that a random surfer is at that entity type with random jumps (q component) or by navigation through relationships ($1-q$ component). Therefore, the resulting importance of the entity types correspond to the stationary probabilities of the Markov chain, given by:

$$I_{ER}(e) = \frac{q}{|\mathcal{E}|} + (1-q) \sum_{e' \in conn(e)} \frac{I_{ER}(e')}{|conn(e')|}$$

In our extension to it we add a new component to the formula in order to jump not only to the connected entity types but also to the virtually connected ones through the navigation relationships uncovered in the schema rules. The definition is now:

$$I^+_{ER}(e) = \frac{q}{|\mathcal{E}|} + (1-q)\left(\sum_{e' \in conn(e)} \frac{I^+_{ER}(e')}{|conn(e')|} + \sum_{e'' \in rconn(e)} \frac{I^+_{ER}(e'')}{|rconn(e'')|}\right)$$

3.5 BEntityRank

This is a variation of the previous method specifying that the probability of randomly jumping to each entity type is not the same for each entity type, but it depends on the number of its attributes [5,6]. The higher the number of attributes, the higher the probability to randomly jump to that entity type. That is:

$$I_{BER}(e) = q\frac{attr(e)}{|\mathcal{A}|} + (1-q) \sum_{e' \in conn(e)} \frac{I_{BER}(e')}{|conn(e')|}$$

Our extension is in the same way as in EntityRank but taking into account the definition of the attributes component of BEntityRank. The definition is:

$$I^+_{BER}(e) = q\frac{attr(e)}{|\mathcal{A}|} + (1-q)\left(\sum_{e' \in conn(e)} \frac{I^+_{BER}(e')}{|conn(e')|} + \sum_{e'' \in rconn(e)} \frac{I^+_{BER}(e'')}{|rconn(e'')|}\right)$$

3.6 CEntityRank

Finally, the method that we call CEntityRank (m_4 in [6]) follows the same idea than EntityRank and BEntityRank, but including the generalization relationships. Each generalization between ascendants and descendants is viewed as a bidirectional transition, as in the case of associations. Formally:

$$I_{CER}(e) = q_1\frac{attr(e)}{|\mathcal{A}|} + q_2 \sum_{e' \in gen(e)} \frac{I_{CER}(e')}{|gen(e')|} + (1-q_1-q_2) \sum_{e'' \in conn(e)} \frac{I_{CER}(e'')}{|conn(e'')|}$$

One more time, our extension includes the uncovered navigations of the schema rules as bidirectional transitions for the random surfer. The new definition is:

$$\begin{aligned} I^+_{CER}(e) = {} & q_1\frac{attr(e)}{|\mathcal{A}|} + q_2 \sum_{e' \in gen(e)} \frac{I^+_{CER}(e')}{|gen(e')|} \\ & + (1-q_1-q_2)\left(\sum_{e'' \in conn(e)} \frac{I^+_{CER}(e'')}{|conn(e'')|} + \sum_{e''' \in rconn(e)} \frac{I^+_{CER}(e''')}{|rconn(e''')|}\right) \end{aligned}$$

4 Experimental Evaluation

We have implemented the six methods described in the previous section, in both the original and the extended versions. We have then evaluated the methods using two distinct case studies: the osCommerce [9] and the UML metaschema. The original methods have been evaluated with the input knowledge they are able to process: the entity types, attributes, associations and generalization/specialization relationships of the structural schemas.

For the osCommerce, the extended versions have been evaluated with the complete structural schema, and the complete behavioural schema (including event types and their pre/post conditions). The osCommerce schema comprises 346 entity types (of which 261 are event types), 458 attributes, 183 associations, 204 general constraints and derivation rules and 220 pre- and post conditions. For the UML metaschema there is no behavioral schema and therefore we have only used the complete structural schema. The version of the UML metaschema we have used comprises 293 entity types, 93 attributes, 377 associations, 54 derivation rules and 116 general constraints. In the following, we summarize the two main conclusions we have drawn from the study of the result data.

4.1 Correlation between the Original and the Extended Versions

Figure 2 shows, for each method, the results obtained in the original and the extended versions for the osCommerce. The horizontal axis has a point for each of the 85 entity types of the structural schema, ordered descendently by their importance in the original version. The vertical axis shows the importance computed in both versions. The importance has been normalized such that the sum of the importances of all entity types in each method is 100.

As shown in Fig. 2(f) the highest correlation between the results of both versions is for the CEntityRank (r=0.931), closely followed by the BEntityRank (r=0.929). The lowest correlation is for the Weighted Simple Method (r=0.61).

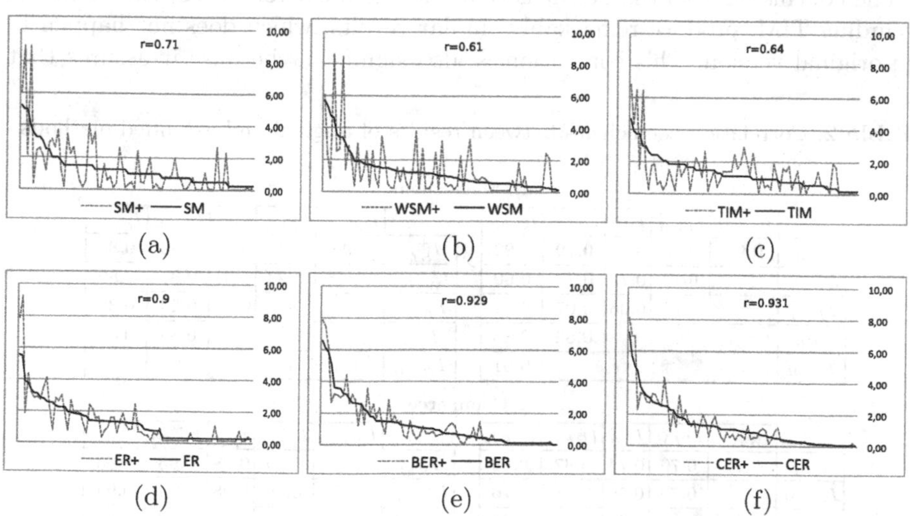

Fig. 2. Comparison between base and extended importance-computing methods once applied to the osCommerce schema

Similar results are obtained for the UML metamodel. In this case the correlation between the two versions of the Weighted Simple Method is 0.84 and that of the CEntityRank is 0.95.

The conclusion from this result is that the method that produces more similar results in both versions is the CEntityRank, followed by the BEntityRank. The conclusion is significant because it implies that if we have to compute the importance of the entity types of a schema, but we only have its attributes, associations and generalization/specialization relationships, the original method that gives results more similar to those that would be obtained in the extended method is the CEntityRank, followed by the BEntityRank. We tend to believe that these are the methods of choice when one wants to compute the relative importance of entity types taking into account the whole schema, but only a fragment of it is available (or only a fragment of it can be processed with the available tools).

This conclusion contrasts with the results reported in [6], which, based on subjective evaluations given by evaluators, concludes that the method that gives the best results is the Simple Method. However, Fig. 2(a) shows that the result given by that method considerably changes when the whole schema knowledge is taken into account.

4.2 Variability of the Original and the Extended Versions

Table 2 shows the correlation between each pair of methods (separately, originals and extended versions), in both case studies. It can be seen that, if we exclude the Transitive Inheritance Method (TIM) because it gives the worst results, the correlation in the original versions of the methods ranges from 0.59 to 0.98, while in the extended versions the range is from 0.83 to 0.99.

The conclusion from this result is that the extended versions of the methods, excluding TIM, produce remarkably similar results, which does not happen in the original version. This conclusion is also significant because it assures that

Table 2. Correlation coefficients between results of original and extended methods

UML Metaschema

	I_{WSM}	I_{TIM}	I_{ER}	I_{BER}	I_{CER}
I_{SM}	0.98	0.15	0.82	0.79	0.92
I_{WSM}		0.16	0.73	0.77	0.90
I_{TIM}			0.06	0.07	0.11
I_{ER}				0.82	0.83
I_{BER}					0.91

	I^+_{WSM}	I^+_{TIM}	I^+_{ER}	I^+_{BER}	I^+_{CER}
I^+_{SM}	0.99	0.26	0.93	0.83	0.86
I^+_{WSM}		0.27	0.91	0.85	0.89
I^+_{TIM}			0.25	0.24	0.30
I^+_{ER}				0.84	0.84
I^+_{BER}					0.91

osCommerce

	I_{WSM}	I_{TIM}	I_{ER}	I_{BER}	I_{CER}
I_{SM}	0.97	0.79	0.74	0.87	0.88
I_{WSM}		0.79	0.59	0.78	0.76
I_{TIM}			0.40	0.54	0.61
I_{ER}				0.94	0.94
I_{BER}					0.97

	I^+_{WSM}	I^+_{TIM}	I^+_{ER}	I^+_{BER}	I^+_{CER}
I^+_{SM}	0.99	0.79	0.98	0.93	0.93
I^+_{WSM}		0.79	0.98	0.94	0.94
I^+_{TIM}			0.78	0.73	0.83
I^+_{ER}				0.94	0.93
I^+_{BER}					0.97

the use of the Simple Method (extended version) whose computational cost is very low, and on the other hand it allows the incremental recalculation of the importance of entity types when the schema changes, produces "good-enough" results.

5 Conclusions and Further Work

The visualization and the understanding of large conceptual schemas require the use of specific methods. These methods generate indexed, clustered, summarized or focused schemas that are easier to visualize and understand. Almost all of these methods require computing the importance of each entity type in the schema. We have argued that the objective importance of an entity type in a schema should be related to the amount of knowledge that the schema defines about it. There are several proposals of metrics for entity type importance. All of them are mainly based on the amount of knowledge defined in the schema, but -surprisingly- they only take into account the fragment of that knowledge consisting on the number of attributes, associations and specialization/generalization relationships. A complete conceptual schema also includes cardinalities, general constraints, derivation rules and the specification of events, all of which contribute to the knowledge of entity types.

We have analyzed the influence of that additional knowledge on a representative set of six existing metrics. We have developed extended versions of each of those metrics. We have evaluated both versions of those methods in two large real-world schemas. The two main conclusions are: (1) Among the original versions of the methods, the methods of choice are those based on the link analysis following the same approach than Google's PageRank; and (2) The extended versions of most methods produce remarkably similar results, which does not happen in the original version.

We plan to continue this work in two main directions. The first, is the experimentation with other large industrial schemas to check whether the above conclusions may have a larger experimental basis. The second, is the extension of the work to other existing metrics.

Acknowledgements. Thanks to the anonymous referees and the people of the GMC group for their useful comments to previous drafts of this paper. This work has been partly supported by the Ministerio de Ciencia y Tecnologia under TIN2008-00444 project, Grupo Consolidado.

References

1. Olivé, A., Cabot, J.: A research agenda for conceptual schema-centric development. In: Krogstie, J., Opdahl, A.L., Brinkkemper, S. (eds.) Conceptual Modelling in Information Systems Engineering, pp. 319–334. Springer, Heidelberg (2007)
2. Lindland, O.I., Sindre, G., Sølvberg, A.: Understanding quality in conceptual modeling. IEEE Software 11(2), 42–49 (1994)

3. Castano, S., Antonellis, V.D., Fugini, M.G., Pernici, B.: Conceptual schema analysis: Techniques and applications. ACM Trans. Database Syst. 23(3), 286–332 (1998)
4. Moody, D.L., Flitman, A.: A Methodology for Clustering Entity Relationship Models – A Human Information Processing Approach. In: Akoka, J., Bouzeghoub, M., Comyn-Wattiau, I., Métais, E. (eds.) ER 1999. LNCS, vol. 1728, pp. 114–130. Springer, Heidelberg (1999)
5. Tzitzikas, Y., Hainaut, J.L.: How to tame a very large ER diagram (using link analysis and force-directed drawing algorithms). In: Delcambre, L.M.L., Kop, C., Mayr, H.C., Mylopoulos, J., Pastor, Ó. (eds.) ER 2005. LNCS, vol. 3716, pp. 144–159. Springer, Heidelberg (2005)
6. Tzitzikas, Y., Kotzinos, D., Theoharis, Y.: On ranking rdf schema elements (and its application in visualization). J. UCS 13(12), 1854–1880 (2007)
7. Yu, C., Jagadish, H.V.: Schema summarization. In: Dayal, U., Whang, K.Y., Lomet, D.B., Alonso, G., Lohman, G.M., Kersten, M.L., Cha, S.K., Kim, Y.K. (eds.) VLDB, pp. 319–330. ACM, New York (2006)
8. Olivé, A.: Conceptual Modeling of Information Systems. Springer, Heidelberg (2007)
9. Tort, A., Olivé, A.: The osCommerce Conceptual Schema. Universitat Politècnica de Catalunya (2007), `http://guifre.lsi.upc.edu/OSCommerce.pdf`
10. Object Management Group (OMG): Unified Modeling Language (UML) Superstructure Specification, version 2.2 (February 2009)
11. Object Management Group (OMG): Object Constraint Language Specification (OCL), version 2.0 (May 2006)
12. Baroni, A.L.: Formal definition of object-oriented design metrics. Master's thesis, Vrije Universiteit Brussel (2002)
13. Brin, S., Page, L.: The anatomy of a large-scale hypertextual web search engine. In: Computer Networks and ISDN Systems, pp. 107–117. Elsevier Science Publishers B. V., Amsterdam (1998)

Preface to ETheCoM 2009

Markus Kirchberg[1] and Klaus-Dieter Schewe[2]

[1] Institute for Infocomm Research, A*STAR, Singapore
Markus.Kirchberg@ieee.org
[2] Information Science Research Centre, Palmerston North, New Zealand
kdschewe@acm.org

The aim of the first international workshop on "Evolving Theories of Conceptual Modelling" (ETheCoM) was to bring together researchers with an interest in theoretical foundations of conceptual modelling. The emphasis is on evolving theories that address mathematical and logical underpinnings of new developments in conceptual modelling, e.g. addressing service-oriented software systems, personalisation of Information Systems and services, network-centric and web-based applications, biomedical applications, games and entertainment, etc. We were interested in precise semantics, in particular with respect to constraints, and in the usage of such semantics for the reasoning and inferencing about model properties.

The call for papers solicited submissions addressing theories of concepts, mathematical semantics of service-oriented systems, integrity constraints maintenance and dependency theory, theoretical foundations of personalisation of Information Systems and services, formal semantics of network-centric and web-based applications, formal methods for data and knowledge base design, reasoning about data and knowledge base dynamics, logical and mathematical models for novel application areas, adaptivity for personalised data and knowledge bases, formal information integration in data and knowledge bases, knowledge discovery in data and knowledge bases, formal linguistics for data and knowledge bases and others.

A total number of 16 full papers was submitted to the workshop. Each paper was reviewed by three or four members of the international programme committee, and we finally selected the five best rated submissions for presentation at the workshop. The workshop programme was further complemented by two invited presentations by Bernhard Thalheim from Christian-Albrechts-University Kiel, Germany, and Edward Hermann Haeusler from Pontifícia Universidade Católica do Rio de Janeiro, Brazil.

We wish to thank all authors of submitted papers and all workshop participants, without whom the workshop would not have been possible. In particular, we would like to thank Edward Hermann Haeusler and Bernhard Thalheim for their willingness to give invited presentations at the workshop. We are grateful to the members of the programme committees for their timely expertise in carefully reviewing the submissions. Finally, our thanks go to the organisers of ER 2009 and the workshop chairs Carlos Heuser and Günther Pernul for giving us the opportunity to organize this workshop.

C.A. Heuser and G. Pernul (Eds.): ER 2009 Workshops, LNCS 5833, p. 33, 2009.

Is It Important to Explain a Theorem? A Case Study on UML and $\mathcal{ALCQI}$

Edward Hermann Haeusler and Alexandre Rademaker

PUC-Rio

{hermann,arademaker}@inf.puc-rio.br

1 Introduction

Description Logics (DL) are quite well-established as underlying logics for KR. $\mathcal{ALC}$ is a basic description logic. ER and UML are among the most used semi-formal artifacts in computer science. The DL-community has shown that one needs to go a bit further to reason on ER and UML models. $\mathcal{ALCQI}$ is able to express most of the features involved in an ER and UML modeling. DL-Lite would also be taken for doing this, although it might be more verbose.

When we define a theory, from ER or UML models, the reasoner should provide understandable explanations in order to facilitate the process of evolving the theory towards its validation or extension. There are some works on explanation in DL, we cite [1,2,3,4] among them. They rely on the proof system implemented by reasoner. Tableaux and Sequent Calculus (SC) are the main proof systems used for. Natural Deduction (ND) is a proof system that tries to naturally represent human mathematical/formal reasoning, at least Gentzen[1]aimed this. Prawitz improved it, on top of Jaskowski's work, characterizing proofs without detours, called *Normal Proofs.* They correspond to Analytic Tableaux and cut-free SC proofs. Howard, after Curry, observed that the procedure that yields normal proofs from non-normal is related to the way typed λ-terms are evaluated. This is the Curry-Howard isomorphism, existing between algorithms and ND proofs. The typed λ-term associated to a ND deduction is taken as its computational content. We believe that the computational content of ND helps in choosing it as the basis to generate adequate explanations on theoremhood in a theory.

We discuss why ND is the most adequate structure to explain theorems and then use a ND for $\mathcal{ALCQI}$ to explain reasoning on an UML model cited by the DL-community. A ND for the core logic $\mathcal{ALC}$ is shown and then extended to $\mathcal{ALCQI}$. In the following we discuss ND Analytic Tableaux and Sequent Calculus as a basis to explanation generation. In section 4 we compare the use of these system in providing explanation on UML reasoning.

[1] First I wished to construct a formalism that comes as close as possible to actual reasoning. Thus arose a "calculus of natural deduction" [5].

C.A. Heuser and G. Pernul (Eds.): ER 2009 Workshops, LNCS 5833, pp. 34–44, 2009.

2 Proofs and Explanations

A deduction of a proposition α from a set of hypothesis Γ is essentially a mean of convincing that Γ entails α. When validating a theory, represented by a set of logical formulas, we use to test entailments, by submitting them to a theorem prover. Each test may result in either a proof/deduction or an "invalid" answer. Depending on the logic being used, the "invalid" answer may not be always provided[2]. Here we are interested in tests providing a false positive answer, that is, the prover shows a deduction/proof for an assertion that must be invalid in the considered Theory. This is one of the main reasons to explain a theorem when validating a Theory, provide explanation on why a false positive is entailed. Another reason to provide explanations on a theorem, has to do with providing explanation on why some assertion is a true positive. This latter use is concerned to certification, in this case the proof/deduction itself serves as a certification document. This article does not take into account educational uses of theorem provers, and their resulting theorems, since explanations in these cases are more demanding.

We compare Analytic Tableaux (AT)[6], Sequent Calculus (SC)[7] and Natural Deduction (ND)[8] as presented in the respective cited references. Because of the lack of space we do not show the set of rules for each system. They are quite well-known and this may not prejudice reader's understanding. In this section we consider the propositional logic (Minimal, Intuitionistic and Classical, as defined in [8]).

Sequent Calculus seems to be the oldest among the three systems here considered. Gentzen decided to move from ND to SC in order to detour from technical problems faced by in his syntactical proof of the consistency of Arithmetic in 1936. As mentioned by Prawitz[8], SC can be understood as a meta-calculus for the deducibility relation in ND. A consequence of this is that ND can represent in only one deduction of α from $\gamma_1, \ldots, \gamma_n$ many SC proofs of the sequent $\gamma_1, \ldots, \gamma_n \Rightarrow \alpha$. Gentzen made SC formally state rules that were implicit in ND, such as the structural rules. We advice the reader that the SC used here (see [7]) is a variation of Gentzen's calculus designed with the goal of having, in each inference rule, any formula occurring in a premise as a subformula of some formula occurring in the conclusion. This subformula property facilitates the implementation of a prover based on this very system.

Consider a normal ND deduction of α from $\gamma_1, \ldots, \gamma_k$, and, a deduction of γ_i (for some $i = 1, k$) from $\delta_1, \ldots, \delta_n$. Using the latter to prove γ_i in the former yields a (possibly non-normal) deduction of α from $\gamma_1, \ldots, \delta_1, \ldots, \delta_n, \ldots, \gamma_k$. This can be done in SC by applying a cut rule between the proofs of the corresponding sequents $\delta_1, \ldots, \delta_n \Rightarrow \gamma_i$ and $\gamma_1, \ldots, \gamma_k \Rightarrow \alpha$ yielding a proof of the sequent $\gamma_1, \ldots, \delta_1, \ldots, \delta_n, \ldots, \gamma_k \alpha$. The new ND deduction can be normalized, in the former case, and the cut introduced in the latter case can be eliminated. In the case of AT, the fact that they are closed by *modus ponens* implies that closed ATs for $\delta \rightarrow \gamma$ and $\gamma \rightarrow \alpha$ entails the existence of a closed AT for $\delta \rightarrow \alpha$.

[2] Theoremhood testing is undecidable for some logics.

The use of cuts, or equivalently, lemmas may reduce the size of a derivation. However, the relevant information conveyed by a deduction or proof in any of this systems has to firstly consider normal deductions, cut-free proofs and analytic Tableaux. They are the most representative formal objects in each of these systems as a consequence of the subformula property, holding in ND too.

Consider the following two derivations (a) and (b) in Sequent Calculus. They both correspond to the same Natural Deduction derivation that is showed in (c). (a) and (b) only differ in the order of rule applications (1) and (2). In ND there is no such distinction . In this example, this order of application is irrelevant in terms of explanation, although it is not for the prover's implementation. Rule applications (3) and (4) can also be permuted producing more (distinct) SC proofs of the same sequent, while their ND corresponding derivation is the same. The pattern represented by this ND deduction is close to what one expects from an argument drawing a conclusion from any conjunction that it contains. The dual pattern of drawing disjunctions from any of their components is also present in this example. This example shows how SC proofs carry more information than the need for a meaningful explanation. Concerning the AT system, Smullyan noted that its proofs correspond to SC proofs by considering sequents formed by positively signed formulas ($T\alpha$) at the antecedent and negatively signed ones ($F\alpha$) appearing at the succedent. A Block AT is defined then by considering AT expansion rules in the form of inference rules. In this way, our example in SC would carry the same content useful for explanation carried by the SC proofs. (d) below shows the Block AT associated to the first proof in SC below. Note that AT also carries out the order of rule applications as one of its inherent features. We must note that different SC proofs and its corresponding AT proofs, as the one shown, are represented, all of them, by the sole ND in (c).

$$\dfrac{\dfrac{\dfrac{\dfrac{A_1, A_2, A_3 \Rightarrow A_4, A_2, A_5}{A_1, A_2, A_3 \Rightarrow A_4, A_2 \vee A_5}\,(1)}{A_1, A_2 \wedge A_3 \Rightarrow A_4, A_2 \vee A_5}\,(2)}{A_1, A_2 \wedge A_3 \Rightarrow A_4 \vee (A_2 \vee A_5)}\,(3)}{A_1 \wedge (A_2 \wedge A_3) \Rightarrow A_4 \vee (A_2 \vee A_5)}\,(4)$$

(a)

$$\dfrac{\dfrac{\dfrac{\dfrac{A_1, A_2, A_3 \Rightarrow A_4, A_2, A_5}{A_1, A_2 \wedge A_3 \Rightarrow A_4, A_2, A_5}}{A_1, A_2 \wedge A_3 \Rightarrow A_4, A_2 \vee A_5}}{A_1, A_2 \wedge A_3 \Rightarrow A_4 \vee (A_2 \vee A_5)}}{A_1 \wedge (A_2 \wedge A_3) \Rightarrow A_4 \vee (A_2 \vee A_5)}$$

(b)

$$\dfrac{\dfrac{\dfrac{\dfrac{A_1 \wedge (A_2 \wedge A_3)}{A_2 \wedge A_3}}{A_2}}{A_2 \vee A_5}}{A_4 \vee (A_2 \vee A_5)}$$

(c)

$$\dfrac{\dfrac{\dfrac{\dfrac{T(A_1 \wedge (A_2 \wedge A_3)), F(A_4 \vee (A_2 \vee A_5)), TA_1, T(A_2 \wedge A_3), FA_4, F(A_2 \vee A_5), TA_2, TA_3, FA_2, FA_5}{T(A_1 \wedge (A_2 \wedge A_3)), F(A_4 \vee (A_2 \vee A_5)), TA_1, T(A_2 \wedge A_3), FA_4, F(A_2 \vee A_5), TA_2, TA_3}}{T(A_1 \wedge (A_2 \wedge A_3)), F(A_4 \vee (A_2 \vee A_5)), TA_1, T(A_2 \wedge A_3), FA_4, F(A_2 \vee A_5)}}{T(A_1 \wedge (A_2 \wedge A_3)), F(A_4 \vee (A_2 \vee A_5)), TA_1, T(A_2 \wedge A_3)}}{T(A_1 \wedge (A_2 \wedge A_3)), F(A_4 \vee (A_2 \vee A_5))}$$

(d)

This example is carried out in Minimal Logic. For Classical reasoning, an inherent feature of most DLs, including $\mathcal{ALC}$, the above scenario changes. Any classical proof of the sequent $\gamma_1, \gamma_2 \Rightarrow \alpha_1, \alpha_2$ corresponds a ND deduction of $\alpha_1 \vee \alpha_2$ from γ_1, γ_2, or, of α_1 from $\gamma_1, \gamma_2, \neg\alpha_2$, or, of α_2 from $\gamma_1, \gamma_2, \neg\alpha_1$, or, of $\neg\gamma_1$ from $\neg\alpha_1, \gamma_2, \neg\alpha_2$, and so on. In Classical[3] logic, each SC may represent more than

[3] Intuitionistic Logic and Minimal Logic have similar behaviours concerning the relationship between their respective systems of ND and SC.

one deduction, since we have to choose which formula will be the conclusion in the ND side. We recall that it still holds that to each ND deduction there is more than one SC proof. In order to serve as a good basis for explanations of classical theorems we choose ND as the most adequate. Note that we are not advocating that the prover has to produce ND proofs directly. An effective translation to a ND might be provided. Of course there must be a ND for the logic involved, together with a proof of Normalization. In the following we present a ND for $\mathcal{ALC}$. Because of our space limits, we cannot show an example illustrating the Classical case. The interested reader can prove the sequent $A_1 \wedge A_2 \rightarrow B \Rightarrow (A_2 \rightarrow B) \vee ((D \wedge A_1) \rightarrow E)$ and compare what is obtained with the corresponding ND deduction of $(A_2 \rightarrow B) \vee ((D \wedge A_1) \rightarrow E)$ from $A_1 \wedge A_2 \rightarrow B$. In section 4 an example illustrating the use of theorem to explain reasoning on UML models is accomplished by proofs in ND , SC and AT.

3 A Natural Deduction for $\mathcal{ALC}$and $\mathcal{ALCQI}$

In this section we present a Natural Deduction (ND) system for $\mathcal{ALC}$, named $N_{\mathcal{ALC}}$. We briefly discuss the motivation and the basic considerations behind the design of $N_{\mathcal{ALC}}$. We sketch completeness , soundness and normalization theorems. The complete proofs of these theorems can be found in http://tecmf.inf.puc-rio.br/reports/DLND.

$\mathcal{ALC}$ is a basic description language [9] and its syntax of concept descriptions, denoted as C, is described as follows:

$$C ::= \bot \mid \top \mid A \mid \neg C \mid C_1 \sqcap C_2 \mid C_1 \sqcup C_2 \mid \exists R.C \mid \forall R.C$$

where A stands for atomic concepts and R for atomic roles. $\top$ can be taken as $\neg\bot$.

The semantics of concept descriptions is defined in terms of an *interpretation* $\mathcal{I} = (\Delta^{\mathcal{I}}, .^{\mathcal{I}})$. The domain $\Delta^{\mathcal{I}}$ of $.^{\mathcal{I}}$ is a non-empty set of individuals and the interpretation function $.^{\mathcal{I}}$ maps each atomic concept A to a set $A^{\mathcal{I}} \subseteq \Delta^{\mathcal{I}}$ and for each atomic role a binary relation $r^{\mathcal{I}} \subseteq \Delta^{\mathcal{I}} \times \Delta^{\mathcal{I}}$. The interpretation function $.^{\mathcal{I}}$ is extended to concept descriptions inductive as follows:

$$\begin{array}{c}
\top^{\mathcal{I}} = \Delta^{\mathcal{I}} \quad \bot^{\mathcal{I}} = \emptyset \quad (\neg C)^{\mathcal{I}} = \Delta^{\mathcal{I}} \setminus C^{\mathcal{I}} \\
(C \sqcap D)^{\mathcal{I}} = C^{\mathcal{I}} \cap D^{\mathcal{I}} \quad (C \sqcup D)^{\mathcal{I}} = C^{\mathcal{I}} \cup D^{\mathcal{I}} \\
(\exists R.C)^{\mathcal{I}} = \{a \in \Delta^{\mathcal{I}} \mid \exists b.(a,b) \in R^{\mathcal{I}} \wedge b \in C^{\mathcal{I}}\} \\
(\forall R.C)^{\mathcal{I}} = \{a \in \Delta^{\mathcal{I}} \mid \forall b.(a,b) \in R^{\mathcal{I}} \rightarrow b \in C^{\mathcal{I}}\}
\end{array}$$

The $N_{\mathcal{ALC}}$ presented in Figure 1 is based on the extension of the $\mathcal{ALC}$ language in which concepts are decorated by two lists of labels, possible empty, the left-side and right-side list.

$$L_L ::= R, L_L \mid \emptyset \qquad L_R ::= R, L_R \mid R(L_L), L_R \mid \emptyset \qquad C ::= {}^{L_L}C^{L_R}$$

where R stands for atomic role names, $\emptyset$ for an empty list, L_L (L_R) for left-side (right-side) list of roles and $R(L_L)$ is an *skolemized* label expression. Only the right-side list of labels can contain skolemized label expressions. That is, labels

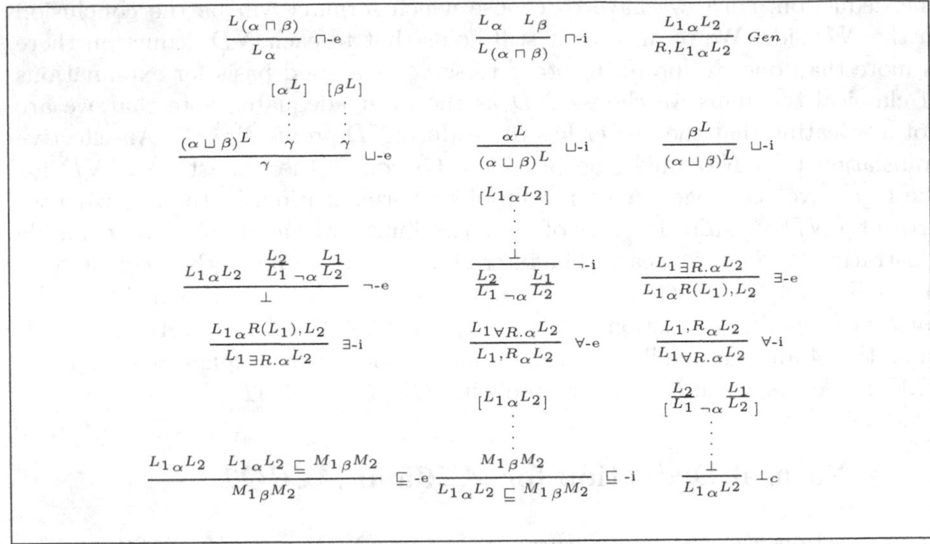

Fig. 1. The Natural Deduction system for $\mathcal{ALC}$

are nothing but role names. We say that a labeled $\mathcal{ALC}$ concept is *consistent* if it has an $\mathcal{ALC}$ concept equivalent. Let $L = \exists R_2.\forall Q_2.\exists R_1.\forall Q_1$ be the roles prefix of $\exists R_2.\forall Q_2.\exists R_1.\forall Q_1.\alpha$. In ${}^{Q_2,Q_1}\alpha^{R_1(Q_2),R_2}$, the prefix L is split into two lists. The left-side list holds the universal quantified roles ($\forall$) and the right-side list holds the existential quantified roles ($\exists$). Each existential quantified role are made dependent from the list of universal quantified roles appearing before it in L[4]. Labels are syntactic artifacts of our system, which means that labeled concepts and its equivalent $\mathcal{ALC}$ have the same semantics.

Consider ${}^{L_1}\alpha^{L_2}$; the notation ${}^{\frac{L_2}{L_1}}\alpha^{\frac{L_1}{L_2}}$ denotes the exchanging of the universal roles occurring in L_1 for the existential roles occurring in L_2 in a consistent way such that the skolemization is dually placed. This is used to express the negation of labeled concepts. If $\beta \equiv \neg\alpha$ the formula ${}^{\frac{Q}{R}}\beta^{\frac{R}{Q(R)}}$ is the negation of ${}^{R}\alpha^{Q(R)}$.

$N_{\mathcal{ALC}}$ was designed to be extended to DLs with role contructors and subsumptions. This is one of the main reasons to use roles-as-labels in its formulation.

Despite the use of labelled formulas, the main non-standard feature of $N_{\mathcal{ALC}}$ is the fact that it is defined on two kind of formulas, namely *concept descriptions* and *subsumptions*. If $\Delta \vdash \Psi$ is an inference rule involving only concept descriptions then it states that providing any DL-interpretation for the premise concepts, if a is an individual belonging to the intersection of interpreted concepts in Δ then it also belongs to the interpreted conclusion. A subsumption $\Phi \sqsubseteq \Psi$ has no concept associated to. It is a truth-value statement, its true depends on whether the interpretation of Ψ includes the corresponding interpretation of Ψ.

[4] When the existential quantified role does not have quantified role before, we write R instead of $R(\emptyset)$ in the right-side list of labels.

The fact that DL has no concept internalizing $\sqsubseteq$ imposes quite particular features on the form of the normal proofs in $N_{\mathcal{ALC}}$.

In rule $\sqsubseteq$-i, ${}^{L_1}\alpha^{L_2} \sqsubseteq {}^{L_1}\beta^{L_2}$ depends only on the assumption ${}^{L_1}\alpha^{L_2}$ and no other. The proviso to rule *Gen* application is that the premise ${}^{L_1}\alpha^{L_2}$ does not depend on any hypothesis. In $\bot_c$-rule, ${}^{L_1}\alpha^{L_2}$ has to be different from $\bot$. Deriving, in $N_{\mathcal{ALC}}$ axioms and rules of the standard axiomatization of $\mathcal{ALC}$ entails.

Theorem 1. *$N_{\mathcal{ALC}}$ is complete regarding the standard semantics of $\mathcal{ALC}$.*

Definition 1. *Let $\Omega = (\mathcal{C}, \mathcal{S})$ be a tuple composed by a set of labeled concepts $\mathcal{C} = \{\alpha_1, \ldots, \alpha_n\}$ and a set of subsumption $\mathcal{S} = \{\gamma_1 \sqsubseteq \delta_1, \ldots, \gamma_k \sqsubseteq \delta_k\}$. We say that an interpretation $\mathcal{I} = (\Delta^{\mathcal{I}}, .^{\mathcal{I}})$ satisfies Ω and write $\mathcal{I} \models \Omega$ whenever: (i) $\mathcal{I} \models \mathcal{C}$, which means $\bigcap_{\alpha \in \mathcal{C}} \alpha^{\mathcal{I}} \neq \emptyset$; and (ii) $\mathcal{I} \models \mathcal{S}$, which means that for all $\gamma_i \sqsubseteq \delta_i \in \mathcal{S}$, we have $\gamma_i^{\mathcal{I}} \subseteq \delta_i^{\mathcal{I}}$.*

We use $\Omega \vdash F$ if exists a deduction Π with conclusion F (concept or subsumption) and all the hypothesis in Ω. This lemma is used to prove theorem 2.

Lemma 1. *Let Π be a deduction in $N_{\mathcal{ALC}}$ of F with all hypothesis in $\Omega = (\mathcal{C}, \mathcal{S})$, then (i) if F is a concept then $\mathcal{S} \models \bigsqcap_{A \in \mathcal{C}} A \sqsubseteq F$; and (ii) if F is a subsumption $A_1 \sqsubseteq A_2$ then $\mathcal{S} \models \bigsqcap_{A \in \mathcal{C}} A \sqcap A_1 \sqsubseteq A_2$.*

Theorem 2. *$N_{\mathcal{ALC}}$ is sound regarding the standard semantics of $\mathcal{ALC}$. if $\Omega \vdash \gamma$ then $\Omega \models \gamma$.*

In proving normalization for $N_{\mathcal{ALC}}$, the reductions for obtaining a normal proof in classical propositional logic also apply to. We follow Prawitz [8] approach incremented by Seldin's [10] permutation rules for the classical absurdity $\bot_c$. Permutation rules move applications of $\bot_c$-rule downwards the conclusion. After this transformation we end up with a proof having in each *branch* at most one $\bot_c$-rule application as the last rule of it. Normalization considers the fragment $\{\neg, \forall, \sqcap, \sqsubseteq\}$. $N^-_{\mathcal{ALC}}$ is $N_{\mathcal{ALC}}$ restricted to it. Any $\mathcal{ALC}$ formula can be rewritten in an equivalent one in $N^-_{\mathcal{ALC}}$.

Proposition 1. *The $N_{\mathcal{ALC}}$ $\sqcup$-rules and $\exists$-rules are derived in $N^-_{\mathcal{ALC}}$.*

We use Prawitz's [8] terminology: formula-tree, deductions or derivations, rule application, minor and major premises, *threads* and so on. Some terminologies have different meaning in our system, we present them in the sequel.

A *branch* in a $N_{\mathcal{ALC}}$ deduction is an initial part $\alpha_1, \alpha_2, \ldots, \alpha_n$ of a thread such that α_n is either (i) the first formula occurrence in the thread that is a minor premise of an application of $\sqsubseteq$-e or (ii) the last formula occurrence of a thread (the end-formula of the deduction) if there is no such premise in it.

Lemma 2 (Moving $\bot_c$ downwards on branches). *If $\Omega \vdash_{N^-_{\mathcal{ALC}}} \alpha$ then there is a deduction Π in $N^-_{\mathcal{ALC}}$ of α from Ω, such that, each branch in Π has at most one application of $\bot_c$-rule, which is the last rule in it.*

A *maximal formula* is a formula occurrence that is consequence of an intro-rule and the major premise of an elim-rule. Let Π be a deduction of α from Ω

containing a maximal formula occurrence F. We say that Π' is a reduction of Π at F if we obtain Π' by removing F using the reductions below. Since F clearly can not be atomic, each reduction refers to a possible principal sign of F. If the principal sign of F is L, then Π' is said to be a L-reduction of Π. In each case, one can easily verify that the Π' obtained is still a deduction of α from Ω.

$\sqcap$-reduction	$\forall$-reduction
$\dfrac{\dfrac{\begin{matrix}\Pi_1 & \Pi_2\\ {}^{L}\alpha & {}^{L}\beta\end{matrix}}{{}^{L}(\alpha \sqcap \beta)}}{{}^{L}\alpha} \quad \rhd \quad \begin{matrix}\Pi_1\\ {}^{L}\alpha\end{matrix}$	$\dfrac{\dfrac{\begin{matrix}\Pi_1\\ {}^{L_1,R}\alpha^{L_2}\end{matrix}}{{}^{L_1}\forall R.\alpha^{L_2}}}{{}^{L_1,R}\alpha^{L_2}} \quad \rhd \quad \begin{matrix}\Pi_1\\ {}^{L_1,R}\alpha^{L_2}\end{matrix}$
$\neg$-reduction	$\sqsubseteq$ -reduction
$\dfrac{\dfrac{\begin{matrix}[{}^{L_1}\alpha^{L_2}]\\ \Pi_1\\ \bot\end{matrix}}{{}^{\frac{L_2}{L_1}}\neg\alpha^{\frac{L_1}{L_2}}} \quad \begin{matrix}\Pi_2\\ {}^{L_1}\alpha^{L_2}\end{matrix}}{\bot} \quad \rhd \quad \begin{matrix}\Pi_2\\ [{}^{L_1}\alpha^{L_2}]\\ \Pi_1\\ \bot\end{matrix}$	$\dfrac{\begin{matrix}\Pi_1\\ \alpha\end{matrix} \quad \dfrac{\begin{matrix}[\alpha]\\ \Pi_2\\ \beta\end{matrix}}{\alpha \sqsubseteq \beta}}{\beta} \quad \rhd \quad \begin{matrix}\Pi_1\\ [\alpha]\\ \Pi_2\\ \beta\end{matrix}$

Lemma 3 (Eliminating maximal $\sqsubseteq$-formulas). *If Π is a deduction of α from Ω, in $N^{-}_{\mathcal{ALC}}$ then there is a deduction Π', in $N^{-}_{\mathcal{ALC}}$ of α from Ω without any maximal $\sqsubseteq$-formulas, i.e., maximal formulas with $\sqsubseteq$ as principal sign.*

A $N^{-}_{\mathcal{ALC}}$ deduction is *normal* when it does not have maximal formulas. Let Π be a deduction in $N^{-}_{\mathcal{ALC}}$. By Lemma 3 we obtain Π' without any maximal $\sqsubseteq$-formulas. Lemma 2 reduces the number of applications of $\bot_c$-rule on each branch and move them downwards to its end. Thus, any deduction in $N^{-}_{\mathcal{ALC}}$ can be normalized in one having no maximal $\sqsubseteq$-formula and at most one $\bot_c$-rule application per branch, the last one. Normal forms facilitate explanations, their pattern are the same for valid subsumptions in any $\mathcal{ALC}$ theory.

Theorem 3 (normalization of $N_{\mathcal{ALC}}$). *If $\Omega \vdash_{N^{-}_{\mathcal{ALC}}} \alpha$, then there is a normal deduction in $N^{-}_{\mathcal{ALC}}$ of α from Ω.*

The sub-formula principle is corollary of Normalization. It states that in a proof of a subsumption $\alpha \sqsubseteq \beta$ from a set Δ of subsumptions, every concept-formula is either a subformula-concept occurring in the conclusion or in some of the subsumptions belonging to Δ.

Since theories must be closed under generalizations, we introduce the following rules in order to reflect this closure, possibly with R or Q omitted.

$$\frac{\alpha \sqsubseteq \beta}{{}^{R}\alpha^{Q} \sqsubseteq {}^{R}\alpha^{Q}}$$

One of the main goals of this article is to show how $N_{\mathcal{ALC}}$ facilitates the reasoning explanation from formal $\mathcal{ALC}$ proofs. To illustrate this in real cases, we will need to move to a more expressive DL. In fact, since the results of [11,12,13,14,2] we know that in order to express ER or UML modeling and reasoning, we have to use $\mathcal{ALCQI}$. It is $\mathcal{ALC}$ with number restrictions and inverse roles.

$$C ::= \bot \mid A \mid \neg C \mid C_1 \sqcap C_2 \mid C_1 \sqcup C_2 \mid \exists R.C \mid \forall R.C \mid \leq nR.C \mid \geq nR.C$$
$$R ::= P \mid P^{-}$$

where A stands for atomic concepts and R for atomic roles. Some of the above operators can be mutually defined: (i) $\bot$ for $A \sqcap \neg A$; (ii) $\top$ for $\neg\bot$; (iii) $\geq kR.C$ for $\neg(\leq k-1R.C)$; (iv) $\leq kR.C$ for $\neg(\geq k+1R.C)$; (v) $\exists R.C$ for $\geq 1R.C$. An $\mathcal{ALCQI}$ theory is a finite set of inclusion assertions of the form $C_1 \sqsubseteq C_2$. The semantics of $\mathcal{ALCQI}$ constructors and theory is analogous to that of $\mathcal{ALC}$.

$$(P^-)^{\mathcal{I}} = \{(a,a') \in \Delta^{\mathcal{I}} \times \Delta^{\mathcal{I}} \mid (a',a) \in P^{\mathcal{I}}\}$$
$$(\leq kR.C)^{\mathcal{I}} = \{a \in \Delta^{\mathcal{I}} \mid \#\{a' \in \Delta^{\mathcal{I}} \mid (a,a') \in R^{\mathcal{I}} \wedge a' \in C^{\mathcal{I}}\} \leq k\}$$

4 An Example on UML and ALCQI

In [11], *DLs* are used to formalize UML diagrams. It uses two DL languages: $\mathcal{DLR}_{ifd}$ or $\mathcal{ALCQI}$. The diagram and its formalization on Fig. 3, are from [11]. We use examples of *DL* deductions in the above mentioned paper on page 84, using $N_{\mathcal{ALCQI}}$ to reasoning on the $\mathcal{ALCQI}$ KB. The idea is to exemplify how one can obtain from $N_{\mathcal{ALCQI}}$ proofs, a more precise and direct explanation.

The first example concerns a refinement of a multiplicity. That is, from reasoning on the diagram, one can deduce that the class `MobileCall` participates on the association `MobileOrigin` with multiplicity 0..1, instead of the 0..* presented in the diagram. The proof on $N_{\mathcal{ALCQI}}$ is as follows, where we abbreviate the class names for their first letters, for instance, `Origin` (`O`), `MobileCall` (`MC`), `call` (`c`) and so on. Note that $\neg \geq 2\texttt{c}^-.\texttt{MO}$ is actually an abbreviation for $\leq 1\texttt{c}^-.\texttt{MO}$.

$$\dfrac{\dfrac{\dfrac{[\geq 2\ \texttt{c}^-.\texttt{MO}]^2 \quad \dfrac{\texttt{MO} \sqsubseteq \texttt{O}}{\geq 2\ \texttt{c}^-.\texttt{MO} \sqsubseteq\ \geq 2\ \texttt{c}^-.\texttt{O}}}{\geq 2\ \texttt{c}^-.\texttt{O}} \qquad \dfrac{\dfrac{\dfrac{[\texttt{MC}]^1 \quad \texttt{MC} \sqsubseteq \texttt{PC}}{\texttt{PC}} \quad \texttt{PC} \sqsubseteq\ \geq 1\ \texttt{c}^-.\texttt{O} \sqcap \leq 1\ \texttt{c}^-.\texttt{O}}{\geq 1\ \texttt{c}^-.\texttt{O} \sqcap \leq 1\ \texttt{c}^-.\texttt{O}}}{\leq 1\ \texttt{c}^-.\texttt{O}}}{\dfrac{\bot}{\neg \geq 2\ \texttt{c}^-.\texttt{MO}}\ 2}}{\texttt{MC} \sqsubseteq \neg \geq 2\ \texttt{c}^-.\texttt{MO}}\ 1$$

To exemplify deductions on diagrams, an incorrect generalization between two classes was introduced. The generalization asserts that each `CellPhone` is a `FixedPhone`, which means the introduction of the new axiom `CellPhone` $\sqsubseteq$ `FixedPhone` in the KB. From that improper generalization, several undesirable properties could be drawn.

The first conclusion about the modified diagram is that `Cellphone` is now inconsistent. The $N_{\mathcal{ALCQI}}$ proof below explicits that from the newly introduced axiom and from the axiom `CellPhone` $\sqsubseteq$ `¬FixedPhone` in the KB, one can conclude that `CellPhone` is now inconsistent.

$$\dfrac{\dfrac{\dfrac{\texttt{Cell} \sqsubseteq \neg\texttt{Fixed} \quad [\texttt{Cell}]^1}{\neg\texttt{Fixed}} \qquad \dfrac{\texttt{Cell} \sqsubseteq \texttt{Fixed} \quad [\texttt{Cell}]^1}{\texttt{Fixed}}}{\bot}}{\texttt{Cell} \sqsubseteq \bot}\ 1$$

The second conclusion is that in the modified diagram, `Phone` $\equiv$ `FixedPhone`. Note that we have only to show that `Phone` $\sqsubseteq$ `FixedPhone` since `FixedPhone` $\sqsubseteq$ `Phone` is an axiom already in the original KB. We can conclude from the proof

$$m \geq n\ \frac{{}^{L,R_m}\alpha^{M}}{{}^{L,R_n}\alpha^{M}}\ Q_{+} \qquad m \leq n\ \frac{{}^{L,R_n}\alpha^{M}}{{}^{L,R_m}\alpha^{M}}\ Q_{-} \qquad m \leq n\ \frac{{}^{L,R^m}\alpha^{M}}{{}^{L,R^n}\alpha^{M}}\ Q^{-}$$

$$m \geq n\ \frac{{}^{L,R^n}\alpha^{M}}{{}^{L,R^m}\alpha^{M}}\ Q^{+} \qquad \frac{{}^{L}\alpha^{R(L),M}}{{}^{L,R_1}\alpha^{M}}\ \mathrm{Q}\exists \qquad n \geq 1\ \frac{{}^{L,R_n}\alpha^{M}}{{}^{L}\alpha^{R(L),M}}\ \mathrm{Q}\exists$$

$$n \geq 1\ \frac{{}^{L,R_n}\alpha^{M}}{{}^{L}\alpha^{R(L),M}}\ \exists\mathrm{Q} \qquad \frac{{}^{L}\alpha^{R(L),M}}{{}^{L,R_1}\alpha^{M}}\ \exists\mathrm{Q} \qquad \frac{{}^{L_1}\!\leq nR.\alpha^{L_2}}{{}^{L_1,R^n}\alpha^{L_2}}\ \leq$$

$$\frac{{}^{L_1,R^n}\alpha^{L_2}}{{}^{L_1}\!\leq nR.\alpha^{L_2}}\ \leq \qquad \frac{{}^{L_1}\!\geq nR.\alpha^{L_2}}{{}^{L_1,R_n}\alpha^{L_2}}\ \geq \qquad \frac{{}^{L_1,R_n}\alpha^{L_2}}{{}^{L_1}\!\geq nR.\alpha^{L_2}}\ \geq$$

$$\frac{{}^{L_1}\alpha^{M_1,R} \sqsubseteq {}^{L_2}\beta^{M_2}}{{}^{L_1}\alpha^{M_1} \sqsubseteq {}^{R^-,M_1}\beta^{M_2'}}\ \mathrm{inv}$$

Fig. 2. The ND system $N_{\mathcal{ALCQI}}$ for $\mathcal{ALCQI}$ (only rules that extend $N_{\mathcal{ALC}}$)

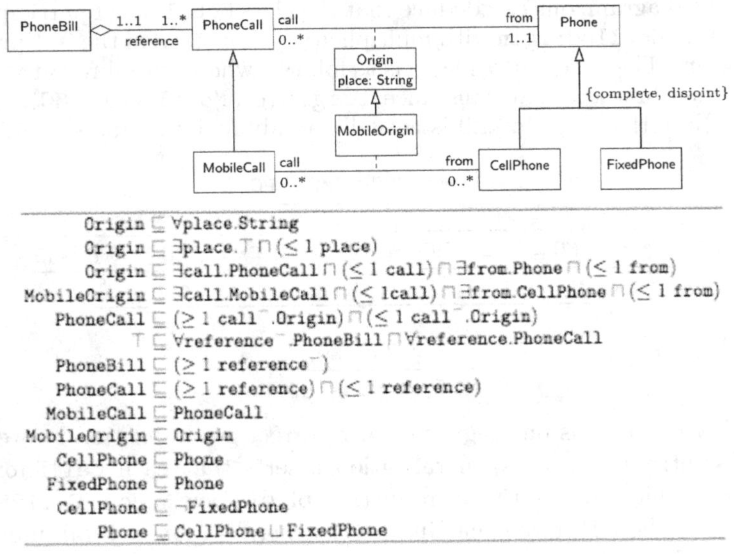

Fig. 3. $\mathcal{ALCQI}$ KB and its corresponding UML diagram

below that `Phone` $\sqsubseteq$ `FixedPhone` is not a direct consequence of `CellPhone` being inconsistent, as stated in [11], but mainly as a direct consequence of the newly introduced axiom and a cases analysis over the possible subtypes of `Phone`.

$$\dfrac{\dfrac{[\texttt{Phone}]^1 \qquad \texttt{Phone} \sqsubseteq \texttt{Cell} \sqcup \texttt{Fixed}}{\texttt{Cell} \sqcup \texttt{Fixed}} \qquad \dfrac{[\texttt{Cell}] \qquad \texttt{Cell} \sqsubseteq \texttt{Fixed}}{\texttt{Fixed}} \qquad [\texttt{Fixed}]}{\dfrac{\texttt{Fixed}}{\texttt{Phone} \sqsubseteq \texttt{Fixed}}\ 1}$$

Below it is shown the above discussed subsumption proved in SC.

$$\dfrac{\dfrac{\dfrac{\dfrac{\dfrac{\mathtt{MO} \Rightarrow \mathtt{O}}{\geq 2\ call^{-}.\mathtt{MO} \Rightarrow \geq 2\ call^{-}.\mathtt{O}}}{\mathtt{MC}, \geq 2\ call^{-}.\mathtt{MO} \Rightarrow \geq 2\ call^{-}.\mathtt{O}} \quad \dfrac{\dfrac{\mathtt{MC} \Rightarrow \mathtt{PC} \quad \mathtt{PC} \Rightarrow \geq 1\ call^{-}.\mathtt{O} \sqcap \leq 1\ call^{-}.\mathtt{O}}{\mathtt{MC} \Rightarrow \geq 1\ call^{-}.\mathtt{O} \sqcap \leq 1\ call^{-}.\mathtt{O}}}{\mathtt{MC}, \geq 2\ call^{-}.\mathtt{MO} \Rightarrow \geq 1\ call^{-}.\mathtt{O} \sqcap \leq 1 call^{-}.\mathtt{O}}}{\mathtt{MC}, \geq 2\ call^{-}.\mathtt{MO} \Rightarrow \geq 1\ call^{-}.\mathtt{O} \sqcap \leq 1 call^{-}.\mathtt{O} \sqcap \geq 2 call^{-}.\mathtt{O}}}{\mathtt{MC}, \geq 2\ call^{-}.\mathtt{MO} \Rightarrow \bot}}{\mathtt{MC} \Rightarrow \neg \geq 2\ call^{-}.\mathtt{MO}}$$

We prove the same $\mathtt{MC} \sqsubseteq \neg \geq 2\ \mathtt{call}^{-}.\mathtt{MO}$ subsumption in AT. First translate the subsumption problem to a satisfiability problem. $C \equiv \mathtt{MC} \sqcap \geq 2\ \mathtt{call}^{-}.\mathtt{MO}$ is already in the NNF (negation normal form), and so, we are ready to the AT algorithm. We must try to construct a finite interpretation $\mathcal{I}$ such that $C^{\mathcal{I}} \neq \emptyset$. $\mathcal{I}_0$ is the initial version of the interpretation. By $\sqcap$ rule, we get $\mathcal{I}_1$. Than, by $\geq n\ R.C$ we get $\mathcal{I}_2$. $\mathcal{I}_3$ is obtained by using the theory axioms $\mathtt{MO} \sqsubseteq \mathtt{O}$ and $\mathtt{MC} \sqsubseteq \mathtt{PC}$ and $\mathcal{I}_4$ by using the theory axiom $\mathtt{PC} \sqsubseteq \geq 1\ call^{-}.\mathtt{O} \sqcap \leq 1\ call^{-}.\mathtt{O}$. Next, $\mathcal{I}_5$ by $\sqcap$ rule. Interpretation $\mathcal{I}_5$ now contains a contradiction, the individual a is required to have at most one sucessor of type $\mathtt{O}$ in the role $call^{-}$. Nevertheless, b and c are also required to be of type $\mathtt{O}$ and sucessors of a in role $call^{-}$, vide $\mathcal{I}_3$ and $\mathcal{I}_2$. This shows that C is unsatisfiable, and thus $\mathtt{MC} \sqsubseteq \neg \geq 2\ \mathtt{call}^{-}.\mathtt{MO}$.

$$\{(\mathtt{MC} \sqcap \geq 2\ \mathtt{call}^{-}.\mathtt{MO})(a)\} \qquad (\mathcal{I}_0)$$

$$\mathcal{I}_0 \cup \{\mathtt{MC}(a), (\geq 2\ call^{-}.\mathtt{MO})(a)\} \qquad (\mathcal{I}_1)$$

$$\mathcal{I}_1 \cup \{call^{-}(a,b), call^{-}(a,c), \mathtt{MO}(b), \mathtt{MO}(c)\} \qquad (\mathcal{I}_2)$$

$$\mathcal{I}_2 \cup \{\mathtt{O}(b), \mathtt{O}(c), \mathtt{PC}(a)\} \qquad (\mathcal{I}_3)$$

$$\mathcal{I}_3 \cup \{(\geq 1\ call^{-}.\mathtt{O} \sqcap \leq 1\ call^{-}.\mathtt{O})(a)\} \qquad (\mathcal{I}_4)$$

$$\mathcal{I}_4 \cup \{(\geq 1\ call^{-}.\mathtt{O})(a), (\leq 1\ call^{-}.\mathtt{O})(a)\} \qquad (\mathcal{I}_5)$$

5 Conclusion

We presented a ND system for $\mathcal{ALC}$ and $\mathcal{ALCQI}$ and showed, by means of some examples, how it can be useful to explain formal facts on theories obtained from UML models. Instead of UML, ER could also be used according a similar framework. Regarding the examples used in this article and the explanations obtained, it is worthwhile noting that the Natural Deduction proofs obtained are quite close to the natural language explanation provided by the authors of the article whose the examples come from. Because of lack of space we cannot provide the respective excerpts for a comparison. This article shows that ND deduction systems are better than Tableaux and Sequent Calculus as structures to be used in explaining theorem when validating theories in the presence of false positives.

References

1. McGuinness, D.: Explaining Reasoning in Description Logics. PhD thesis, Rutgers University (1996)
2. Calvanese, D., et al: DL-Lite: Practical reasoning for rich DLs. In: Proc. of the DL 2004 (2004)

3. Borgida, A., Franconi, E., Horrocks, I., McGuinness, D., Patel-Schneider, P.: Explaining $\mathcal{ALC}$ subsumption. In: Proc. of the Int. Workshop on DLs, pp. 33–36 (1999)
4. Liebig, T., Halfmann, M.: Explaining subsumption in $\mathcal{ALEHF}_{R+}$ tboxes. In: DL 2005, pp. 144–151 (2005)
5. Gentzen, G.: Untersuchungen über das logische Schließen. Math. Z. 39 (1935)
6. Smullyan, R.: First-Order Logic. Springer, Heidelberg (1968)
7. Takeuti, G.: Proof Theory. North-Holland, Amsterdam (1975)
8. Prawitz, D.: Natural deduction: a proof-theoretical study. Almqvist & Wiksell (1965)
9. Baader, F., et al.: The Description Logic Handbook: theory, implementation, and applications. Cambridge University Press, Cambridge (2003)
10. Seldin, J.: Normalization and excluded middle. I. Studia Logica 48(2), 193–217 (1989)
11. Berardi, D., Calvanese, D., De Giacomo, G.: Reasoning on UML class diagrams. Artificial Intelligence 168(1-2), 70–118 (2005)
12. Calvanese, D., Lenzerini, M., Nardi, D.: Description logics for conceptual data modeling. In: Logics for Databases and Information Systems, pp. 229–263. Kluwer, Dordrecht (1998)
13. Calvanese, D., et al.: Information integration: Conceptual modeling and reasoning support. In: Proc. of, CoopIS 1998 (1998)
14. Calvanese, D., et al.: Conceptual modeling for data integration. In: Borgida, A.T., et al. (eds.) Mylopoulos Festschrift. LNCS, vol. 5600, pp. 173–197. Springer, Heidelberg (2009)

Towards a Theory of Conceptual Modelling

Bernhard Thalheim

Christian Albrechts University Kiel, Department of Computer Science
Olshausenstr. 40, D-24098 Kiel, Germany
thalheim@is.informatik.uni-kiel.de

Abstract. Conceptual modelling is a widely applied practice and has led to a large body of knowledge on constructs that might be used for modelling and on methods that might be useful for modelling. It is commonly accepted that database application development is based on conceptual modelling. It is however surprising that only very few publications have been published on a *theory of conceptual modelling*.

Modelling is typically supported by languages that are well-founded and easy to apply for the description of the application domain, the requirements and the system solution. It is thus based on a *theory of modelling constructs*. At the same time, modelling incorporates a description of the application domain and a prescription of requirements for supporting systems. It is thus based on methods of *application domain gathering*. Modelling is also an engineering activity with engineering steps and engineering results. It is thus *engineering*. The first facet of modelling has led to a huge body of knowledge. The second facet is considered from time to time in the scientific literature. The third facet is underexposed in the scientific literature.

This paper aims in developing principles of conceptual modelling. They cover modelling constructs as well as modelling activities as well as modelling properties and as well as management of models. We first clarify the notion of conceptual modelling. Principles of modelling may be applied and accepted or not by the modeler. Based on these principles we can derive a theory of conceptual modelling that combines foundations of modelling constructs, application capture and engineering.

A general theory of conceptual modelling is far too comprehensive and far too complex. It is not yet visible how such a theory can be developed. This paper therefore aims in introducing a framework and an approach to a general theory of conceptual modelling. We are however in urgent need of such a theory. We are sure that this theory can be developed and use this paper for the introduction of the main ingredients of this theory.

1 Introduction

The main purpose of conceptual modelling is classically understood as the elicitation [3,6,10] of a high-quality conceptual schema of a (software, information, workflow, ...) system. This understanding mainly concentrates on the result of conceptual modelling and hinders the development of a general theory of conceptual modelling. Modelling is based on languages which might be sophisticated

C.A. Heuser and G. Pernul (Eds.): ER 2009 Workshops, LNCS 5833, pp. 45–54, 2009.

[3] and well understood [10] like the ER modelling language or might be fuzzy with lazy semantics development like the UML. Let us analyse the complexity of the modelling task and then let us draw some conclusions for the modelling "act".

The Three Dimensions of Conceptual Modelling. Conceptual modelling is often only discussed on the basis of modelling constructs and illustrated by some small examples. It has however three fundamental dimensions:

1. *Modelling language constructs* are applied during conceptual modelling. Their syntactics, semantics and pragmatics must be well understood.
2. *Application domain gathering* allows to understand the problems to be solved, the opportunities of solutions for a system, and the requirements and architecture that might be prescribed for the solution that has been chosen.
3. *Engineering* is oriented towards encapsulation of experiences with design problems pared down to a manageable scale.

The first dimension is handled and well understood in the literature. Except few publications, e.g. [1], the second dimension has not yet got a sophisticated and well understood support. The third dimension has received much attention by data modelers [8] but did not get through to research literature. It must therefore be our goal to combine the three dimensions into a holistic framework.

Implications for a Theory of Conceptual Modelling. The three dimensions of conceptual modelling must be integrated into a framework that supports the relevant dimension depending on the modelling work progress. The currently most difficult dimension is the engineering dimension. Engineering is inherently concerned with failures of construction, with incompleteness both in specification and in coverage of the application domain, with compromises for all quality dimensions, and with problems of technologies currently at hand.

At the same time, there is no universal approach and no universal language that cover all *aspects of an application*, that have a well-founded *semantics* for all constructions, that reflect any *relevant facet in applications*, and that *support engineering*. Models are at different levels of abstraction and *particularisation*. We therefore are going to develop *a number of different models* that reflect different aspects of the system that is under development. [11] introduces *model suites* as a set of models with explicit *associations* among the models, with explicit *controllers* for maintenance of coherence of the models, with application schemata for their explicit *maintenance and evolution*, and tracers for establishment of their *coherence*. Model suites and multi-model specification increases the complexity of modelling. Interdependencies among models must be given in an explicit form. Models that are used to specify different views of the same problem or application must be used consistently in an integrated form. Changes within one model must be propagated to all dependent models. Each singleton model must have a well-defined semantics as well as a number of representations for display of model content. The representation and the model must be tightly coupled. Changes within any model, must either be refinements of previous models or explicit revisions of such models. The change management must support rollback to earlier versions of the model suite. Model suites may have different versions.

Quality Assessment, Control and Improvement. According to [4] the result of conceptual modelling depends on *information available* about the UoD; on information about the UoD, regarded as *not relevant* for the concept or conceptual model at hands, and therefore abandoned or renounced; on the *philosophical background* to be applied in the modelling work; on the *additional knowledge* included by the modeler, e.g. knowledge primitives, conceptual 'components', selected logical or mathematical presuppositions, mathematical structures, etc.; on the *collection of problems* that may be investigated in this environment; on the *ontology* (or better language with its lexical semantics [7]) used as a basis of the conceptualization process; in the *epistemological theory*, which directs how ontology should be applied in recognizing and formulating concepts, conceptual models or theories, and in constructing information, data, and knowledge, on different levels of abstraction; on the *the purpose and goal of the conceptual modelling work*; on the collection of *methods for conceptual modelling*; on the *process of the practical concept formation and modelling work*; and finally on the *knowledge and skill of the person making modelling*, as well as those of the people giving information for the modelling work.

Quality properties are

- *static qualities* of models such as the development quality (pervasiveness, analysability, changeability, stability, testability, privacy of the models, ubiquity), internal quality (accuracy, suitability, interoperability, robustness, self-contained, independence), and quality of use (understandability, learnability, operability, attractiveness, appropriateness), and
- *dynamic qualities* within a selected development approach (executability, refinement quality, scope restriction, effect preservation, context explicitness, completion tracking).

The information system modelling process is intentionally or explicitly ruled by a number of development strategies, development steps, and development policies. Modelling steps lead to new specifications to which quality criteria can be applied. Typical quality criteria are completeness and correctness in both the syntactical and semantical dimensions. We assume that at least these four quality criteria are taken into consideration.

Outline and Tasks of the Paper. The paper aims in introducing a theory of conceptual modelling. We first describe modelling as a task. Then we show that a theory of conceptual modelling consists of a number of sub-theories that can be developed separately and that can be integrated. This approach leads to a framework for method development for conceptual modelling. We may develop a number of properties that allow to judge on the quality of the models. It is not our intention to present a fully developed theory. Instead we propose a path for the development of such a theory.

It has not been our intention to survey the modelling literature. This would be far too large already for the conceptual modelling research. Good sources to this research are [3,6,10].

Sample Application: *Traffic light control.* Given a crossroad, e.g. consisting of two streets (north-south, east-west) with an intersection and of traffic lights that direct traffic. We assume at the first glance that traffic lights might switch from red to green and from green to red. We also might assume that both opposite cross lights show the same colour. Classical approaches typically start with a model for each cross light. Next the interdependence among the state changes is either modelled through integrity constraints or through implicit modelling constructs. The well-known Petri net solution is depicted in Figure 1.

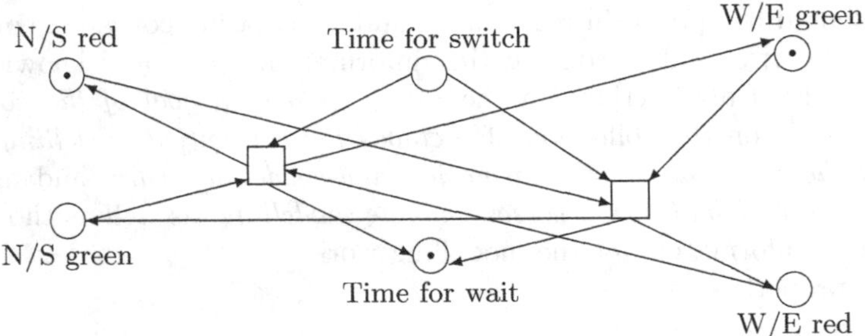

Fig. 1. Traffic control based on Petri nets

This model neither scales nor has a good development, internal, or dynamic quality. The extension to yellow colour is an intellectual challenge as well the extension to more flexible directing. The main reason for the poor quality and the conceptual and implementation inadequacy is its wrong attitude, wrong scope, wrong abstraction, and wrong granularity.

2 Modelling

The Notion of Model in Science Theory. Information system models are typically representations (how specified) of certain application solutions (origin, whereof) for a community of users (whom), for certain application goals and intentions (for what), within a certain time span (when), and with certain restrictions (normal, exception and forbidden cases).

A model represents subjects or things

- Based on an *analogy* of structuring, functionality, or behaviour,
- Considering certain *application purposes*, and
- Providing a simple handling or *service* or consideration of the things under consideration.

The model definition given is one [9] of many options. A model has typically a *model capacity*:

· The model provides some understanding of the original;
· The model provides an explanation of demonstration through auxiliary information and thus makes original subject easier or better to understand;
· The model provides an indication and facilities for making properties viewable;
· The model allows to provide variations and support optimisation;
· The model support verification of hypotheses within a limited scope;
· The model supports construction of technical artifacts;
· The model supports control of things in reality;
· The model allows a replacement of things of reality and acts as a mediating means.

Typically these functions are used simultaneously and in competition. Therefore *to model* means different activities at the same time: to plan or form after a pattern or shape, to make into an organization (as an army, government, or parish), to produce a representation or simulation to model a problem, and to construct or fashion in imitation of a particular model.

This competition of meanings results in a number of *problems* of conceptual modelling such as competing attitudes and profiles of modelers, varieties of styles of specification, multi-model reasoning, and integration into a general coherent model ensemble (or model suite). Therefore we face the *"grand challenge of harmonisation"*.

Models are typically given by the triple (original, image mapping) that is extended by properties of the image, of the mapping, of the system under consideration, that are based on a common modelling "culture" or understanding, and depends on the aim of the model. Therefore we envision that modelling can be considered as an art similar to the 'art of programming'.

The "Act" of Modelling. *Modelling* typically means the construction of models which can be used for detection of insights or for presentation of perceptions of systems. Modelling is typically based on languages and thus has a semiotic foundation.

The act of modelling consists of

1. a selection and construction of an appropriate model depending on the task and purpose and depending on the properties we are targeting and the context of the intended system and thus of the language appropriate for the system,
2. a workmanship on the model for detection of additional information about the original and of improved model,
3. an analogy conclusion or other derivations on the model and its relationship to the real world, and
4. a preparation of the model for its use in systems, to future evolution and to change.

The Modelling Gap. Modelling is inherently incomplete, biased and ruled by scoping by the initiators of a project, by restricting attention to parts of the application that are currently under consideration and ruling out any part of the application that will never be under consideration. This intentional restriction is typically not communicated and directly results in the "modelling gap" [5]. Additionally, modelling culture results in different models and different understandings.

Incompleteness of specifications is caused by incomplete knowledge currently available, incomplete coverage of the specification or by inability to represent the knowledge in the application. Incompleteness may be considered as the main source of the modelling gap beside culture and skills of modelers. The application is either partially known, or only partially specified, or cannot be properly specified. To overcome the problems of specifications we may either use

- *negated specifications* that specify those cases which are not valid for the application,
- *robust specifications* that cover the main cases of the applications, or
- *approximative specifications* that cover the application on the basis of control parameters and abstract from order parameters.

Principles of Abstraction. The development of a model is *the* result of modelling. It relates things $\mathcal{D}$ under consideration with concepts $\mathcal{C}$. This relationship $\mathcal{R}$ is characterised by restrictions ρ to its applicability, by a modality θ or rigidity of the relationship, and by the confidence Ψ in the relationship. The model is agreed within a group $\mathcal{G}$ and valid in a certain world $\mathcal{W}$. Stachowiak [9] defined three characteristic properties of models: the *mapping* property (have an original), *truncation* property (the model lacks some of the ascriptions made to the original), and *pragmatic* property (the model use is only justified for particular model users, tools of investigation, and period of time). We can additionally consider the *extension* property. The property allows that models represent judgments which are not observed for the originals. In computing, for example, it is often important to use executable models. Finally, the *distortion* property is often used for improving the physical world or for inclusion of visions of better reality.

These principles result typically result in *forgetful mappings* from the origin to the model.

Sample Application. First we decide whether the analogy to real-life is based on the behaviour of the entire system or on the combined behaviour of the behaviour of components. This distinction directly implies a choice between a model that represents the entire application as one system and the components as its elements (*local-as-view model*) and a model that combines local models to a global one (*global-as-view model*). All conceptual solutions known in literature use the global-as-view model and results in very complex models.

We might prefer the local-as-view approach. States reflect the entire state of the crossroad, i.e. *NSredEWgreen, NSredEWred, NSgreenEWred.* The last state reflects that the north-south direction is open and the east-weast direction is closed. We might add the state *NSredEWred* for representation of the exception state and the state *NSnothingEWnothing* for the start and the end state. The state *NSgreenEWgreen* is a conflict state and thus not used for the model.

3 Constituents of a Theory of Conceptual Modelling

Next we highlight main constituents of a theory of conceptual modelling. It is surprising that literature mainly covers only the first one.

Theory of Modelling Concepts. Modelling concepts are elements of a certain language for the representation (r) of things (t) under consideration, with restrictions for their applicability (a), with a rigidity or modality (m), with a confidence (c) on their validity, based on a common understanding of a group (g) within their world (w) or culture. We therefore may represent the result of modelling by a tuple (r, t, a, m, c, g, w). The group may use its reference models or modes. The theory of conceptual modelling constructs for object-relational database applications can be based entirely on the extended ER model [10].

Theory of Modelling Activities. Modelling activities are based on modelling acts. Modelling is a specific form and we may thus develop workflows of modelling activities. These workflows are based on work steps [10] such as 'decompose' or 'extend', abstraction and refinement acts, validation and verification, equivalences of concepts, transformation techniques, pragmatistic solutions and last but not least the domain-specific solutions and languages given by the application and implementation domains.

Theory of Properties of Modelling Activities. It is often neglected that models have their properties. We therefore need a reasoning, evaluation, control and management facility for providing an insight into the model quality itself.

Model Management. The development of a holistic model that covers all but all aspects of an application overburdens modelling languages and overloads cognitive skills of modelers. Therefore, we may separate different aspects and concerns and model those reduced tasks to separate models. These models must however be harmonised and integrated. Therefore, modelling must allow one to reason on model ensembles and their coherence.

Goals and Portfolio of Modelling. Models are different for different purposes. We may develop a model for analysis of an application domain, for construction of a system, for communicating about an application, for assessment, and for governance. These different purposes result in different goals and task portfolios.

Results of Modelling. The conceptual schemata are typically considered to be *the* result of conceptual modelling. We may however have different equivalent representations of the same schema, a documentation of the entire development process and the reasons for development decisions, on the scope and restrictions, and last but not least on the bindings among the schemata. The last result offers an opportunity for evolution.

The Fundamental Structural Relations. The five fundamental structural relations used for construction abstraction are aggregation/participation, generalisation/specialisation, exhibition/characterisation, classification and instantiation, and separation between introduction and utilisation.

Aggregation/participation characterizing which object consists of which object or resp. which object is part of which object. Aggregation is based on constructors such as sets, lists, multisets, trees, graphs, products etc. It may include naming. *Generalizeation/specialization* characterizing which object generalizes which object or resp. which object specializes which object. Hierarchies may be defined through different classifications and taxonomies. So, we may have a different hierarchy for each point of view. Hierarchies are built based on inheritance assumptions. So, we may differentiate between generalization and specialization in dependence on whether characterization are not or are inherited and on whether transformation are or are not applicable. *Exhibition/characterization* specifying which object exhibits which object or resp. which object is characterized by which object. Exhibitions may be multi-valued depending of the data type used. They may be qualitative or quantitative. *Classification/instantiation* characterizing which object classifies which object or resp. which object is an instance of which object. *Introduction/utilisation* allows to distinguish between an introduction of an object, the shared or exclusive utilisation and the finalisation of an object.

Sample Application. The local-as-view model in Figure 2 is based on a *two-layer architecture* that uses a global schema and local view schemata. We explicitly specify properties and binding among the global and local schemata, e.g. master-slave binding. State changes and the pedestrian calls are not recorded after they have been issued. The scheduler is based on this schema and might use workflow diagrams, trigger rules or ASM rules [2] for specification of BPMN diagrams. We can use a generic pattern approach that supports extensions, e.g. for kinds of states and kinds of state changes. Typical rules are the following:

ChangeAction := getState; choosePossibleStateChange(state);
apply(possibleStateChange(state)
AlarmAction := `on` alarm changeStateToErrorState
Clock := `on` tick observeWhetherChangeRequired
NormalAction := `if` change = true then ChangeAction
PedestrianCall := `on` callAtPoint(cp) ChangeNextStepIssuedAt(cp).

In a similar form we specify views for local display.

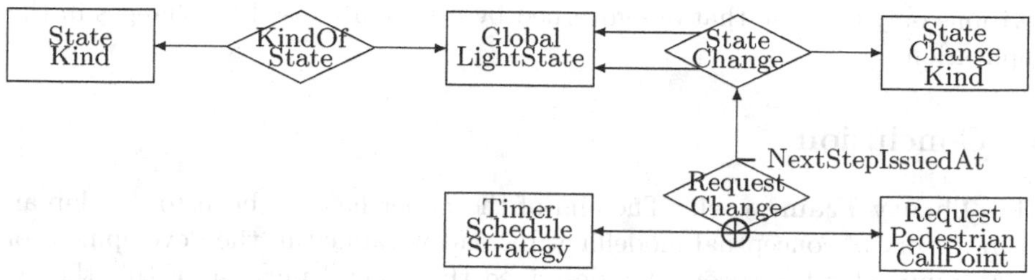

Fig. 2. The traffic light support database schema

4 Application of the Framework to Conceptual Modelling

Methods of Conceptual Modelling. The theory of conceptual modelling is based on a small number of methods. The main methods are *abstraction, modularisation generalisation/refinement, transformations, selection and application of modelling styles,* and *separation of concern*. Abstraction and refinement are well understood. Modularisation is based on an architectural decomposition of a large model into components and a development of a linking or binding scheme for the separated components. Typically, conceptual modelling only considers one transformation technique for the mapping among layers, e.g. from conceptual to logical schemata. The mapping technique and the mapping rules may however vary depending on the goals. Separation of concern allows to provide a clear understanding of parts of the application.

Properties of Conceptual Modelling. We are interested in a general guidance for the entire modelling process and in a management of all models that supports coherence among the models. The theory of conceptual modelling discussed above should support a selection of a *general modelling strategy*. Layered conceptual modelling is one of the well-known strategies. It is based on modularisation, abstraction and refinement. We decompose a system into components and support a consideration of models in various abstractions and details.

The theory of conceptual modelling may also be used for a selection and development of an assembly of modelling styles. Typical well-known styles [10] are inside-out refinement, top-down refinement, bottom-up refinement, modular refinement, and mixed skeleton-driven refinement. These different kinds of refinement styles allow one to derive *plans* for refinement and and *primitives* for refinement.

The introduction of strategies and styles allow to provide a general support for maintenance of results and qualities achieved so far and for restricting the scope of a change in modelling. We therefore may develop a strategy that has a number of *properties* such as monotonicity of accepted results, incrementality of changes depending only on the most recent solution, finiteness of checks for any criterion requested for the modelling process, application domain consistency without additional reference to later steps, and conservativeness that restricts revisions only to those that are governed by errors or caused by changes in the application.

5 Conclusion

The Theory Framework. The aim of the paper has not been to develop an entire theory of conceptual modelling. Instead we aimed in the development of a programme for the theory. We described the general purpose of this theory, demonstrated how different paradigms can be selected, and showed which scope, modelling acts, modelling methods, modelling goals and modelling properties might be chosen for this theory.

Future Work. The programme requires far more work. The theory needs a variable taxonomy that allows a specialisation to languages chosen for a given application domain, must be based on a mathematical framework that allows to prove properties, must be flexible for coping with various modelling methodologies, must provide an understanding of the engineering of modelling, and finally should be supported by a meta-CASE tool that combines existing CASE to to a supporting workbench.

References

1. Bjørner, D.: Domain engineering. COE Research Monographs, vol. 4. Japan Advanced Institute of Science and Technology Press, Ishikawa (2009)
2. Börger, E., Thalheim, B.: A method for verifiable and validatable business process modeling. In: Börger, E., Cisternino, A. (eds.) Advances in Software Engineering. LNCS, vol. 5316, pp. 59–115. Springer, Heidelberg (2008)
3. Chen, P.P., Akoka, J., Kangassalo, H., Thalheim, B. (eds.): Conceptual modeling: current issues and future directions. LNCS, vol. 1565. Springer, Heidelberg (1999)
4. Kangassalo, H.: Approaches to the active conceptual modelling of learning. In: Chen, P.P., Wong, L.Y. (eds.) ACM-L 2006. LNCS, vol. 4512, pp. 168–193. Springer, Heidelberg (2007)
5. Kaschek, R.: Konzeptionelle Modellierung. PhD thesis, University Klagenfurt, Habilitationsschrift (2003)
6. Olivé, A.: Conceptual modeling of information systems. Springer, Berlin (2007)
7. Schewe, K.-D., Thalheim, B.: Semantics in data and knowledge bases. In: Schewe, K.-D., Thalheim, B. (eds.) SDKB 2008. LNCS, vol. 4925, pp. 1–25. Springer, Heidelberg (2008)
8. Simsion, G.: Data modeling - Theory and practice. Technics Publications, LLC, New Jersey (2007)
9. Stachowiak, H.: Modell. In: Seiffert, H., Radnitzky, G. (eds.) Handlexikon Zur Wissenschaftstheorie, pp. 219–222. Deutscher Taschenbuch Verlag GmbH & Co. KG, München (1992)
10. Thalheim, B.: Entity-relationship modeling – Foundations of database technology. Springer, Berlin (2000)
11. Thalheim, B.: The conceptual framework to multi-layered database modelling. In: Prc. EJC, Maribor, Slovenia, pp. 118–138 (2009)

Assessing Modal Aspects of OntoUML Conceptual Models in Alloy

Alessander Botti Benevides, Giancarlo Guizzardi, Bernardo F.B. Braga, and João Paulo A. Almeida

Ontology and Conceptual Modeling Research Group (NEMO), Computer Science Department, Federal University of Espírito Santo (UFES), Av. Fernando Ferrari, n 514, Goiabeiras, Vitória, ES, Brazil
{abbenevides,gguizzardi,bfbbraga}@inf.ufes.br, jpalmeida@ieee.org

Abstract. Assessing the quality of conceptual models is key to ensure that conceptual models can be used effectively as a basis for understanding, agreement and construction of information systems. This paper proposes an approach to assess conceptual models defined in OntoUML by transforming these models into specifications in the logic-based language Alloy. These Alloy specifications include the modal axioms of the theory underlying OntoUML, allowing us to validate the modal meta-properties of the OntoUML types and part-whole relations.

1 Introduction

John Mylopoulos [1] defines conceptual modeling as "the activity of formally describing some aspects of the physical and social world around us for purposes of understanding and communication". In this view, a conceptual model is a means to represent what modelers (or stakeholders represented by modelers) perceive in some portion of the physical and social world, *i.e.*, a means to express their conceptualization [2] of a certain universe of discourse.

If conceptual models are to be used effectively as a basis for understanding, agreement, and, perhaps, construction of an information system, conceptual models should express as accurately as possible a modeler's intended conceptualization. More specifically, the model should ideally describe all states of affairs that are deemed *admissible* and rule out those deemed *inadmissible* according to the conceptualization [2].

As argued for in [2], the quality of a conceptual modeling language can be assessed by considering the extent to which the language supports the definition of models that capture this intended conceptualization. This concern has justified the revision of a portion of UML into a conceptual modeling language named OntoUML. This revision enables modelers to make finer-grained distinctions between different types of classes and different types of part-whole relations according to the UFO foundational ontology [2].

More specifically, this revision introduces modal meta-properties for object classifiers that enable one to distinguish between rigid, semi-rigid and anti-rigid

C.A. Heuser and G. Pernul (Eds.): ER 2009 Workshops, LNCS 5833, pp. 55–64, 2009.

classifiers (and therefore distinguish properties that apply necessarily to objects from those that apply contingently) as well as meta-properties for part-whole relations to distinguish between mandatory, essential, inseparable and immutable parts, and immutable wholes [2].

Although the language revision impacts on a modeler's ability to express the intended conceptualization accurately, one would certainly be naive to assume that modelers make no mistakes while constructing the models and that they fully understand the theory that supports the language. These cases could lead to ill-defined conceptual models, which may be: (i) syntactically incorrect; (ii) syntactically correct, but unsatisfiable; (iii) syntactically correct, satisfiable, but invalid according to the intended conceptualization.

Previous efforts in addressing the assessment of OntoUML models have focussed on syntactic correctness and led to the specification of OntoUML's syntactical constraints as OCL expressions on the language's metamodel [3] and the building of a graphical editor [3] that is capable of automatic syntax verification. In this paper, we go beyond syntax verification and aim at addressing the satisfiability and validity of OntoUML models. More specifically, we discuss an approach based on formal specifications in the logic-based language Alloy [4] to generate instances of an OntoUML model with the purpose of showing that the model is satisfiable and improving the modeler's confidence in the validity of the model.

In our approach, the Alloy specification is fed into the Alloy Analyzer to generate a branching-time Kripke world structure [5] that reveals the possible dynamics of object creation, classification, association and destruction. Each world in this structure is an instance of the OntoUML model and represents a snapshot of the objects and relations that exist in that world. This world structure is necessary since the meta-properties characterizing most of the ontological distinctions in UFO are modal in nature. (For example, the definition of a "rigid" classifier states that it applies necessarily to its instances in all worlds in which they exist.) We have specified UFO's modal axioms in Alloy to guarantee that the generated world structure satisfies these axioms by construction. Therefore, the sequence of possible snapshots in this world structure will support claims of model satisfiability and improve our confidence on claims of validity.

This paper is further structured as follows. Section 2 presents the modal aspects of UFO's object types and part-whole relations and presents our running example; section 3 briefly describes the suitable Kripke world structures and their representation in Alloy; section 4 exemplifies a world structure generated by the Alloy Analyzer; section 5 discusses related work and, finally, section 6 presents our final considerations.

2 OntoUML Concepts

Due to space limitations, we concentrate here on a fragment of the Unified Foundation Ontology (UFO) [2], with a specific focus on those distinctions that are spawned by variations in meta-properties of a modal nature. UFO's main

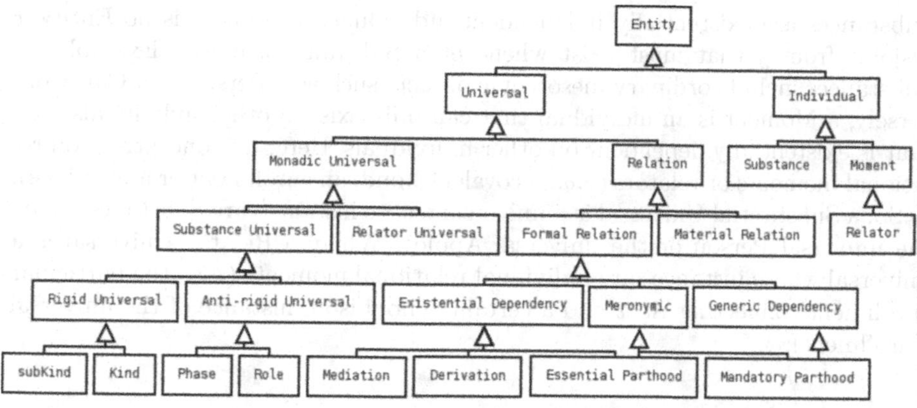

Fig. 1. Excerpt of UFO taxonomy [2]

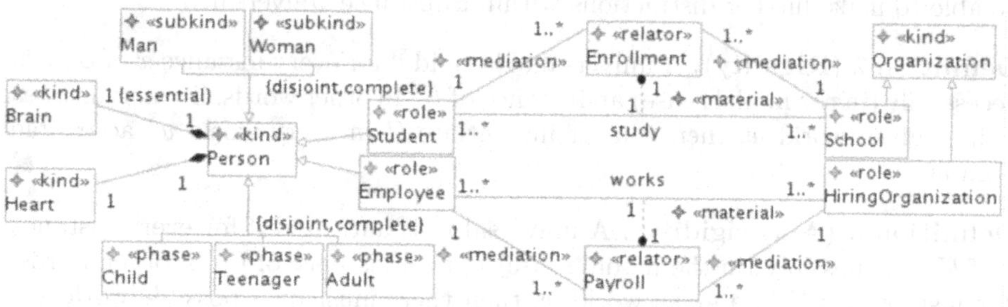

Fig. 2. Running example

categories are depicted in Fig. 1 below and are briefly discussed in the remainder of this section by using a running example depicted in Fig. 2. Since OntoUML is a modeling language whose metamodel is designed to be isomorphic to the UFO ontology, the leaf ontological distinctions in Fig. 1 appear as modeling primitives in the language (see stereotyped classes and relationships in Fig. 2).

2.1 Substances and Moments

UFO is based on a fundamental distinction between Individuals and Universals (roughly instances and types, respectively) and, within the category of individuals, it differentiates between Substances and Moments. The distinction between Substances and Moments is based on the formal notion of existential dependence, a modal notion that can be briefly defined as follows:

Definition 1 (existential dependence): an individual x is *existentially dependent* on another individual y iff, as a matter of necessity, y must exist whenever x exists. In other words, in every world w, if x exists in w then y must also exist in w. ■

Substances are existentially independent individuals, *i.e.*, there is no Entity x disjoint from y that must exist whenever a Substance y exists. Examples of Substances include ordinary mesoscopic objects such as a Person or a Car. Conversely, a Moment is an individual that can only exist in other individuals, *i.e.*, that is existentially dependent on other individuals. Here, we concentrate on relational moments or relators (*e.g.*, a covalent bond, an enrollment or a marriage).

So, a Substantial Universal is a universal whose instances are Substances (*e.g.*, the universal Person or the universal Apple). While, a Relator Universal is a universal whose instances are individual relational moments (*e.g.*, the particular enrollment connecting *Alex* and a certain School is an instance of the universal Enrollment).

2.2 Substance Universals

We need to define some additional modal notions (rigidity and anti-rigidity) to be able to make further distinctions within Substance Universal.

Definition 2 (Rigidity): A universal U is rigid if for every instance x of U, x is necessarily (in the modal sense) an instance of U. In other words, if x instantiates U in a given world w, then x must instantiate U in every world w' accessible from w. ■

Definition 3 (Anti-rigidity): A universal U is anti-rigid if for every instance x of U, x is possibly (in the modal sense) not an instance of U. In other words, if x instantiates U in a given world w, then there must be a possible world w', accessible from w, in which x does not instantiate U. ■

Substantial Universals that are rigid are named Kinds and subKinds. These universals define a stable backbone, a taxonomy of rigid universals instantiated by a given individual (the Kind being the unique top-most rigid universal instantiated by an individual).

Within the category of anti-rigid substantial universals we have a further distinction between Phases and Roles. Both Phases and Roles are specializations of rigid universals (Kinds/subKinds). However, they are differentiated w.r.t. their specialization conditions. For the case of Phases, the specialization condition is always an intrinsic one. For instance, in Fig. 2, a child is a Person whose age is within a certain range. For Roles, in contrast, their specialization condition is a relational one: a Student is a Person who is enrolled in (has a study relation to) a School, *etc.* Formally speaking, this distinction is based on a meta-property named Relational Dependence:

Definition 4 (Relational Dependence): A type T is relationally dependent on another type P via relation R iff in every world w, for every instance x of T there is an instance y of P in that world such that x and y are related via R in w. ■

Finally, as discussed in [2], Phases (in contrast to Roles) are always defined in a partition set. For instance, in Fig. 2, the universals Child, Teenager and Adult define a phase partition for the Kind Person. As consequence, we have that in an each world w, every Person is either a Child, a Teenager or an Adult in w and never more than one of these. Additionally, if x is a Child (Teenager, Adult) in w, there is always a possible world w', accessible from w, in which x will not be a Child, in which case he will be either a Teenager or an Adult.

In summary, in the example of Fig. 2, these model distinctions are exemplified by contrasting the (Kind) universal Person, the (Role) universal Student and the (Phase) universal Teenager.

2.3 Relator Universals and Relations

In order to represent the relation between Student and Person, one should model Student as a Role played by Person in a certain context, where he is enrolled in a School. Analogously, one should model School as a Role played by an Organization when providing educational services to a Student. This context is materialized by the Material Relation *study* (represented as the «material» stereotype in OntoUML), which is in turn, derived from the existence of the Relator Universal Enrollment («relator»). In other words, we can say that a particular student x studies at a particular school y iff there is an Enrollment z that mediates x and y. This situation is illustrated in Fig. 2. The formal relations of mediation in this model represents the existential dependence of the relator on its bearers [2].

Once more, we concentrate here on two different types of part-whole relations that are distinguished based on modal meta-properties, in particular, the previously defined notions of Existential and Relational Dependence. As one can observe contrasting the Definitions 1 and 4, the former is a relation between two individuals, whilst the latter is a relation between types. An Essential Parthood relation is a parthood relation that implies existential dependence. Contrariwise, a Mandatory Parthood relation is one that implies (generic) relational dependence (where the relation R defined in Definition 4 is a relation of formal parthood). These two types of relations are exemplified in Fig. 2 by the relations Brain-Person and Heart-Person, respectively. An Essential Parthood relation between the universals Brain and Person implies that: for every x instance of Person there is an individual y instance of Brain such that x cannot exist without having that specific Brain as a part (*i.e.*, y cannot change from world to world). The mandatory parthood between the universals Heart and Person instead implies that: for every x instance of Person, x cannot exist without a (generic) instance of Heart as a part.

3 Representing Modality in Alloy

Our approach is based on the transformation of OntoUML into Alloy. The product of this transformation is an Alloy specification that can be fed into the Alloy Analyzer to generate a Kripke structure and its associated worlds that together respect UFO's (modal) axioms. This allow us to show that an OntoUML model is semantically consistent (satisfiable). Furthermore, we believe that the analysis of a well-chosen set of these structures (*e.g.*, structures that exhibit important behavior of model's instances) can improve the modeler's confidence in the validity of the model.

We represent modality explicitly in the generated Alloy specification. This means that this specification reifies the notion of Kripke world structure and explicitly manipulates the accessibility relations between worlds. This is necessary to specify UFO's modal axioms, given that no notion of modality is built-in in Alloy.

Our intention is to represent a possible worlds structure in which the accessibility relations between worlds represents the common sense temporal structure. In our ordinary language, we are able to talk about the present, the past, the possible future, and the facts that could have happened, but accidentally did not (*i.e.*, the counterfactuals). So, in this work, we consider a Kripke structure that is able to handle all these notions.

The transformation we have implemented maps OntoUML classes to Alloy signatures and OntoUML binary relationships to Alloy ternary relations declared inside signatures. For example, an OntoUML relation R that relates A to B is mapped to an Alloy relation $R = < A, B, World >$, where the third field denotes the worlds in which the relationship exists. The characteristics of the OntoUML classes (*e.g.*, rigidity, anti-rigidity) and the ones of the relationships (*e.g.*, cardinality constraints, shareability, existential dependency and disjointness) are mapped to constraints in Alloy.

Listing 1 shows an excerpt of the Alloy specification that was automatically generated from the model depicted in Fig. 2. This specification illustrates the representation of the OntoUML class Person in Alloy.

Listing 1. Alloy model excerpt

```
1 sig Person_Set in Concept { Person: some World }
2 {
3    Person in existsIn
4    all w1: World | w1 in Person => (all w2:
          w1.access | (w2 in existsIn) => (w2 in
          Person)) -- Rigidity
5    some w: World | w in this.Child -- Phase
6    some w: World | w in this.Teenager -- Phase
7    some w: World | w in this.Adult -- Phase
8    ⋮
9 }
```

This excerpt shows the specification of the Rigidity of the Kind Person (line 4) and the possibility of a Person to be an instance of Child, Teenager or Adult (lines 5, 6 and 7, respectively).

4 Generating Instances

Fig. 3 shows the Kripke structure and associated worlds that are generated by the Alloy Analyzer for the model shown in Fig. 2. This structure demonstrates the satisfiability of the model and exemplifies some of its dynamic aspects, assisting the user in the validation process.

In the following, we will briefly explain the sequence of events pictured in this Kripke structure: (i) in the first moment (PastWorld2), there is a child (*Alex*) and an organization; (ii) in the second moment (PastWorld1), *Alex* studies in that organization, which plays the role of a school, and there is a second organization; (iii) in the third moment (PastWorld0), *Alex* becomes a teenager, still studying in the same school, but now also an employee (trainee) of the second organization, which plays the role of a hiring organization; (iv) in a counterfactual moment just after PastWorld0, *Alex* has undergone a heart transplant and becomes a healthy adult who works for the same organization; he no longer studies; (v) in the current moment (CurrentWorld), *Alex* is dead; (vi) FutureWorld1 depicts the possibility of both organizations continuing to exist in the future, while (vii) FutureWorld0 depicts the possibility that one of them no longer exists.

This Kripke structure exemplifies some important constraints like the rigidity of the Kind Person exemplified by *Alex* (he never ceases to be an instance of Person while he exists); the anti-rigidity of the Phases Child, Teenager and Adult (for every world w in which *Alex* is in one of these Phases, there is a world w', accessible from w, in which *Alex* is not in that Phase); the anti-rigidity of the Roles Student and Employee (for every world w in which *Alex* plays one of these Roles, there is a world w', accessible from w, in which *Alex* does not play that Role); the relational dependence of the Roles Student and Employee (*Alex* can only play these Roles while related to a school by an enrollment (in the case he is a student), or related to a hiring organization by a payroll (in the case he is an employee)); as well as some well known conceptual modeling primitives, such as abstractness (of Person) (instances of Person have to be instances of Man or Woman), disjointness and completeness (of Man and Woman; and Child, Teenager and Adult).

Also, this Kripke structure illustrates the existential dependence of a person to his/her brain (*Alex* never changes his brain), depicted by the "essential" tag in the relationship between Person and Brain (Fig. 2). One can notice that in the counterfactual world, *Alex* changed his heart (maybe he underwent a heart transplant that saved his life). This behaviour is totally acceptable, as *Alex* is generically dependent on the «kind» Heart.

5 Related Work

Several approaches in literature aim at assessing whether conceptual models comply with their intended conceptualizations. Although many approaches (*e.g.*, [6] and [7]) focus on analysis of behavioural UML models, we are primarily concerned with structural models and thus refrain further analysis of behavioural-focused work.

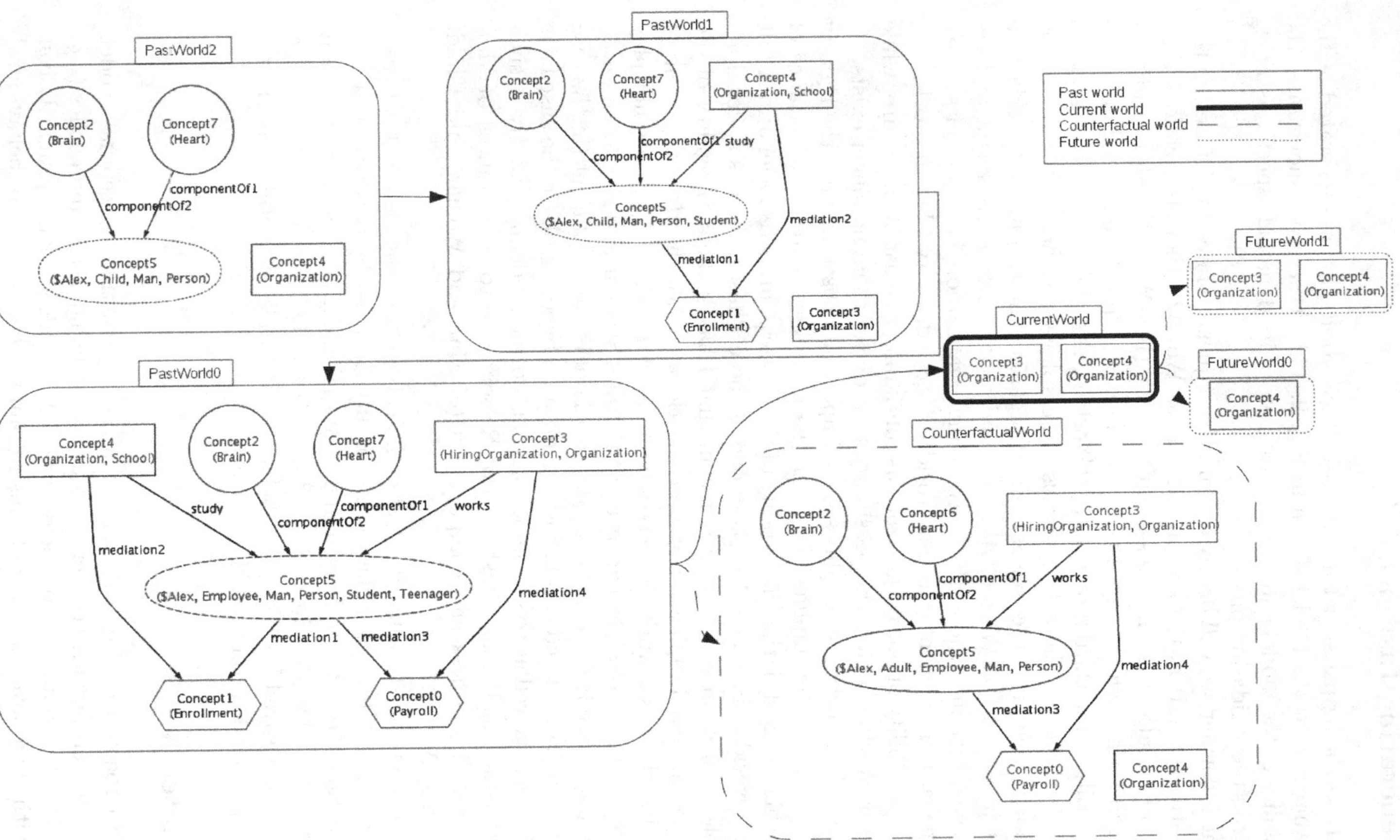

Fig. 3. Visual representation of an example world structure for the OntoUML conceptual model pictured in Fig. 2

A prominent example is the USE (UML Specification Environment) tool proposed in [8]. The tool is able to indicate whether instances of a UML class diagram respect constraints specified in the model through OCL. Differently from our approach, which is based on the automatic creation of example world structures, in USE the modeler must specify sequences of snapshots in order to gain confidence on the quality of the model (either through the user interface or by specifying sequences of snapshots in a tool-specific language called ASSL, A Snapshot Sequence Language). Since no modal meta-property of classifiers is present in UML, this tool does not address modal aspects and validates constraints considering only a sole snapshot.

Finally, the approaches of [9] and [10] are similar to ours in that they translate UML class diagrams to Alloy. However, both of them translate all classes into Alloy signatures, which suggests that no dynamic classification is possible in these approaches. Similarly to our approach, [10] implements a model transformation using model-driven techniques to automatically generate Alloy specifications, while [9] relies on manual translation to Alloy. Similar to USE, [9] focuses on analysis and constraint validation on single snapshots. [10] introduces a notion of state transition but still does not address the modal aspects of classes since these are not part of UML.

6 Final Considerations

A mature approach to conceptual modeling requires modelers to gain confidence on the quality of the models they produce, assessing whether these models express as accurately as possible an intended conceptualization. This paper contributes to that goal, by providing tools to validate the modal properties of a conceptual model in OntoUML.

Following a model-driven approach, we have defined and automated a transformation of OntoUML models into Alloy specifications. The generated Alloy specifications are fed into the Alloy Analyzer to create world structures that show the possible dynamics of object creation, classification, association and destruction as defined in the model. The snapshots in this world structure confront a modeler with states-of-affairs that are deemed admissible by the model. This enables modelers to detect unintended states-of-affairs and take the proper measures to rectify the model. We believe that the example world structures support a modeler in the validation process, especially since it reveals how state-of-affairs evolve in time and how they may eventually evolve (revealing alternative scenarios implied by the model.)

The generated Alloy specification is correct by construction such that it reflects UFO's modal axioms. As a consequence, any world structure created by the Alloy Analyzer respects UFO's modal axioms and shows that the model is satisfiable. If the Alloy Analyzer fails to find an example world structure, this may indicate unsatisfiability, although no guarantee of unsatisfiability is given. This is a consequence of Alloy's choices to cope with tractability. For instance, Alloy searches for example structures within a restricted context, *i.e.*, a given finite maximum number of elements.

As future work, we intend to incorporate support for domain constraints in our approach, *e.g.*, including OCL constraints in an OntoUML model. This will require transforming these constraints into Alloy in order to guarantee that the constraints are satisfied in all instances generated by the Analyzer.

Further, we intend to work on methodological support for the validation process, proposing guidelines for modelers to select relevant world structures. We will aim for an interactive approach in which a modeler can select which of the alternative scenarios to consider. We believe that this may help pruning the branches in the world structure keeping the size of this structure manageable.

Ideally, by exploring visualization techniques, we could use the instances generated by Alloy as example scenarios to be exposed to the stakeholders of the conceptual model (such as domain experts) in order to validate whether their conceptualization has been captured accurately by the modeler.

Acknowledgments. This work has been supported by CNPq, FAPES (INFRA-MODELA) and FACITEC (MODELA). The authors also thank Kyriakos Anastasakis for his support and suggestions.

References

1. Mylopoulos, J.: Conceptual Modeling and Telos. In: Conceptual Modeling, Databases, and CASE: An Integrated View of Information Systems Development. Wiley, Chichester (1992)
2. Guizzardi, G.: Ontological foundations for structural conceptual models. PhD thesis, University of Twente, Enschede, The Netherlands, Enschede (October 2005)
3. Benevides, A.B., Guizzardi, G.: A model-based tool for conceptual modeling and domain ontology engineering in ontouml. In: Filipe, J., Cordeiro, J. (eds.) ICEIS 2009. LNBIP, vol. 24, pp. 528–538. Springer, Heidelberg (2009)
4. Jackson, D.: Software abstractions: logic, language, and analysis. MIT Press, Cambridge (2006)
5. Hughes, G.E., Cresswell, M.J.: A Companion to Modal Logic. Routledge and Kegan Paul, London (1985)
6. Beato, M.E., Barrio-Solórzano, M., Cuesta, C.E.: UML automatic verification tool (TABU). In: SAVCBS 2004 Specification and Verification of Component-Based Systems at ACM SIGSOFT 2004/FSE-12 (2004)
7. Schinz, I., Toben, T., Mrugalla, C., Westphal, B.: The rhapsody uml verification environment. In: SEFM 2004: Proceedings of the Software Engineering and Formal Methods, Second International Conference, Washington, DC, USA, pp. 174–183. IEEE Computer Society, Los Alamitos (2004)
8. Gogolla, M., Büttner, F., Richters, M.: Use: A uml-based specification environment for validating uml and ocl. Science of Computer Programming 69, 27–34 (2007)
9. Massoni, T., Gheyi, R., Borba, P.: A uml class diagram analyzer. In: 3rd International Workshop on Critical Systems Development with UML, affiliated with 7th UML Conference, pp. 143–153 (2004)
10. Maintainers: UML2Alloy. Project website: `http://www.cs.bham.ac.uk/~bxb/UML2Alloy`

First-Order Types and Redundant Relations in Relational Databases

Flavio A. Ferrarotti[1,2,*], Alejandra L. Paoletti, and José M. Turull Torres[3]

[1] Yahoo! Research Latin America, Santiago, Chile
flavio@dcc.uchile.cl

[2] Departamento de Ingeniería Informática, Facultad de Ingeniería, Universidad de Santiago de Chile

[3] School of Engineering and Advanced Technology, College of Sciences, Massey University, Wellington, New Zealand

Abstract. Roughly, we define a *redundant relation* in a database instance (dbi) as a k-ary relation R such that there is a first-order query which evaluated in the reduced dbi, gives us R. So, we can eliminate that relation R as long as the equivalence classes of the relation of equality of the first-order types for all k-tuples in the dbi are not altered. It turns out that in a fixed dbi, the problem of deciding whether a given relation in the dbi is redundant is *decidable*, though intractable. We then study redundant relations with a restricted notion of equivalence so that the problem becomes *tractable*.

1 Introduction

From a conceptual point of view it is desirable for a model of computation of queries to be *representation independent*. This means, roughly, that queries to databases (in the present work we will refer to *database instances* simply as *databases*) which represent the "same" reality should evaluate to the "same" result. In mathematical terms, Chandra and Harel [3] captured the previous concept by asking queries to isomorphic databases to evaluate to the same result. The principle of preservation of isomorphisms has an important consequence if we consider a single database, namely the preservation of automorphisms. That is, considering a fixed database, two elements with the same "structural" properties should be considered as undistinguishable. By structural properties we roughly mean the way in which the two elements are related to all other elements in the database, by means of the different relations according to the schema. The same is also true for tuples of elements, i.e., two tuples with the same "structural" properties should be considered as undistinguishable. To formalize this concept we can make use of the model theoretic notion of type. The notion of type of a tuple is a topic which has been deeply studied in the context of finite model theory [4,11], but which has not received the same attention in the context of database theory. Roughly, if $\mathcal{L}$ is a logic, the $\mathcal{L}$ *type of a tuple* of length k in a given database is the set of $\mathcal{L}$ formulas with up to k free variables which

* Corresponding author.

C.A. Heuser and G. Pernul (Eds.): ER 2009 Workshops, LNCS 5833, pp. 65–74, 2009.

are satisfied by that tuple in the database. As databases are finite structures, it follows that two arbitrary tuples have the same first-order type if and only if they are commutable by some automorphism. So, two arbitrary tuples have the same "structural" properties and should be considered undistinguishable, if and only if, they have the same first-order type.

Redundant storage of information can lead to a variety of practical problems on the updating, insertion and deletion of data. This anomaly is usually known as the redundancy problem and has been studied extensively in the field of databases. Traditionally, the redundancy problem is studied by considering a particular class of properties, the functional dependencies, that are supposed to be satisfied by all instances of a given database. By taking a quite different approach, we will make use of the model theoretic concept of type to study the redundancy problem. Specifically, we initiate in this work the study of a sort of redundancy problem revealed by what we call redundant relations. Roughly, we define a *redundant relation* as a relation R such that there is a first-order query which evaluated in the *reduced database*, gives us R. So, we can eliminate that relation R as long as the equivalence classes of the relation of equality of the first-order types for all k-tuples in the database are not altered. In practical terms, this means that we do not lose information if we eliminate such redundant relation from a database. It turns out that in a fixed database of some relational schema, the problem of deciding whether a given relation in the database is redundant is *decidable*, though intractable. We then study redundant relations with a restricted notion of equivalence so that the problem becomes *tractable*.

We also give the construction of a formula in polynomial time which, provided that R is a redundant relation in the database, will evaluate to R in the reduced database.

Note that the problem of deciding whether a given relation (schema) is redundant in a given class of databases is clearly not decidable in the general case.

The outcome of this research can be of a great relevance to applications like census databases, where we have a huge and stable database instance of a very large schema, and where by eliminating redundant relations we can save an important amount of space and time in the evaluation of queries. We aim to follow this research towards defining a kind of normal form for database instances.

2 Preliminaries

We define a *relational database schema*, or simply *schema*, as a set of relation symbols with associated arities. We do not allow constraints in the schema, and we do not allow constant symbols neither. If $\sigma = \langle R_1, \ldots, R_s \rangle$ is a schema with arities $r_1, \ldots, r_s$, respectively a *database instance* or simply *database* over the schema σ, is a structure $\mathcal{I} = \langle D^{\mathcal{I}}, R_1^{\mathcal{I}}, \ldots, R_s^{\mathcal{I}} \rangle$ where $D^{\mathcal{I}}$ is a finite set which contains exactly all elements of the database, and for $1 \leq i \leq s$, $R_i^{\mathcal{I}}$ is a relation of arity r_i, i.e., $R_i^{\mathcal{I}} \subseteq (D^{\mathcal{I}})^{r_i}$. We will often use $\mathrm{dom}(\mathcal{I})$ instead of $D^{\mathcal{I}}$. We will use $\simeq$ to denote isomorphism. A *k-tuple* over a database $\mathcal{I}$, with $k \geq 1$, is a tuple of length k formed with elements from $\mathrm{dom}(\mathcal{I})$. We will denote a k-tuple of $\mathcal{I}$ as $\bar{a}_k$, and also as $\bar{a}$. We use $\mathcal{B}_\sigma$ to denote the class of all databases of schema σ.

Computable Queries: In this paper, we will consider *total* queries only. Let σ be a schema, let $r \geq 1$, and let R be a relation symbol of arity r. A *computable query of arity r and schema* σ ([3]), is a total recursive function $q^r : \mathcal{B}_\sigma \to \mathcal{B}_{\langle R \rangle}$ which preserves isomorphisms such that for every database $\mathcal{I}$ of schema σ, $\mathrm{dom}(q(\mathcal{I})) \subseteq \mathrm{dom}(\mathcal{I})$. We denote the class of computable queries of schema σ as $\mathcal{CQ}_\sigma$, and $\mathcal{CQ} = \bigcup_\sigma \mathcal{CQ}_\sigma$.

Finite Model Theory and Databases: As usual in finite model theory, we will regard a logic as a language, that is, as a set of formulas (see [5]). We will only consider signatures, or vocabularies, which are purely *relational*. We will always assume that the signature includes a symbol for equality. We consider *finite* structures only. Consequently, if $\mathcal{L}$ is a logic, the notion of *equivalence* between structures or databases, denoted as $\equiv_\mathcal{L}$, will be related to only finite structures. If $\mathcal{L}$ is a logic and σ is a signature, we will denote as $\mathcal{L}_\sigma$ the class of formulas from $\mathcal{L}$ with signature σ. A *database schema* will be regarded as a *relational signature*, and a *database instance* of some schema σ as a finite and relational *σ-structure*. By $\varphi(x_1, \ldots, x_r)$ we denote a formula of some logic whose free variables are *exactly* $\{x_1, \ldots, x_r\}$. Let $\mathrm{free}(\varphi)$ be the set of free variables of the formula φ. If $\varphi(x_1, \ldots, x_k) \in \mathcal{L}_\sigma$, $\mathcal{I} \in \mathcal{B}_\sigma$, $\bar{a}_k = (a_1, \ldots, a_k)$ is a k-tuple over $\mathcal{I}$, let $\mathcal{I} \models \varphi(x_1, \ldots, x_k)[a_1, \ldots, a_k]$ denote that φ is TRUE, when interpreted by $\mathcal{I}$, under a valuation v where for $1 \leq i \leq k$ $v(x_i) = a_i$. Then we consider the set of all such valuations as follows: $\varphi^\mathcal{I} = \{(a_1, \ldots, a_k) : a_1, \ldots, a_k \in \mathrm{dom}(\mathcal{I}) \wedge \mathcal{I} \models \varphi(x_1, \ldots, x_k)[a_1, \ldots, a_k]\}$. Sometimes, we use the same notation when the set of free variables of the formula is *strictly* included in $\{x_1, \ldots, x_k\}$. We denote as FO^k with some integer $k \geq 1$ the fragment of first-order logic (FO) where only formulas whose variables are in $\{x_1, \ldots, x_k\}$ are allowed. In this setting, FO^k itself is a logic. This logic is obviously *less expressive* than FO. We denote as C^k the logic which is obtained by adding to FO^k *counting quantifiers*, i.e., all existential quantifiers of the form $\exists^{\geq m} x$ with $m \geq 1$. Informally, $\exists^{\geq m} x(\varphi)$ means that there are at least m *different* elements in the database which satisfy φ.

Types: Given a database $\mathcal{I}$ and a k-tuple $\bar{a}_k$ in $\mathrm{dom}(\mathcal{I})^k$, we would like to consider *all* properties of $\bar{a}_k$ in the database $\mathcal{I}$ including the properties of every component of the tuple and the properties of all different sub-tuples of $\bar{a}_k$. Therefore, we use the notion of *type*. Let $\mathcal{L}$ be a logic. Let $\mathcal{I}$ be a database of some schema σ and let $\bar{a}_k = (a_1, \ldots, a_k)$ be a k-tuple over $\mathcal{I}$. The $\mathcal{L}$ *type* of $\bar{a}_k$ in $\mathcal{I}$, denoted $tp^\mathcal{L}_\mathcal{I}(\bar{a}_k)$, is the set of formulas in $\mathcal{L}_\sigma$ with free variables *among* $\{x_1, \ldots, x_k\}$ such that every formula in the set is TRUE when interpreted by $\mathcal{I}$ for any valuation which assigns the i-th component of $\bar{a}_k$ to the variable x_i, for every $1 \leq i \leq k$. In symbols $tp^\mathcal{L}_\mathcal{I}(\bar{a}_k) = \{\varphi \in \mathcal{L}_\sigma : \mathrm{free}(\varphi) \subseteq \{x_1, \ldots, x_k\} \wedge \mathcal{I} \models \varphi[a_1, \ldots, a_k]\}$. The following is a well known result.

Proposition 1. *For every schema σ and for every pair of (finite) databases $\mathcal{I}$, $\mathcal{J}$ of schema σ the following holds: $\mathcal{I} \equiv_{FO} \mathcal{J}$ iff $\mathcal{I} \simeq \mathcal{J}$.*

Although types are infinite sets of formulas, due to results of A. Dawar [4] and M. Otto [10], a *single* FO^k (C^k) formula is equivalent to the FO^k (C^k) type of a tuple over a given database. The equivalence holds for all databases of the same schema.

Proposition 2. *([4,10]): For every schema σ, for every database $\mathcal{I}$ of schema σ, for every $k \geq 1$, for every $1 \leq l \leq k$, and for every l-tuple $\bar{a}_l$ over $\mathcal{I}$, there is an FO^k formula $\chi \in tp_{\mathcal{I}}^{FO^k}(\bar{a}_l)$ and a C^k formula $\phi \in tp_{\mathcal{I}}^{C^k}(\bar{a}_l)$, such that for any database $\mathcal{J}$ of schema σ and for every l-tuple $\bar{b}_l$ over $\mathcal{J}$, $\mathcal{J} \models \chi[\bar{b}_l]$ iff $tp_{\mathcal{I}}^{FO^k}(\bar{a}_l) = tp_{\mathcal{J}}^{FO^k}(\bar{b}_l)$ and $\mathcal{J} \models \phi[\bar{b}_l]$ iff $tp_{\mathcal{I}}^{C^k}(\bar{a}_l) = tp_{\mathcal{J}}^{C^k}(\bar{b}_l)$.*

Moreover, such formulas χ and ϕ can be built inductively for a given database. If an FO^k formula χ (C^k formula ϕ, respectively) satisfies the condition of Proposition 2, we call χ an *isolating formula* for $tp_{\mathcal{I}}^{FO^k}(\bar{a}_l)$ (ϕ an *isolating formula* for $tp_{\mathcal{I}}^{C^k}(\bar{a}_l)$, respectively).

Remark 1. Isolating formulas for the FO types of k-tuples can be built in a similar way to that used to build the isolating formulas for FO^k types and C^k types. Considering the formulas $\varphi_{\bar{u}}^m(\bar{x})$, defined in Theorem 2.2.8 in [5], as we are dealing with finite structures there will always be an m big enough such that for all σ-structures $\mathcal{B}$ and k-tuples $\bar{v}$ over $\text{dom}(\mathcal{B})^k$ we have that $\mathcal{B} \models \varphi_{\mathcal{A},\bar{u}}^m[\bar{v}]$ iff $tp_{\mathcal{A}}^{FO}(\bar{u}) = tp_{\mathcal{B}}^{FO}(\bar{v})$, and that is the *isolating formula* for the FO type of $\bar{u}$ in $\mathcal{A}$. It is well known (see [5]) that $n+1$ is a value of m big enough to build the isolating formula for an arbitrary k-tuple in a given database of size n. The size of these formulas is exponential in n. However, for FO types there are other isolating formulas, built from the so called *diagram* of the database, which are of size polynomial in n (see Fact 1 below).

3 Databases with Redundant Relations

The fundamental observation which leads to our definition of redundant relation is that, as the FO types of all k-tuples in a database $\mathcal{A}$ describe all FO properties which are satisfied by the tuples of arity up to k in $\mathcal{A}$, every FO query of arity up to k will be equivalent in $\mathcal{A}$ to the disjunction of some of the FO isolating formulas for the FO types for k-tuples in $\mathcal{A}$. Thus, we could eliminate a relation $R^{\mathcal{A}}$ of arity k from $\mathcal{A}$ as long as the relationship among the FO types of the different k-tuples in $\mathcal{A}$ is not altered.

Definition 1. *Let σ be a relational schema, let $\mathcal{A}$ be a database of schema σ, and let R_i be a given relation symbol in σ. We denote as $\sigma - R_i$ the schema obtained by eliminating from σ the relation symbol R_i, FO_σ and $FO_{\sigma-R_i}$ the set of formulas of FO over the schemas σ and $\sigma - R_i$, respectively, and $\mathcal{A}|_{\sigma-R_i}$ the* reduced database *of schema $\sigma - R_i$ obtained by eliminating the relation $R_i^{\mathcal{A}}$ from $\mathcal{A}$. We say that R_i, a k-ary relation in σ, is a* redundant relation *in the database $\mathcal{A}$ if for all k-tuples $\bar{u}$ and $\bar{v}$ in $\text{dom}(\mathcal{A})^k$, $tp_{\mathcal{A}}^{FO_\sigma}(\bar{u}) = tp_{\mathcal{A}}^{FO_\sigma}(\bar{v})$ iff $tp_{\mathcal{A}|_{\sigma-R_i}}^{FO_{\sigma-R_i}}(\bar{u}) = tp_{\mathcal{A}|_{\sigma-R_i}}^{FO_{\sigma-R_i}}(\bar{v})$, i.e., the equivalence classes induced by the FO_σ types of the k-tuples in $\text{dom}(\mathcal{A})^k$ coincide with the equivalence classes induced by the $FO_{\sigma-R_i}$ types of k-tuples in $\text{dom}(\mathcal{A}|_{\sigma-R_i})^k$.*

Let's see an example of a database with a redundant relation.

Example 1. Below, we show two complete binary trees $\mathcal{G}_1$ and $\mathcal{G}_2$. They can be seen as databases of schema $\tau = \langle E, C \rangle$ with E a binary relation symbol interpreted as the edge relation and C a unary relation symbol interpreted as the set of black nodes.

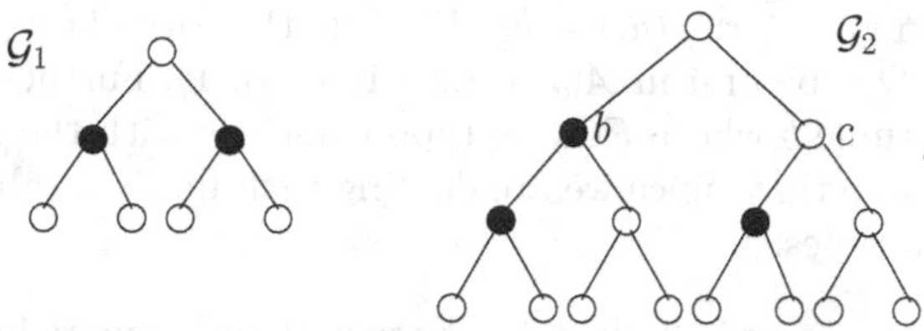

Clearly, if we consider the FO types for tuples of arity 1 in a complete binary tree of depth n then we have $n+1$ different types, because all nodes of the same depth have the same FO type. That is, a node in a complete binary tree cannot be distinguished by any FO formula from another node at the same depth in the tree, therefore, nodes of the same depth can be exchanged by an automorphism of the tree. This fact points out that in our complete binary tree $\mathcal{G}_1$, the relation $C^{\mathcal{G}_1}$ is a redundant relation, i.e., for every elements $u, v \in \text{dom}(\mathcal{G}_1)$, $tp^{FO_\tau}_{\mathcal{G}_1}(u) = tp^{FO_\tau}_{\mathcal{G}_1}(v)$ iff $tp^{FO_{\tau-C}}_{\mathcal{G}_1|_{\tau-C}}(u) = tp^{FO_{\tau-C}}_{\mathcal{G}_1|_{\tau-C}}(v)$. On the other hand, this is not the case for the tree $\mathcal{G}_2$ as the relation $C^{\mathcal{G}_2}$ allows us to distinguish, for levels two and three, some nodes from the others in the same level. So it is not longer the case that all nodes in the same level have the same FO type. Take for instance the nodes b and c in $\mathcal{G}_2$. Let $\varphi_b(x) \equiv \exists y(E(y,x) \wedge \neg\exists z(E(z,y))) \wedge C(x)$ and let $\varphi_c(x) \equiv \exists y(E(y,x) \wedge \neg\exists z(E(z,y))) \wedge \neg C(x)$. Then, $\mathcal{G}_2 \models \varphi_b(x)[b]$ but $\mathcal{G}_2 \not\models \varphi_b(x)[c]$ and $\mathcal{G}_2 \models \varphi_c(x)[c]$ but $\mathcal{G}_2 \not\models \varphi_c(x)[b]$. Clearly, $tp^{FO_\tau}_{\mathcal{G}_2}(b) \neq tp^{FO_\tau}_{\mathcal{G}_2}(c)$ while $tp^{FO_{\tau-C}}_{\mathcal{G}_2|_{\tau-C}}(b) = tp^{FO_{\tau-C}}_{\mathcal{G}_2|_{\tau-C}}(c)$.

We will next show that there is, for every redundant relation $R^{\mathcal{A}}$ in a database $\mathcal{A}$ of schema σ, an FO formula ϕ_R of vocabulary $\sigma - R$ such that if ϕ_R is evaluated in the reduced database $\mathcal{A}|_{\sigma-R}$, it defines the relation $R^{\mathcal{A}}$.

Fact 1. *Let R be a relation symbol in σ of arity r, let $R^{\mathcal{A}}$ be a redundant relation in the database $\mathcal{A}$, let $\bar{a}$ be an r-tuple in $R^{\mathcal{A}}$, and let $\bar{b}$ be an r-tuple in* dom($\mathcal{A}$)*. Then, there is a formula $\psi_{\bar{a}}(z_1, \ldots, z_r)$ of $FO_{\sigma-R}$ such that $\mathcal{A}|_{\sigma-R} \models \psi_{\bar{a}}(\bar{z})[\bar{b}]$ iff $tp^{FO_{\sigma-R}}_{\mathcal{A}|_{\sigma-R}}(\bar{a}) = tp^{FO_{\sigma-R}}_{\mathcal{A}|_{\sigma-R}}(\bar{b})$. And, hence, if $\mathcal{A}|_{\sigma-R} \models \psi_{\bar{a}}(\bar{z})[\bar{b}]$ then $\bar{b} \in R^{\mathcal{A}}$.*

Proof. Following [5] we build $\psi_{\bar{a}}(z_1, \ldots, z_r)$ by using the diagram of $\mathcal{A}|_{\sigma-R}$. Assume $|\text{dom}(\mathcal{A}|_{\sigma-R})| = n$. Let $v : \{x_1, \ldots, x_n\} \rightarrow \text{dom}(\mathcal{A}|_{\sigma-R})$ be an injective valuation such that $v(x_{i_1}) = a_1, \ldots, v(x_{i_r}) = a_r$, where $1 \leq i_1, \ldots, i_r \leq n$. Let $\Theta = \{\alpha | \alpha$ has the form $P(x_{i_1}, \ldots, x_{i_k})$ where $1 \leq i_1, \ldots, i_k \leq n$, and $P \in \sigma - R$ with arity $k \geq 1\}$ and let

$$\psi_{\bar{a}}(z_1,\ldots,z_r) \equiv \exists x_1 \ldots x_n \big(\bigwedge\{\alpha | \alpha \in \Theta, (\mathcal{A}|_{\sigma-R}, v) \models \alpha\} \wedge \\ \bigwedge\{\neg\alpha | \alpha \in \Theta, (\mathcal{A}|_{\sigma-R}, v) \models \neg\alpha\} \wedge \bigwedge_{1 \le i < j \le n} (x_i \neq x_j) \wedge \\ \forall x_{n+1}(x_{n+1} = x_1 \vee \cdots \vee x_{n+1} = x_n) \wedge z_1 = x_{i_1} \wedge \ldots \wedge z_r = x_{i_r} \big)$$

The following facts complete the proof. Clearly, a given tuple $\bar{b} = (b_1, \ldots, b_r)$ satisfies $\psi_{\bar{a}}(z_1, \ldots, z_r)$ iff there exists an automorphism f in $\mathcal{A}|_{\sigma-R}$ which maps $\bar{a}$ onto $\bar{b}$, i.e., for $1 \le i \le r$, $f(a_i) = b_i$. That is, the formula $\psi_{\bar{a}}$ is an isolating formula for the FO type of $\bar{a}$ in $\mathcal{A}|_{\sigma-R}$ (see Remark 1). Furthermore, as $R^{\mathcal{A}}$ is redundant, every tuple $\bar{b}$ whose $FO_{\sigma-R}$ type coincides with the $FO_{\sigma-R}$ type of $\bar{a}$, is also in $R^{\mathcal{A}}$. Note that, since we are dealing with finite databases, FO types are automorphism types. □

Though in our example we include a database with only one redundant relation, databases may contain several redundant relations. The important fact to have into account in such cases is that in analyzing redundant relations we must consider one relation at a time.

The following proposition shows that, given a redundant relation $R^{\mathcal{A}}$ in a database $\mathcal{A}$ of schema σ, there is an FO formula ϕ_R of vocabulary $\sigma - R$ such that if ϕ_R is evaluated in the reduced database $\mathcal{A}|_{\sigma-R}$, it defines the relation $R^{\mathcal{A}}$, and that such formula can be build in polynomial time.

Proposition 3. *Let $\mathcal{A}$ be a database of schema σ, and let $R^{\mathcal{A}} = \{\bar{a}^1_k, \ldots, \bar{a}^n_k\}$ be a redundant relation of arity k and cardinality n in $\mathcal{A}$. Then, the following FO formula $\phi_R(x_1,\ldots,x_k) \equiv \psi_1(x_1,\ldots,x_k) \vee \ldots \vee \psi_n(x_1,\ldots,x_k)$ where, for $1 \le i \le n$, ψ_i is the formula described in Fact 1 for the k-tuple $\bar{a}^i_k$, defines the relation $R^{\mathcal{A}}$ when evaluated in the reduced database $\mathcal{A}|_{\sigma-R}$, i.e., $\phi_R^{\mathcal{A}|_{\sigma-R}} = R^{\mathcal{A}}$. Furthermore, there is an algorithm which builds the formula ϕ_R in polynomial time.*

Remark 2. If we omit in the previous proposition the condition of $R^{\mathcal{A}}$ being a redundant relation, then the relation $\phi_R^{\mathcal{A}|_{\sigma-R}}$ would include not only the tuples in $R^{\mathcal{A}}$, but also all the tuples which are commutable by an automorphism with some tuple in $R^{\mathcal{A}}$.

Note that given an FO formula φ_q which expresses an arbitrary query q over a database $\mathcal{A}$ of schema σ, it can be translated in a straightforward way to a formula φ'_q of schema $\sigma - R$ which expresses the same query q over the reduced database $\mathcal{A}|_{\sigma-R}$. By Proposition 3, a redundant relation $R^{\mathcal{A}}$ of arity k in $\mathcal{A}$ can be expressed by an FO formula $\phi_R(x_1,\ldots,x_k)$ in $\mathcal{A}|_{\sigma-R}$. Therefore, every arbitrary query q which is expressed by an FO formula φ_q in which the relation symbol R occurs, could be expressed in the reduced database $\mathcal{A}|_{\sigma-R}$ using the formula $\phi_R(x_1,\ldots,x_k)$. That is, every atomic formula formed with the relation symbol R in φ_q can be replaced in φ'_q by the formula $\phi_R(x_1,\ldots,x_k)$ in the database $\mathcal{A}|_{\sigma-R}$. We only need to take care of the appropriate re-naming of variables in ϕ_R. In general, we can say that given a logic $\mathcal{L}$ and a formula φ_q in that logic that expresses an arbitrary query q over a database $\mathcal{A}$ of schema σ, it can be translated to a formula φ'_q in the same logic of schema $\sigma - R$ which expresses the same query q over the reduced database $\mathcal{A}|_{\sigma-R}$ provided that the formula ϕ_R can be expressed in the logic $\mathcal{L}$.

4 Computing Redundant Relations

Proposition 4. *The following problems are decidable: (i) Given a schema σ, a relation symbol $R \in \sigma$ of arity k, for some $k \geq 1$, and a database $\mathcal{A}$ of schema σ, to decide whether $R^{\mathcal{A}}$ is a redundant relation in $\mathcal{A}$. (ii) Given a schema σ and a database $\mathcal{A}$ of schema σ, to decide whether there is any relation symbol R in σ such that $R^{\mathcal{A}}$ is a redundant relation in $\mathcal{A}$.*

Proof. (sketch). We use the formulas $\psi_{\bar{a}}$ of Fact 1. We denote by $\psi_{\mathcal{A},\bar{a}}$ the formula built following that fact for the database $\mathcal{A}$. The following algorithm decides (i).
redundant := True; $m := |\text{dom}(\mathcal{A})| + 1$;
For every $\bar{u} \in \text{dom}(\mathcal{A})^k$ {
  Build $\psi_{\mathcal{A},\bar{u}}(\bar{x})$; Build $\psi_{\mathcal{A}|_{\sigma-R},\bar{u}}(\bar{x})$;
  For every $\bar{v} \in \text{dom}(\mathcal{A})^k$ {
    If $\neg(\mathcal{A} \models \psi_{\mathcal{A},\bar{u}}(\bar{x})[\bar{v}] \leftrightarrow \mathcal{A}|_{\sigma-R} \models \psi_{\mathcal{A}|_{\sigma-R},\bar{u}}(\bar{x})[\bar{v}])$ then {
    "If it is not the case that $\bar{u}$ and $\bar{v}$ have the same *FO* type both in $\mathcal{A}$ and in $\mathcal{A}|_{\sigma-R}$"
      redundant := False; Return redundant } } };
Return redundant;
As relational database schemas have a finite number of relation symbols. We can decide (ii) by simply checking, using the previous algorithm, whether for some relation symbol R in σ, $R^{\mathcal{A}}$ is a redundant relation. □

Unfortunately, the algorithm we gave in the proof of Proposition 4 to decide whether a given relation is redundant in a given database, has exponential time complexity. Note that while the formulas $\psi_{\mathcal{A},\bar{a}}$ of the previous proposition and Fact 1 can be built in polynomial time, their evaluation on a given database takes time $\mathcal{O}(n^n)$, since we must consider all valuations on the n variables of the formulas to that end. It is very unlikely that there is a polynomial time algorithm for this problem since it is equivalent to deciding isomorphism. In this section we attack this problem by restricting: (a) the *class of queries* to a sub-class of $\mathcal{CQ}$ in such a way that determining whether a relation is redundant regarding only such sub-class of queries is in P, (b) the *class of databases* to classes where deciding whether a relation is redundant in a database which belongs to the class is in P.

Definition 2. *Let $\mathcal{L}$ be a sub-logic of FO, let σ be a relational schema, let $\mathcal{A}$ be a database of schema σ, let $\mathcal{K}$ be a class of computable queries, let $r \geq 1$, and let R be an r-ary relation symbol in σ. We say that $R^{\mathcal{A}}$ is a $(\mathcal{K}, \mathcal{L})$-*redundant relation *in the database $\mathcal{A}$ if there is an $\mathcal{L}$ formula $\phi_R(x_1, \ldots, x_r)$ such that, for every $q \in \mathcal{K}, q(\mathcal{A}) = q(\langle \mathcal{A}|_{\sigma-R}, \phi_R^{\mathcal{A}|_{\sigma-R}} \rangle)$, where $\langle \mathcal{A}|_{\sigma-R}, \phi_R^{\mathcal{A}|_{\sigma-R}} \rangle$, of schema σ, denotes the reduced database $\mathcal{A}|_{\sigma-R}$ augmented with the relation defined by the formula ϕ_R in $\mathcal{A}|_{\sigma-R}$.*

As a consequence of this definition and Proposition 3, we get the following.

Fact 2. *Let R be a relation symbol in σ of arity r. The relation $R^{\mathcal{A}}$ is $(\mathcal{CQ}, FO)$-redundant in a database $\mathcal{A}$ iff it is redundant in the sense of Definition 1.*

4.1 Subclasses of Queries

We will examine next $(\mathcal{K}, \mathcal{L})$-redundant relations for well studied classes of computable queries which are strictly included in $\mathcal{CQ}$. The classes we will consider here characterize the expressive power of some variations of the *reflective relational machine* (RRM) developed in [1]. In [12] a strict hierarchy was defined in $\mathcal{CQ}$, in terms of the preservation of equivalence in FO^k. We denote the whole hierarchy as $\mathcal{QCQ}^\omega$. For every natural k, the layer denoted as $\mathcal{QCQ}^k$ was proved to be a semantic characterization of the computation power of the RRM of [1] if we restrict to k the number of different variables which can be used in any FO query generated during a computation (denoted by RRM^k). A variation of RRM called *reflective counting machine* (RCM) was defined in [13] together with a characterization of its expressive power through a hierarchy denoted as $\mathcal{QCQ}^{C^\omega}$. For every natural k, we denote as $\mathcal{QCQ}^{C^k}$ the layer of the hierarchy $\mathcal{QCQ}^{C^\omega}$ which consists of those queries that preserve equivalence in C^k. The RCM with variable complexity k (RCM^k) is defined as a variant of the RRM^k in which the dynamic queries are formulas in the logic C^k, instead of FO^k. For every natural k, the layer denoted as $\mathcal{QCQ}^{C^k}$ characterizes exactly the expressive power of the RCM^k. The following fact is a direct consequence of Definition 2 and the fact that the $\mathcal{QCQ}^k$ and $\mathcal{QCQ}^{C^k}$ classes preserve equality of FO^k types and C^k types, respectively, in the set of k-tuples of a database.

Fact 3. *Let $k \geq 1$, (i) a relation $R^{\mathcal{A}}$ of arity $1 \leq r \leq k$ is $(\mathcal{QCQ}^k, FO^k)$-redundant in a database $\mathcal{A}$ iff for all r-tuples $\bar{u}$ and $\bar{v}$ in* $\mathrm{dom}(\mathcal{A})^r$, *$tp_{\mathcal{A}}^{FO^k_\sigma}(\bar{u}) = tp_{\mathcal{A}}^{FO^k_\sigma}(\bar{v})$ iff $tp_{\mathcal{A}|_{\sigma-R}}^{FO^k_{\sigma-R}}(\bar{u}) = tp_{\mathcal{A}|_{\sigma-R}}^{FO^k_{\sigma-R}}(\bar{v})$, (ii) a relation R of arity $1 \leq r \leq k$ is $(\mathcal{QCQ}^{C^k}, C^k)$-redundant in a database $\mathcal{A}$ iff for all r-tuples $\bar{u}$ and $\bar{v}$ in* $\mathrm{dom}(\mathcal{A})^r$, *$tp_{\mathcal{A}}^{C^k_\sigma}(\bar{u}) = tp_{\mathcal{A}}^{C^k_\sigma}(\bar{v})$ iff $tp_{\mathcal{A}|_{\sigma-R}}^{C^k_{\sigma-R}}(\bar{u}) = tp_{\mathcal{A}|_{\sigma-R}}^{C^k_{\sigma-R}}(\bar{v})$.*

By a result of Grohe [7], equivalence in FO^k and C^k is complete for polynomial time. Then we can check in P, C^k equivalence as well as FO^k equivalence between every two extensions of a database with any given pair of tuples. So, we have the following important proposition.

Proposition 5. *Given a schema σ, a relation symbol $R \in \sigma$ and a database $\mathcal{A}$ of schema σ. To decide whether $R^{\mathcal{A}}$ is a $(\mathcal{QCQ}^k, FO^k)$-redundant relation in $\mathcal{A}$, as well as to decide whether $R^{\mathcal{A}}$ is a $(\mathcal{QCQ}^{C^k}, C^k)$-redundant relation in $\mathcal{A}$, is in P.*

The importance of this proposition lies on that, for $k \geq 2$, the classes $\mathcal{QCQ}^{C^k}$ capture a relevant portion of the class $\mathcal{CQ}$ of computable queries. Following [8] though using a slightly different perspective, we define the notion of equality of queries *almost everywhere*, as follows:

$$\mu_{(q=q')} = \lim_{n\to\infty} \frac{|\{\mathcal{I} \in \mathcal{B}_\sigma : \mathrm{dom}(\mathcal{I}) = \{1,\ldots,n\} \wedge q(\mathcal{I}) = q'(\mathcal{I})\}|}{|\{\mathcal{I} \in \mathcal{B}_\sigma : \mathrm{dom}(\mathcal{I}) = \{1,\ldots,n\}\}|}$$

where q, q' are computable queries of schema σ. If $\mathcal{C}$ is a class of finite structures,

$$\mu_{\mathcal{C}} = \lim_{n\to\infty} \frac{|\{\mathcal{I} \in \mathcal{B}_\sigma : dom(\mathcal{I}) = \{1,\ldots,n\} \wedge \mathcal{I} \in \mathcal{C}\}|}{|\{\mathcal{I} \in \mathcal{B}_\sigma : dom(\mathcal{I}) = \{1,\ldots,n\}\}|}$$

Let's consider the following result from [2] and [9].

Proposition 6. *([2] and [9]) There is a class $\mathcal{C}$ of graphs with $\mu_{\mathcal{C}} = 1$ such that for all graphs $\mathcal{I}, \mathcal{J} \in \mathcal{C}$ we have $\mathcal{I} \simeq \mathcal{J}$ iff $\mathcal{I} \equiv_{C^2} \mathcal{J}$. Moreover, for all $\mathcal{I} \in \mathcal{C}$ and $a, b \in \mathrm{dom}(\mathcal{I})$, there is an automorphism mapping a to b iff $tp_{\mathcal{I}}^{C^2}(a) = tp_{\mathcal{I}}^{C^2}(b)$.*

Then it follows that, for *every* computable query q there is a query q' in $\mathcal{QCQ}^{C^2}$ such that $\mu_{(q=q')} = 1$, i.e., such that q' coincides with q over *almost all* databases. Furthermore, there is a big amount of relevant queries, which are not expressible in relational calculus (or FO), that belong to the lower levels of the $\mathcal{QCQ}^{\omega}$ and $\mathcal{QCQ}^{C^{\omega}}$ hierarchies. (i) Assume we have a database with a ternary relation R such that a tuple (a, b, c) is in R iff the supplier a supplies part b to project c. Then, the query "suppliers who supply the biggest number of different parts supplied by any supplier in the database" is in the class $\mathcal{QCQ}^{C^3}$. (ii) The property of the graph being *regular of even degree*, or equivalently of having an *Eulerian cycle*, is in the class $\mathcal{QCQ}^{C^2}$. (iii) *Graph connectivity* is in $\mathcal{QCQ}^3$. (iv) The problem of determining whether the cardinality of the domain of a database is even, belongs to $\mathcal{QCQ}^{C^1}$. (v) The *transitive closure* over graphs is in $\mathcal{QCQ}^3$. (vi) To decide whether a binary relation R is an equivalence relation with an even number of equivalence classes, is in $\mathcal{QCQ}^{C^2}$.

4.2 Subclasses of Databases

Proposition 7. *Let $k \geq 1$ and let $\mathcal{C}$ be a class of databases in which C^k (FO^k) equivalence coincides with isomorphism. Then, the problem of deciding whether a given relation is redundant in a database which belongs to $\mathcal{C}$, is in P, as well as the problem of deciding whether a given database in $\mathcal{C}$ has any redundant relation.*

Some examples of classes where C^k equivalence coincides with isomorphism are: (i) the class of *planar* graphs, where there is a $k \geq 1$ such that C^k equivalence coincides with isomorphism; (ii) for all $k \geq 1$, the class of graphs of *k-bounded tree-width* [6], where C^{k+3} equivalence coincides with isomorphism; (iii) the class of *trees*, where C^2 equivalence coincides with isomorphism. Regarding FO^k, in the class of *linear* graphs [5], FO^2 equivalence coincides with isomorphism and in the class of graphs with *color class size* ≤ 3, FO^3 equivalence coincides with isomorphism. Note that, even if C^k equivalence and FO^k equivalence do not coincide with isomorphism, we have the following result.

Fact 4. *Let C be a class of databases in which isomorphism is decidable in P, then the problem of deciding whether a given relation is redundant in a database which belongs to $\mathcal{C}$, is in P, as well as the problem of deciding whether a given database in $\mathcal{C}$ has any redundant relation.*

The classes of linear graphs, trees, planar graphs and graphs with bounded tree-width are examples of such classes where isomorphism is decidable in P.

References

1. Abiteboul, S., Papadimitriou, C., Vianu, V.: Reflective Relational Machines. Information and Computation 143, 110–136 (1998)
2. Babai, L., Erdös, P., Selkow, S.: Random Graph Isomorphism. SIAM Journal on Computing 9, 628–635 (1980)
3. Chandra, A.K., Harel, D.: Computable Queries for Relational Data Bases. Journal of Computer and System Sciences 21(2), 156–178 (1980)
4. Dawar, A.: Feasible Computation Through Model Theory. Ph.D. thesis, University of Pennsylvania, Philadelphia (1993)
5. Ebbinghaus, H.D., Flum, J.: Finite Model Theory, 2nd edn. Springer, Heidelberg (1999)
6. Grohe, M., Mariño, J.: Definability and Descriptive Complexity on Databases of Bounded Tree-Width. In: Beeri, C., Bruneman, P. (eds.) ICDT 1999. LNCS, vol. 1540, pp. 70–82. Springer, Heidelberg (1998)
7. Grohe, M.: Equivalence in Finite Variable Logics is Complete for Polynomial Time. In: Proceedings of 37th IEEE Symposium on Foundations of Computer Science, pp. 264–273 (1996)
8. Hella, L., Kolaitis, P., Luosto, K.: Almost Everywhere Equivalence of Logics in Finite Model Theory. The Bulletin of Symbolic Logic 2(4), 422–443 (1996)
9. Immerman, N., Lander, E.: Describing Graphs: A First Order Approach to Graph Canonization. In: Selman, A. (ed.) Complexity Theory Retrospective, pp. 59–81. Springer, Heidelberg (1990)
10. Otto, M.: The Expressive Power of Fixed Point Logic with Counting. Journal of Symbolic Logic 61(1), 147–176 (1996)
11. Otto, M.: Bounded Variable Logics and Counting. Springer, Heidelberg (1997)
12. Turull Torres, J.M.: A Study of Homogeneity in Relational Databases. Annals of Mathematics and Artificial Intelligence 33(2), 379–414 (2001); Erratum in Annals of Mathematics and Artificial Intelligence 42, 443–444 (2004)
13. Turull Torres, J.M.: Relational Databases and Homogeneity in Logics with Counting. Acta Cybernetica 17(3), 485–511 (2006)

On Matrix Representations of Participation Constraints

Sven Hartmann[1,*], Uwe Leck[2], and Sebastian Link[3]

[1] Department of Informatics, Clausthal University of Technology, Germany
sven.hartmann@tu-clausthal.de
[2] Department of Mathematics and Computer Science, University of Wisconsin, USA
[3] School of Information Management, Victoria University of Wellington, New Zealand

Abstract. We discuss the existence of matrix representations for generalised and minimum participation constraints which are frequently used in database design and conceptual modelling. Matrix representations, also known as Armstrong relations, have been studied in literature e.g. for functional dependencies and play an important role in example-based design and for the implication problem of database constraints. The major tool to achieve the results in this paper is a theorem of Hajnal and Szemerédi on the occurrence of clique graphs in a given graph.

1 Introduction

Informally, a database relation may be considered as a matrix, where every column contains the data of the same sort and every row contains the data of some object. This approach is very similar to the two-dimensional tables that humans have used to keep track of information for centuries. As an example consider a relation schema (Teacher, Course, Weekday) and the database relation in Figure 1 containing information on classes taught at a university.

Teacher	Course	Weekday
Mary	Java	Mo
John	C++	Tu
John	Python	Tu
Mary	Java	We
Mary	Java	Fr

Fig. 1. A database relation containing information on classes to be taught

Often the data stored in a database relation are not independent from each other. In the example above any two classes on the same course are given by the same teacher. Let Ω denote the set of columns, and let X, Y be non-empty subsets of Ω. Then Y functionally depends on X if any two rows coinciding in the columns of X are also equal in the columns of Y. Further, data entries do

* Corresponding author.

C.A. Heuser and G. Pernul (Eds.): ER 2009 Workshops, LNCS 5833, pp. 75–84, 2009.

not occur arbitrarily often. In the example above every teacher gives between 2 and 3 classes, and for every course there is at most one class per weekday. Given some entry a in the matrix, its degree $\deg(A, a)$ counts how often this entry occurs in the column $A \in \Omega$. Analysing these degrees provides lower and upper bounds on the number of rows that coincide in column A.

Some of the dependencies discussed above may hold by accident. When the database relation is updated they could well be violated. Other dependencies, however, we wish to hold forever, no matter of how the database relation is modified. They reflect the semantics of the real world situation captured by the database. The notion of a database relation itself provides only syntax but does not carry the semantics of the data. Therefore, semantic integrity constraints are used to specify the rules which data have to satisfy in order to reflect the properties of the represented objects in the modelled real world situation. When designing a database system, integrity constraints have been proven useful in ensuring databases with semantically desirable properties, in preventing update anomalies, and in allowing the application of efficient methods for storing, accessing and querying data. Consequently, various classes of integrity constraints have been defined and studied for databases with functional dependencies, multivalued dependencies and inclusion dependencies being the most prominent examples. In addition, properties such as satisfiability and implication have been studied for these constraints. For details we refer e.g. to [21, 25].

In the present paper we study participation constraints which gained much attention in the database design community, but also in conceptual modelling and knowledge representation, cf. [5, 12–14, 16–18, 22, 24, 25].

2 Preliminaries

Let R be a matrix with n columns and s rows and such that no two rows are identical. Let $\Omega = \{C_1, \ldots, C_n\}$ be the n-element set of columns. Further, let $range_i$ contain all entries of R in column C_i. In the context of the relational database model (RDM), the columns are called *attributes*, the elements in $range_i$ are called the values of attribute C_i, the sequence $(C_1, \ldots, C_n)$ is called a *relation schema*, and the matrix R is called a *database relation* over Ω. The rows of R are *tuples* from the cartesian product $range_1 \times \cdots \times range_n$, and each tuple contains the data of one object.

2.1 Participation Constraints

Within this paper, we are mainly concerned with participation constraints. A *participation constraint* is an expression $card^{part}(C_i) = b$ with $b \in \mathbb{N}^\infty$ and $C_i \in \Omega$. This constraint holds in the database relation R if every value $v \in range_i$ appears at most b times in column C_i. For example, the participation constraint $card^{part}(Teacher) = 3$ tells us that every teacher gives at most three classes. We call a participation constraint *finite* if b is finite.

Participation constraints may be easily extended to sets of columns. A *generalised participation constraint* is an expression $card^{part}(X) = b$ with $b \in \mathbb{N}^\infty$

and $\emptyset \neq X \subseteq \Omega$. This constraint holds in a database relation R if there are at most b rows which coincide in each of the columns $C_i \in X$. For example, the generalised participation constraint $card^{part}(\{Course, Weekday\}) = 1$ tells us that every course is taught at most once per weekday. Clearly, a participation constraint $card^{part}(C_i) = b$ corresponds to a generalised participation constraint $card^{part}(X) = b$ with $X = \{C_i\}$. Generalised participation constraints with X containing all but one of the columns are better known as *look-across constraints* or *Chen-style cardinality constraints*, cf. [5]. Note that textbooks on entity-relationship modelling often use the term *cardinality constraints* to refer to either participation constraints or look-across constraints, cf. [13, 17, 25].

In many applications, one is not only interested in upper bounds on the number of occurrences of values but also in lower bounds. A *minimum participation constraint* is an expression $card^{min}(C_i) = a$ with $a \in \mathbb{N}^{\infty}$ and $C_i \in \Omega$. This constraint holds in the database relation R if every value $v \in range_i$ appears at least a times in column C_i. For example, the participation constraint $card^{min}(Teacher) = 2$ tells us that every teacher gives at least two classes.

2.2 The Implication Problem and Closed Constraint Sets

The constraints satisfied by a database relation are usually not independent. A single constraint σ *follows* from a constraint set Σ if σ holds in every database relation R which satisfies Σ. We also say that Σ *implies* σ. Two constraint sets Σ and Σ' are *equivalent* if every constraint in Σ' follows from Σ and vice versa.

For a fixed class $\mathcal{Z}$ of constraints, the *implication problem* for class $\mathcal{Z}$ reads as follows: Given a constraint set $\Sigma \subseteq \mathcal{Z}$ and a single constraint $\sigma \in \mathcal{Z}$, we want to know whether σ follows from Σ. The emergence of the implication problem in database theory is discussed e.g. in [17, 21, 23]. A constraint set Σ is $\mathcal{Z}$*-closed* if it contains every constraint $\sigma \in \mathcal{Z}$ which follows from Σ. Special attention is devoted to the determination of closed constraint sets. Clearly, Σ implies $\sigma \in \mathcal{Z}$ if and only if σ is in the $\mathcal{Z}$-closure of Σ. Thus the characterisation of closed sets in a constraint class $\mathcal{Z}$ completely solves the implication problem for this class.

In the present paper, we are interested in the joint class $\mathcal{P}$ of generalised participation constraints and minimum participation constraints. This extends earlier work on the interaction of ordinary and minimum participation constraints [12, 16, 22, 24].

3 Matrix Representations

Given a database relation R it is often a straightforward task to extract the set $\Sigma(R) \subseteq \mathcal{Z}$ of all constraints from $\mathcal{Z}$ satisfied by R. Clearly, $\Sigma(R)$ must be $\mathcal{Z}$-closed. Conversely, given a constraint set $\Sigma \subseteq \mathcal{Z}$ it is natural to ask whether there is a database relation R such that $\Sigma(R)$ is just the $\mathcal{Z}$-closure of Σ. In this case, R is said to *represent* the constraint set Σ under consideration or to be a $\mathcal{Z}$*-Armstrong relation* for Σ. In this case, R satisfies exactly the logical consequences of Σ among all the constraints in $\mathcal{Z}$.

In view of this property, matrix representations are a popular tool in example-based database design [20]. Armstrong relations satisfy exactly the conditions specified by the database designer. This makes them good examples to represent the real world situation captured by the database. Further, they help the designer to recognise omissions and mistakes in the design. Actually, a major problem that has been noted with the use of automated design tools is to get all necessary design information from the designer into the tool.

Matrix representations have been first studied for functional dependencies. A *functional dependency* is a statement $X \to Y$ where both X and Y are non-empty subsets of Ω. This constraint holds in the database relation R if any two rows coinciding in the columns of X also coincide in the columns of Y. Armstrong [1] observed that closed sets of functional dependencies correspond to closure operations on the set Ω. He proved that every closed set of functional dependencies admits a matrix representation. Demetrovics and Gyepesi [8] proved that in the worst case the minimum size s of an Armstrong relation for a set of functional dependencies satisfies the inequality

$$\frac{1}{n^2}\binom{n}{\lfloor n/2 \rfloor} < s \leq (1 + \tfrac{c}{\sqrt{n}})\binom{n}{\lfloor n/2 \rfloor},$$

for some suitable constant c and $n = |\Omega|$. A functional dependency $X \to \Omega$ is, in particular, called a *key dependency* and X is said to be a *key*. Note that key dependencies are special kinds of generalised participation constraints, namely those ones with $b = 1$. Demetrovics [7] observed that the set of minimal keys is always a Sperner family over the set Ω, that is, minimal keys are mutually inclusion-free. Again, every closed set of key dependencies admits a matrix representation, and [8] shows that in the worst case the minimum size s of an Armstrong relation for a set of key dependencies satisfies the inequality

$$\frac{1}{n^2}\binom{n}{\lfloor n/2 \rfloor} < s \leq 1 + \binom{n}{\lfloor n/2 \rfloor}$$

where $n = |\Omega|$. Matrix representations for functional dependencies have been further studied in the literature, cf. [2, 3, 9, 15, 19]. For a survey on similar results for other constraints, we refer to [21, 25].

Unfortunately, matrix representations are not always possible. Let $n \geq 2$ and consider the empty constraint set Σ which is clearly satisfied by every database relation R over $\Omega = \{C_1, \ldots, C_n\}$. Hence, Σ does not imply any participation constraint $card^{part}(C_1) = b$ with finite b. Conversely, however, each database relation R of size s satisfies the participation constraint $card^{part}(C_1) = s$, which is not a consequence of Σ. In order to be represented by some database relation, Σ must at least imply some finite participation constraint for every $C_i \in \Omega$.

4 Inference Rules

The latter observation again leads to the implication problem. Clearly, we do not want to inspect all possible database relations to decide the implications of a given constraint set. Rather, we are interested in inference rules which help to decide this question. An *inference rule* is an expression $\frac{\Sigma'}{\sigma}\gamma$ where Σ' is a subset of Σ, and γ states some condition on Σ' which has to be satisfied if we want to apply this rule. If Σ contains a subset Σ' satisfying the condition γ, then σ may be *derived* from Σ due to that inference rule. An inference rule is *sound* if Σ implies every constraint σ which may be derived from Σ due to that rule.

We are interested in inference rules which completely describe all the implications of a given constraint set Σ. A *rule system* $\mathcal{R}$ is a set of inference rules. The most prominent example of such a rule system is the Armstrong system for functional dependencies [1]. A set Σ is *syntactically closed* with respect to $\mathcal{R}$ if it contains every constraint σ which may be derived from Σ due to some rule in $\mathcal{R}$. The general problem is to find a rule system $\mathcal{R}$ for the constraint class $\mathcal{Z}$ such that a given set $\Sigma \subseteq \mathcal{Z}$ is $\mathcal{Z}$-closed if and only if it is syntactically closed w.r.t. $\mathcal{R}$. Such a rule system is said to be *sound and complete* for the implication of $\mathcal{Z}$. The Armstrong system for functional dependencies is the most prominent example of a sound and complete rule system.

Let $C_i, C_j \in \Omega$, let X, Y be non-empty subsets of Ω, and let $a, a', b, b' \in \mathbb{N}^{\infty}$. For the class $\mathcal{P}$ of generalised and minimum participation constraints the following inference rules are clearly sound:

$$\frac{}{card^{part}(X) = \infty}, \quad \frac{}{card^{part}(\Omega) = 1}, \quad \frac{}{card^{min}(C_i) = 1},$$

$$\frac{card^{part}(X) = b}{card^{part}(Y) = b}\; X \subset Y, \quad \frac{card^{part}(X) = b}{card^{part}(X) = b'}\; b < b', \quad \frac{card^{min}(C_i) = a}{card^{min}(C_i) = a'}\; a > a',$$

$$\frac{card^{part}(C_i) = b, card^{min}(C_i) = a}{card^{part}(C_i) = 0}\; a > b,$$

$$\frac{card^{part}(C_i) = 0}{card^{part}(C_j) = 0}, \quad \frac{card^{part}(C_i) = 0}{card^{min}(C_i) = \infty}.$$

Note that the last three rules describe situations where Σ is only satisfied by the empty database relation. We call such a constraint set *conflicting*. In the sequel, matrix representations will help us to verify that the rule system above is in fact complete for minimum and generalised participation constraints.

5 Representation Graphs

In the sequel we make use of a nice graph-theoretic analogue of matrix representations, cf. [9]. For every column C_i, we introduce its *representation graph* $\mathcal{G}_i$ whose vertices are the rows of R, and where two vertices r and r' are connected by an edge just when the rows r and r' coincide in column C_i.

By $\mathcal{K}_k$ we denote the *complete graph* on k vertices. In a complete graph any two vertices are connected by an edge. A *clique* of size k in a graph $\mathcal{G}$ is a maximal complete subgraph with k vertices in $\mathcal{G}$. A *clique graph* is a graph where every connected component is a complete graph. Obviously the representation graph

$\mathcal{G}_i$ is a clique graph where each clique corresponds to exactly one value of the attribute C_i. Conversely, suppose we are given a collection $\mathcal{O}$ of subgraphs $\mathcal{G}_i$ of the complete graph $\mathcal{K}_s$ such that each of them is a clique graph. Then it is easy to construct a database relation R of size s whose representation graphs are just the given graphs $\mathcal{G}_i$.

For any non-empty subset $X \subseteq \Omega$, let $\mathcal{G}_X$ denote the intersection of the representation graphs $\mathcal{G}_i$ with $C_i \in X$. This intersection is again a clique graph. The following observation is straightforward.

Proposition 1. *A database relation R satisfies the generalised participation constraint $card^{part}(X) = b$ if and only if the intersection graph $\mathcal{G}_X$ has maximum clique size at most b. A database relation R satisfies the minimum participation constraint $card^{min}(C_i) = a$ if and only if the representation graph $\mathcal{G}_i$ has minimum clique size at least a.*

This explains our interest in collections of clique graphs whose intersections have prescribed clique sizes. In the remainder of this section we assemble a number of lemmata ensuring the existence of such collections. The final lemma in this series will then turn out to be the major tool to establish matrix representations for generalised and minimum participation constraints. In order to prove this final lemma we are going to apply a theorem of Hajnal and Szemerédi [11]. By $\mu\mathcal{K}_k$ we denote the clique graph consisting of μ vertex-disjoint copies of $\mathcal{K}_k$.

Theorem 2 (Hajnal and Szemerédi). *Let $\mathcal{H}$ be a graph with $m = \mu k$ vertices and minimum valency $\delta(\mathcal{H}) \geq m - \mu$. Then $\mathcal{H}$ has a subgraph isomorphic to the clique graph $\mu\mathcal{K}_k$.*

This deep result was first conjectured by Erdős [10] and gives a necessary condition on the occurrence of clique graphs as subgraphs in a given graph $\mathcal{H}$. For a detailed discussion, we refer to Bollobás [4]. Throughout, suppose we are given positive integers k_X for every non-empty subset $X \subseteq \Omega$ such that $k_X \geq k_Y$ whenever $X \subseteq Y$. For simplicity, we write k_j instead of k_{C_j} for every $C_j \in \Omega$.

Lemma 3. *Let $s = \sum_{\emptyset \neq X \subseteq \Omega} k_X$. Then there is a collection of spanning subgraphs $\mathcal{G}_1, \ldots, \mathcal{G}_n$ of $\mathcal{K}_s$ satisfying the following conditions:*

(i) For every j with $1 \leq j \leq n$, the subgraph G_j is a clique graph.
(ii) For every non-empty subset $X \subseteq \Omega$, the intersection graph $\mathcal{G}_X$ has maximum clique size k_X.

Proof. To begin with, we partition the vertex set of $\mathcal{K}_s$ into subsets V_Z where V_Z consists of k_Z vertices and Z runs through all non-empty subsets $Z \subseteq \Omega$. Then, for every $j = 1, \ldots, n$, we choose $\mathcal{G}_j$ to be the clique graph whose components are the complete graphs on the sets V_Z with $j \in Z$ together with the isolated vertices contained in the sets V_Z with $j \notin Z$. Each G_j satisfies the first condition as $k_Z \leq k_j$ holds whenever $j \in Z \subseteq \Omega$. Given some non-empty subset $X \subseteq \Omega$, the intersection graph $\mathcal{G}_X$ is just the clique graph whose non-singleton components are complete graphs on the sets V_Z with $X \subseteq Z$. The inequality $k_Z \leq k_X$ for $X \subseteq Z$ proves $\mathcal{G}_X$ to be of maximum clique size k_X as claimed.

Lemma 4. *Let* $s = \sum_{\emptyset \neq X \subseteq \Omega} \left(k_X - |X| k_X + \sum_{j \in X} k_j \right)$. *Then there is a collection of spanning subgraphs* $\mathcal{G}_1, \ldots, \mathcal{G}_n$ *of* $\mathcal{K}_s$ *satisfying the following conditions:*

(i) For every j *with* $1 \leq j \leq n$, *the subgraph* G_j *is a clique graph such that each of its cliques is of size 1 or* k_j.
(ii) For every non-empty subset $X \subseteq \Omega$, *the intersection graph* $\mathcal{G}_X$ *has maximum clique size* k_X.

Proof. First, we select a subset V' of size $s' = \sum_{\emptyset \neq X \subseteq \Omega} k_X$ among the vertices of $\mathcal{K}_s$. For these vertices we proceed as in the preceding lemma which gives us a collection $\mathcal{O}'$ of graphs $\mathcal{G}'_j$ with vertex set V' satisfying the conditions in the preceding lemma. The remaining vertices not in V' should be partitioned into subsets $V_{j,Z}$ where $V_{j,Z}$ consists of $k_j - k_Z$ vertices, and j, Z runs through all pairs j, Z with $1 \leq j \leq n$ and $j \in Z \subseteq \Omega$. Next, for every $j = 1, \ldots, n$, we have to extend the subgraph $\mathcal{G}'_j$ on vertex set V' to a spanning subgraph $\mathcal{G}_j$ containing all vertices of $\mathcal{K}_s$. For that, we extend the component with vertex set V'_Z in $\mathcal{G}'_j$ to a complete graph on the vertex set $V'_Z \cup V_{j,Z}$ where Z runs through all subsets $Z \subseteq \Omega$ containing j. Due to our choice of the vertex sets V'_Z and $V_{j,Z}$, all the cliques in the resulting clique graph $\mathcal{G}_j$ are of size 1 or k_j as desired. The second condition immediately follows from our construction and the preceding lemma. Note that the intersection graph $\mathcal{G}_X$ is just the intersection graph $\mathcal{G}'_X$ on the vertex set V' augmented by a number of isolated vertices.

Choose λ to be a positive integer such that $\lambda \prod_{j=1}^{n} k_j \geq \sum_{\emptyset \neq X \subseteq \Omega} k_X$.

Lemma 5. *Let* $s = (\lambda + 1) \prod_{j=1}^{n} k_j$. *Then there is a collection of spanning subgraphs* $\mathcal{G}_1, \ldots, \mathcal{G}_n$ *of* $\mathcal{K}_s$ *satisfying the following conditions:*

(i) For every j *with* $1 \leq j \leq n$, *the subgraph* $\mathcal{G}_j$ *is isomorphic to the clique graph* $k_j \mathcal{K}_{s/k_j}$.
(ii) For every non-empty subset $X \subseteq \Omega$, *the intersection graph* $\mathcal{G}_X$ *has maximum clique size* k_X.

Proof. Let V denote the vertex set of $\mathcal{K}_s$. First, we select a subset $V' \subseteq V$ of size $s' = \sum_{\emptyset \neq X \subseteq \Omega} \left(k_X - |X| k_X + \sum_{j \in X} k_j \right)$. For these vertices we proceed as in the preceding lemma which gives us a collection $\mathcal{O}'$ of graphs $\mathcal{G}'_j$ with vertex set V' satisfying the conditions in the preceding lemma. Now, for every $j = 1, \ldots, n$, we have to extend the subgraph $\mathcal{G}'_j$ on vertex set V' to a spanning subgraph $\mathcal{G}_j$ on vertex set V. Assume we have already constructed suitable subgraphs $\mathcal{G}_i$ for $i < j$, and are now going to construct $\mathcal{G}_j$. Let V'' consist of all the isolated vertices in $\mathcal{G}'_j$ and all the vertices in $V - V'$. Put

$$\mu = (\lambda + 1) \prod_{i \neq j} k_i - |\{Z \subseteq \Omega : j \in Z\}|.$$

It is an easy calculation to see that V'' is just of size μk_j. The subgraph of $\mathcal{G}'_j$ induced by $V - V''$ is clearly isomorphic to the clique graph $((\lambda+1) \prod_{i \neq j} k_i - \mu) \mathcal{K}_{k_j}$.

Hence, to ensure condition (i), it essentially remains to arrange the vertices in V'' to cliques of size k_j each. For that, however, we may use neither the edges in the subgraphs $\mathcal{G}_i$, $i < j$, nor the edges in the subgraphs $\mathcal{G}'_i$, $i \geq j$. Let $\mathcal{H}$ be the graph on vertex set V containing all the remaining, i.e. permitted edges for $\mathcal{G}_j$. Further, let $\mathcal{H}''$ be the subgraph of $\mathcal{H}$ induced by the vertex set V''. Every vertex in $\mathcal{H}''$ has valency at least

$$\delta(\mathcal{H}'') \geq |V''| - 1 - \sum_{i \neq j}(k_i - 1) = \mu k_j - 1 - \sum_{i \neq j}(k_i - 1).$$

This allows us to apply the Theorem of Hajnal and Szemerédi which verifies that $\mathcal{H}''$ contains a subgraph with vertex set V'' which is isomorphic to $\mu\mathcal{K}_{k_j}$. Together with the copy of $((\lambda + 1)\prod_{i \neq j} k_i - \mu)\mathcal{K}_{k_j}$ with vertex set $V - V''$ this gives us the subgraph $\mathcal{G}_j$ satisfying condition (i) as desired. Again, condition (ii) immediately follows from our construction and the preceding lemma. Note that the intersection graph $\mathcal{G}_X$ is just the intersection graph $\mathcal{G}'_X$ on the vertex set V' augmented by a number of isolated vertices.

6 Main Results

We are now ready to state our results on matrix representations of generalised and minimum participation constraints. As a consequence we also obtain a characterisation of closed sets in the class $\mathcal{P}$ of generalised and minimum participation constraints.

Theorem 6. *Let Σ be a set of generalised and minimum participation constraints containing some finite participation constraint for every $C_i \in \Omega$. Then Σ may be represented by a database relation R.*

Proof. Let Σ^+ contain Σ and all the consequences of Σ derived by applying the rules in Section 4. If Σ^+ contains a constraint $card^{part}(C_i) = 0$ for some (and thus for all) $C_i \in \Omega$, the empty database relation represents Σ. Otherwise, put $b_X = \min\{b : card^{part}(X) = b \text{ is in } \Sigma^+\}$ for every non-empty $X \subseteq \Omega$, and $a_i = \max\{a : card^{min}(C_i) = a \text{ is in } \Sigma^+\}$. For short, we again write b_i instead of b_{C_i}. By hypothesis, all these values are finite. Two applications of the final lemma in the preceding section will provide representation graphs $\mathcal{G}_1, \ldots, \mathcal{G}_n$ which yield the claimed database relation R. First, we choose $k_X = b_X$ for every non-empty $X \subseteq \Omega$. This gives us a collection $\mathcal{O}^1$ of clique graphs $\mathcal{G}_1^1, \ldots, \mathcal{G}_n^1$. Next, we choose $k_i = a_i$ for every $C_i \in \Omega$ and $k_X = 1$ for every subset $X \subset \Omega$ of size at least 2. This gives us a collection $\mathcal{O}^2$ of clique graphs $\mathcal{G}_1^2, \ldots, \mathcal{G}_n^2$. Afterwards, for every $C_i \in \Omega$, we take $\mathcal{G}_i$ as the vertex-disjoint union of $\mathcal{G}_i^1$ and $\mathcal{G}_i^2$. Due to Proposition 1, it is an easy exercise to check that the database relation R corresponding to the chosen representation graphs in fact represents Σ^+ and, thus, Σ.

Corollary 7. *Let $n \geq 2$. A set Σ of generalised and minimum participation constraints admits a $\mathcal{P}$-Armstrong relation if and only if Σ is conflicting or contains some finite participation constraint for every $C_i \in \Omega$.*

Proof. By virtue of the discussion at the end of Section 3, it suffices to show that Σ does not imply a finite participation constraint for a fixed $C_j \in \Omega$ unless it is conflicting or contains such a constraint. Suppose Σ is not conflicting and contains no finite participation constraint for C_j, but assume Σ implies some constraint $card^{part}(C_j) = b$. Adjoin new participation constraints $card^{part}(C_j) = b + 1$ and $card^{part}(C_i) = a_i$ for every C_i, $i \neq j$, without a finite participation constraint in Σ where a_i is defined as in the proof of the preceding theorem. By this theorem, the augmented constraint set may be represented by a database relation R. Hence, R satisfies Σ, but violates $card^{part}(C_j) = b$.

Corollary 8. *The rule system presented in Section 4 is sound and complete for generalised and minimum participation constraints, that is, a set Σ of generalised and minimum participation constraints is $\mathcal{P}$-closed if and only if Σ is syntactically closed w.r.t. these rules.*

7 Final Remarks

Before closing this paper, there are two remarks called for. The interested reader might wonder why we did not extend the concept of minimum participation constraints to subsets of Ω. This idea seems natural when considering the investigation of generalised participation constraints. The reason for this is twofold. Firstly, one may argue that lower bounds for the occurrence of pairs, triples, etc. of entries are rarely used in database design. Secondly, it turns out that matrix representations for this extended class of constraints are hard to guarantee in many cases. Let $n \geq 3$ and Σ contain the constraints $card^{part}(C_i) = g$ and $card^{min}(C_i) = g$ for every $C_i \in \Omega$, and $card^{part}(X) = 1$ and $card^{min}(X) = 1$ for every two-element subset $X \subseteq \Omega$. A database relation representing Σ is a transversal design $TD(n, g)$ with block size n and group size g. For relevant notions from combinatorial design theory, we refer the interested reader to [6]. Note that a transversal design $TD(n, g)$ corresponds to a set of $n-2$ mutually orthogonal Latin squares of order g. There are still many cases were the existence of transversal designs has not yet been decided. For example, the question whether there exists a $TD(5, 10)$, that is, a set of 3 mutually orthogonal Latin squares of order 10, is still open despite many efforts in the design theory community to settle this case. This, we hope, explains our reservations with extending the concept of minimum participation constraints to subsets of Ω.

Further, one may ask for matrix representations of minimum size. Let $n \geq 3$ and Σ contain the constraints $card^{part}(C_i) = g$ and $card^{min}(C_i) = g$ for every $C_i \in \Omega$, and $card^{part}(X) = 1$ for every two-element subset $X \subseteq \Omega$. If there exists a transversal design $TD(n, g)$, this would be a matrix representation of minimum size for Σ. Hence, again, it appears hard to determine the minimum size of a database relation representing a fairly simple set of generalised and minimum participation constraints. In either case, the known results on key dependencies [8, 15] show that in the worst case the minimum size will be exponential in n.

References

1. Armstrong, W.W.: Dependency structures of database relationship. Inform. Process. 74, 580–583 (1974)
2. Beeri, C., Dowd, M., Fagin, R., Statman, R.: On the structure of Armstrong relations for functional dependencies. J. ACM 31, 30–46 (1984)
3. Bennett, F.E., Wu, L.: On minimum matrix representation of Sperner systems. Discrete Appl. Math. 81, 9–17 (1998)
4. Bollobás, B.: Extremal graph theory. Academic Press, London (1978)
5. Chen, P.P.: The entity-relationship model: towards a unified view of data. ACM Trans. Database Systems 1, 9–36 (1976)
6. Colbourn, C.J., Dinitz, J.H. (eds.): The CRC handbook of combinatorial designs. CRC Press, Boca Raton (1996)
7. Demetrovics, J.: On the equivalence of candidate keys and Sperner systems. Acta. Cybernet. 4, 247–252 (1979)
8. Demetrovics, J., Gyepesi, G.: On the functional dependency and some generalizations of it. Acta. Cybernet. 5, 295–305 (1981)
9. Demetrovics, J., Katona, G.O.H., Sali, A.: Design type problems motivated by database theory. J. Statist. Plann. Inference 72, 149–164 (1998)
10. Erdős, P.: Extremal problems in graph theory. In: A seminar in graph theory, Holt, Rinehart and Winston (1967)
11. Hajnal, A., Szemerédi, E.: Proof of a conjecture of Erdős. In: Combinatorial theory and its applications. North-Holland, Amsterdam (1970)
12. Hartmann, S.: On the implication problem for cardinality constraints and functional dependencies. Ann. Math. Art. Intell. 33, 253–307 (2001)
13. Hartmann, S.: Reasoning about participation constraints and Chen's constraints. In: ADC. CRPIT, vol. 17, pp. 105–113. ACS (2003)
14. Hartmann, S., Link, S.: Numerical keys for XML. In: Leivant, D., de Queiroz, R. (eds.) WoLLIC 2007. LNCS, vol. 4576, pp. 203–217. Springer, Heidelberg (2007)
15. Katona, G.O.H., Tichler, K.: Some contributions to the minimum representation problem of key systems. In: Dix, J., Hegner, S.J. (eds.) FoIKS 2006. LNCS, vol. 3861, pp. 240–257. Springer, Heidelberg (2006)
16. Lenzerini, M., Nobili, P.: On the satisfiability of dependency constraints in entity-relationship schemata. Inform. Systems 15, 453–461 (1990)
17. Levene, M., Loizou, G.: A guided tour of relational databases and beyond. Springer, Heidelberg (1999)
18. Liddle, S.W., Embley, D.W., Woodfield, S.N.: Cardinality constraints in semantic data models. Data Knowledge Eng. 11, 235–270 (1993)
19. Lopes, S., Petit, J.-M., Lakhal, L.: Efficient discovery of functional dependencies and Armstrong relations. In: Zaniolo, C., Grust, T., Scholl, M.H., Lockemann, P.C. (eds.) EDBT 2000. LNCS, vol. 1777, pp. 350–364. Springer, Heidelberg (2000)
20. Mannila, H., Räihä, K.: Design by Example: An application of Armstrong relations. J. Comput. Syst. Sci., 126–141 (1986)
21. Mannila, H., Räihä, K.: The design of relational databases. Addison-Wesley, Reading (1992)
22. McAllister, A.: Complete rules for n-ary relationship cardinality constraints. Data Knowledge Eng. 27, 255–288 (1998)
23. Thalheim, B.: Dependencies in relational databases. Teubner (1991)
24. Thalheim, B.: Fundamentals of cardinality constraints. In: Pernul, G., Tjoa, A.M. (eds.) ER 1992. LNCS, vol. 645, pp. 7–23. Springer, Heidelberg (1992)
25. Thalheim, B.: Entity-relationship modeling. Springer, Berlin (2000)

Toward Formal Semantics for Data and Schema Evolution in Data Stream Management Systems

Rafael J. Fernández-Moctezuma[1], James F. Terwilliger[2], Lois M.L. Delcambre[1], and David Maier[1]

[1] Department of Computer Science
Portland State University, Portland OR 97207, USA
{rfernand,lmd,maier}@cs.pdx.edu
[2] Microsoft Research, Redmond WA 98052, USA
james.terwilliger@microsoft.com

Abstract. Data Stream Management Systems (DSMSs) do not statically respond to issued queries — rather, they continuously produce result streams to standing queries, and often operate in a context where any interruption can lead to data loss. Support for schema evolution in continuous query processing is currently unaddressed. In this work we address evolution in DSMSs by proposing semantics for three evolution primitives: Add Attribute and Drop Attribute (schema evolution), and Alter Data (data evolution). We characterize how a subset of commonly used query operators in a DSMS act on and propagate these primitives.

1 Introduction

Similar to traditional Database Management Systems (DBMSs), Data Stream Management Systems (DSMSs) allow users to declaratively construct and execute queries. Unlike a DBMS, which allows a user to issue a query over a persistent database instance and retrieve a single answer, a DSMS continuously produces results from a *standing* query as new data flows through the system. Data often arrives faster than it can be persisted in a streaming environment, so any time a standing query comes down, data is potentially lost.

Data streams are unbounded in nature, but assuming that their schema remains static is unreasonable. With current DSMS technologies, supporting schema evolution entails bringing down a standing query and instantiating an evolved version of the query against the new schema. This approach may be sufficient for some queries, but it fails in the presence of stateful operators: either state must be propagated from the old query to the evolved query to resume computation, or query answers derived from old state might be lost, in addition to data lost during downtime. We address the problem of schema and data evolution in DSMSs by describing the semantics of a system where queries can automatically respond to evolution and thus continue to actively processing data.

Traditional challenges in DSMSs such as low-latency result production and efficient use of resources have been addressed by adding *markers* in the stream that

C.A. Heuser and G. Pernul (Eds.): ER 2009 Workshops, LNCS 5833, pp. 85–94, 2009.

signal canonical progress. Examples include CTIs in CEDR [2], Heartbeats in Gigascope [3], and Punctuations in NiagaraST [4,13]. We see an opportunity to use embedded markers in a stream to support data and schema evolution. We introduce markers called *accents* into a data stream alongside tuples. The goal is to have a standing query that was instantiated before an accent is issued remain active during and after the processing of the accent. Operators in the query must be aware of and capable of handling accents. The following challenges arise for each operator: (1) Determine whether an operator can adapt to the schema and data changes in an accent, (2) define and implement the processing of each accent, for each operator, minimizing blocking and state accumulation while adapting to the evolution. We support evolution that affects only a subset of the stream.

2 The Evolution Problem Exposed

Our examples use the schema `sensors(ts, s, t)`, `placement(s, l)` to describe a stream of temperature readings from a sensor network. A monotonically increasing attribute `ts` represents timestamps, `s` uniquely identifies a sensor in the network, `l` is a sensor location, and `t` is a reading in Fahrenheit. Consider the following standing query in extended Relational Algebra:

$$\gamma^{AVERAGE}_{\{l,wid\},t}(\mathcal{W}^{5\text{ minutes}}_{ts,wid}(sensors \bowtie_{sensors.s=placement.s} placement))$$

This query joins the two data streams *sensors* and *placement* on *s*, then assigns each resulting tuple a window identifier, *wid*, (for tumbling windows of length 5 minutes) based on the value of the *ts* attribute. Finally, it computes the average temperature per location and time window. We explore several evolution scenarios and how they affect this query.

Example 1. A firmware update changes temperature units. Sensors receive an update and temperature is now reported in Celsius. Note that sensors may not receive the update simultaneously. Consider the example in Figure 1(a), where sensor 2 has implemented the firmware update as of time `ts = 5`, but none of the other sensors have. The content of one window includes tuples with both semantics, and the aggregation in the query produces inaccurate results. Thus, in addition to propagating the desired change through the operators (and the output), we also need to account for subsets of the stream having different semantics. Operators need to know which sensor readings need to be scaled back to Fahrenheit or scaled forward to Celsius.

Example 2. A firmware update extends the schema. Sensors now produce `pressure` information in addition to temperature, resulting in a new schema: `sensors(ts,s,t,p)`. The query shown above can either continue to produce the same results by projecting out the new attribute, or the query can be modified to produce both averages. In general, as a response to evolution, a query may be (1) unaffected, (2) adapted to accommodate the evolution, or (3) have its execution rendered impossible. As with Example 1, the stream may contain a mix of old and new data.

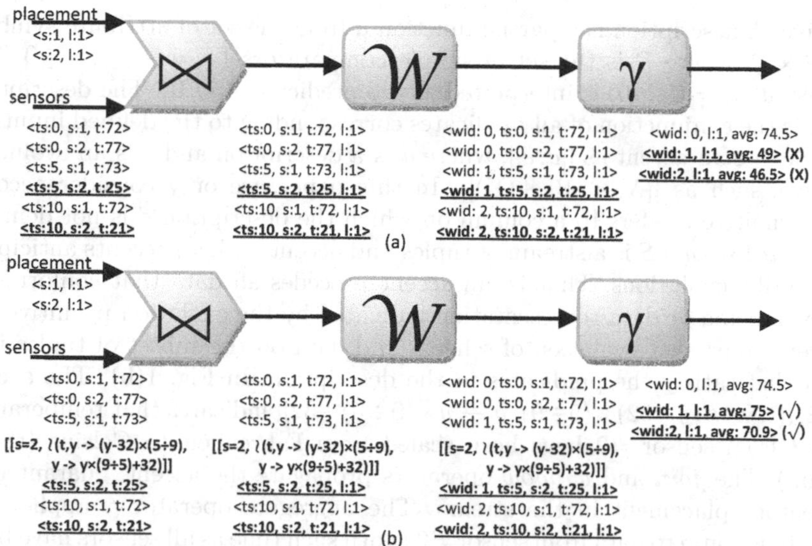

Fig. 1. Query plan with sample tuples from Example 1. (a) A change in domain in temperature readings from sensor 2 (tuples in bold) propagates through the query leading to an erroneous average. (b) Metadata (see Section 3) indicates the shift in domain allowing operators to handle the evolution and compute the correct average.

3 Modeling Streams and Evolution

To support schema flexibility, we redefine streams to have optional attributes in a similar fashion to Lorel [1] or RDF[11]. We define a tuple in a data stream to be a partial function t from the set of attribute symbols $\mathcal{C}$ to a set of attribute values $\mathcal{V}$. Let X be an element in $\mathcal{C}$. If $t(X)$ is defined, the tuple has the value $t(X)$ for attribute X; if $t(X)$ is undefined, it is semantically equivalent to evaluating to $\perp$ (null) for attribute X. We represent a tuple as a set of attribute-value pairs enclosed by angle brackets. The tuple with value 1 for A and value 2 for B, and $\perp$ for all other attributes is <A:1, B:2>.

A *stream* is a (possibly infinite) list T of tuples. The stream has an intrinsic order based on the arrival of tuples into the system. A *finite substream* of stream T is a finite set of tuples that appear consecutively in T.

We consider three evolution primitives in data streams:

- Add Attribute $+(A)$: attribute A is added to tuples in an incoming stream.
- Drop Attribute $-(A)$: attribute A is removed from an incoming stream.
- Alter Data $\wr(A, \alpha, \beta)$: data in attribute A is altered by the invertible and order-preserving function α whose inverse is β. Order-preserving means that if $x < y$, $\alpha(x) < \alpha(y)$. The inverse, β, can be used to return a value in a tuple to the previous domain.

We refer to the actions $+$, $-$, and $\wr$ as *evolution primitives.* To model evolution in a stream, we define an *accent* as a *description* coupled with an evolution primitive. A description is a partial function d from the set of attribute symbols $\mathcal{C}$ to $\mathcal{P} \times \mathcal{V}$, where $\mathcal{P}$ is the set of simple comparators $\{=, \neq, <, \leq, >, \geq\}$. For example, $d(A) = (<, 10)$ is interpreted as the predicate $A < 10$. The description d itself is the conjunction of all predicates corresponding to the defined inputs of d. We denote an accent as $[[d, e]]$, where d is a description and e is an evolution primitive, such as $[[X < 10, +(Y)]]$. In this paper, we only consider accents whose primitive e refers to a column on which the description d is not defined. An *accented stream* $\tilde{S}$ is a stream of tuples and accents, where accents anticipate and describe evolutions. That is, an accent precedes all data that conforms to the new schema or data representation indicated by the evolution primitives.

Accents represent evolution of schema and data on the subset of tuples in a stream that satisfy the predicates in the description. In Fig. 1(b), The accent $[[s = 2, \wr(t, y \rightarrow (y-32)\times(5\div9), y \rightarrow y\times(9\div5)+32]]$ indicates that temperature values t from sensor #2 have been scaled from Fahrenheit to Celsius (the α function). The *join* and *window* operators propagate the accent, guaranteeing anticipatory placement in the stream. The aggregate operator γ applies the inverse function β to data from sensor #2 (until such time as all sensors have been updated) to produce the correct aggregation values. No accent is propagated after γ as there is no evolution beyond that point.

An accent $a = [[d, e]]$ establishes, for tuples that satisfy description d, a relationship between tuples that occur before a versus after it in the stream:

- If e is an Add Attribute primitive $+(X)$, then tuples that follow accent a in the stream and satisfy d may be defined for $t(X)$, while tuples before a are undefined for $t(X)$ (up until an accent with primitive $-(X)$ appears prior to a in the stream, if present).
- If e is a Drop Attribute primitive $-(X)$, then $t(X)$ will be undefined for any tuple t that follows accent a in the stream that satisfies d (up until another Add Attribute primitive occurs, if any).
- If e is an Alter Data primitive $\wr(X, \alpha, \beta)$, then tuples that follow accent a in the stream and satisfy d have their values for attribute X scaled by α. Let t and t' be tuples in the stream that match description d. Recall that order of arrival induces an ordering on tuples. If $t < a < t'$, $t \equiv t'$ if $\forall_{B \in \mathcal{C}, B \neq X}(t(B) = t'(B)) \wedge (t(X) = \beta(t'(X))(\text{or equivalently } \alpha(t(X)) = t'(X)))$.

The primitive $\wr$ describes how to reason about equality over tuple values. For instance, consider the finite substream in Fig. 1(b). Any service operating on the stream, whether it be a query operator or an application reading a query output, may treat the `t` value at `ts = 0` and the `t` value at `ts = 5` for `s = 2` as equal, since $(77 - 32) \times (5/9) = 25$ (converting 77 Fahrenheit to Celsius).

4 Modeling Stream Operator Behavior

Most DSMSs support a variety of query operators that accept streams as input and produce streams as output. For instance, for any finite substream of tuples T as input to the operator $\sigma_{C<V}$, the operator produces as output the finite

substream $\{t \in T | t(C) < V\}$. In this section, we describe the semantics of stream operators when they work on accented streams, in particular how and when accents are output. We also detail the circumstances under which operators fail to support a particular accent, at which point a query would fail to support an evolution. Efficient management of state in DSMSs (for example, in stateful operators) is enabled by markers in the stream [2,3,4,13]. Our description of operators includes state associated with the management of accents. While we do not detail how this state can be purged in this work, we anticipate mechanisms similar to the ones currently used by DSMSs to efficiently manage state.

4.1 Stream Operators

We use the notation $\tilde{S}$ to denote accented streams and $\tilde{S}_O$ to denote an operator's output. We describe how operators react in the presence of accents, and how they propagate accents. Tuple processing occurs on arrival.

Select ($\sigma_{C\theta V}\tilde{S}$)**.** Select operator on attribute C, with comparator θ and value V on stream $\tilde{S}$. The attribute named in the selection condition C is *necessary* to the Select operator's functioning. This means that C must be present in the initial schema of the query. Responses to various accents are as follows:
- $[[d, +(X)]]$: If $X \in C$, the accent is not propagated since the attribute X must already exist, otherwise propagate $[[d, +(X)]]$ to $\tilde{S}_O$.
- $[[d, -(X)]]$: Regardless of description d, if $X \in C$, Select fails to support the evolution, and the query aborts execution. If $X \notin C$, output $[[d, -(X)]]$ to $\tilde{S}_O$.
- $[[d, \wr(X, \alpha, \beta)]]$: If $X \in C$, Select adjusts the selection predicate to be $C\theta\alpha(V)$ for tuples described by d (and $C\theta V$ for all others), and the accent $[[d, \wr(X, \alpha, \beta)]]$ is output to $\tilde{S}_O$. If $X \notin C$, no adjustment is necessary and $[[d, \wr(X, \alpha, \beta)]]$ is output to $\tilde{S}_O$.

Project ($\Pi_C\tilde{S}$)**.** Project tuples on the attribute set C. The project operator may seem trivial at a glance, but in fact is sensitive to accents with a description expressed in terms of attributes projected out ($R = \mathcal{C} - C$; $\mathcal{C}$ is the set of all possible attribute symbols). We define the operation of project assuming no mechanism exists to map descriptions referring to R to descriptions referring to C:
- $[[d, +(X)]]$: Any added attribute X does not match the projection set C, hence no accents with add column primitives need to be output to $\tilde{S}_O$.
- $[[d, -(X)]]$: If d refers only to attributes in C and $X \in C$, output $[[d, -(X)]]$ to $\tilde{S}_O$. Fail if d refers to attributes in R. Otherwise, no accent is output to $\tilde{S}_O$.
- $[[d, \wr(X, \alpha, \beta)]]$: If d refers only to attributes in C and $X \in C$, output $[[d, \wr(X, \alpha, \beta)]]$ to $\tilde{S}_O$. If d refers only to attributes in C and $X \notin C$, no accent is output. Fail if d refers to attributes in R.

Union ($\tilde{S} \cup \tilde{T}$)**.** Union of input streams $\tilde{S}$ and $\tilde{T}$. Because tuples are defined as partial functions, there is no need for there to be union compatibility between S and T. To coordinate across input streams, we assume maintenance of input-related state.

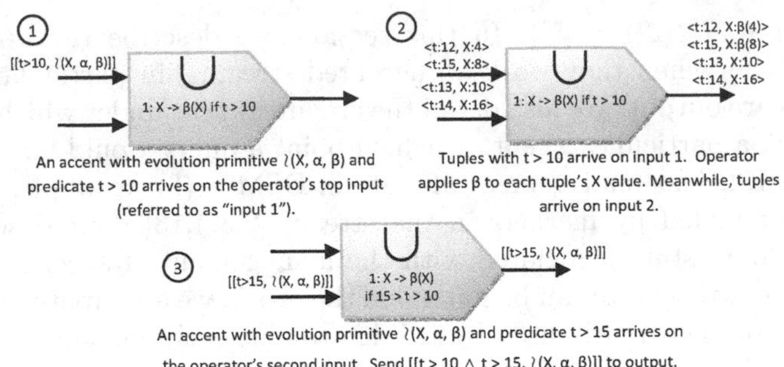

Fig. 2. An example of the Union operator responding to accents on both inputs

- $[[d, +(X)]]$: W.l.o.g., assume the accent is seen on input $\tilde{S}$. If an accent adding attribute X is not in input $\tilde{T}$'s state, add $[[d, +(X)]]$ to state of input $\tilde{S}$ and output $[[d, +(X)]]$ to $\tilde{S}_O$. If an accent $[[f, +(X)]]$ is in input $\tilde{T}$'s state, retrieve its description f. Replace $[[f, +(X)]]$ with $[[f \wedge \neg d, +(X)]]$ in input $\tilde{T}$'s state and output $[[f \wedge d, +(X)]]$ to $\tilde{S}_O$.
- $[[d, -(X)]]$: Similarly to accents with the primitive $+(X)$, Union maintains state, but does not propagate a drop attribute accent until it has been seen on both inputs.
- $[[d, \wr(X, \alpha, \beta)]]$: Similar to the cases when the accent has an Add or Drop Attribute primitive, Union does not propagate an accent with an Alter Data primitive until there is coordination. W.l.o.g., assume the accent is seen on input $\tilde{S}$ and no accent describing an Alter Data on X by α is in input $\tilde{T}$'s state. Any subsequent tuple t seen in input $\tilde{S}$ will have its X value replace with $\beta(t(X))$ in $\tilde{S}_O$, and $[[d, \wr(X, \alpha, \beta)]]$ is added to input $\tilde{S}$'s state. If an accent $[[f, \wr(X, \alpha, \beta)]]$ is in input $\tilde{T}$'s state, retrieve its description f. Replace $[[f, \wr(X, \alpha, \beta)]]$ with $[[f \wedge \neg d, \wr(X, \alpha, \beta)]]$ in input $\tilde{T}$'s state, output $[[f \wedge d, \wr(X, \alpha, \beta)]]$ to $\tilde{S}_O$, and stop altering t. This scenario is illustrated in Fig. 2.

Join ($\tilde{S} \bowtie_{\mathcal{J}(C_{\tilde{S}}, C_{\tilde{T}})} \tilde{T}$). Join input streams $\tilde{S}$ and $\tilde{T}$ using join conditions $\mathcal{J}$, which reference attributes $C_{\tilde{S}}$ and $C_{\tilde{T}}$ from streams $\tilde{S}$ and $\tilde{T}$ respectively. Join's behavior on accents that reference the join attributes are operationally the same as Union, i.e., there needs to be synchronization before propagation, and that synchronization is irrespective of the specific join condition $\mathcal{J}$. Unlike Union, but similar to Select, dropping attributes named in the join condition causes Join to fail.

- $[[d, +(X)]]$: W.l.o.g., assume the accent is seen on input $\tilde{S}$. If an accent adding attribute X is not in input $\tilde{T}$'s state, add $[[d, +(X)]]$ to state of input $\tilde{S}$ and output $[[d, +(X)]]$ to $\tilde{S}_O$. If an accent adding attribute X is in input $\tilde{T}$'s state ($[[f, +(X)]]$), abort execution (fail), since duplicate attributes are not allowed.

- $[[d, -(X)]]$: If $X \in (C_{\tilde{S}} \cup C_{\tilde{T}})$, fail. Otherwise, similarly to accents with the primitive $+(X)$, Join maintains the state of which attributes have been seen so far.
- $[[d, \wr(X, \alpha, \beta)]]$: Join does not propagate an accent with an Alter Data primitive on a join condition column until there is coordination (all others proceed directly to output). W.l.o.g., assume the accent is seen on input $\tilde{S}$ and no accent describing an Alter Data on X by α is in input $\tilde{T}$'s state. Any subsequent tuple t seen in input $\tilde{S}$ will have its X component replaced by $\beta(t(X))$, and $[[d, \wr(X, \alpha, \beta)]]$ is added to input $\tilde{S}$'s state. If an accent $[[f, \wr(X, \alpha, \beta)]]$ is in input $\tilde{T}$'s state, retrieve its description f. Replace $[[f, \wr(X, \alpha, \beta)]]$ with $[[f \wedge \neg d, \wr(X, \alpha, \beta)]]$ in input $\tilde{T}$'s state, output $[[f \wedge d, \wr(X, \alpha, \beta)]]$ to $\tilde{S}_O$, update the join condition s.t. the comparisons are performed on $\alpha(X)$, and stop altering t.

Window ($\mathcal{W}^w_{C,wid}\tilde{S}$). Windowing operator for input stream $\tilde{S}$, argument attribute C, output attribute *wid*, and windowing function $w : C \rightarrow wid$. The Window operator applies a function w to values produced by $t \in \tilde{S}$ to produce a Window ID, most often used for grouping by subsequent aggregate operators. Attributes in C are *essential* to Window's operation.

- $[[d, +(X)]]$: If $X \in C$, the accent is not propagated since the attribute X must already exist, otherwise Window propagates $[[d, +(X)]]$ to $\tilde{S}_O$.
- $[[d, -(X)]]$: Regardless of description d, if $X \in C$, Window fails to support the evolution, and the query aborts execution. If $X \notin C$, Window propagates $[[d, -(X)]]$ to $\tilde{S}_O$.
- $[[d, \wr(X, \alpha, \beta)]]$: If $X \in C$, Window adjusts the windowing function to operate on $\beta(X)$ for tuples described by d, and the accent is not emitted in $\tilde{S}_O$. If $X \notin C$, make no adjustment and output $[[d, \wr(X, \alpha, \beta)]]$ to $\tilde{S}_O$.

Aggregate ($\gamma^f_{\boldsymbol{G},\boldsymbol{E}}S$). Aggregation operator with aggregate function f, grouping attributes G, and exclusion attributes E on input S. This operator performs a grouping operation, similar to the GROUP BY clause in SQL. The operator partitions the input tuples of stream S according to their values in attributes G, and produces the result of the aggregation function for each partition. Unlike the traditional GROUP BY clause (and also unlike the standard stream query aggregate operator as used in Fig. 1), rather than specify the attributes that hold the aggregate data (like the `t` attribute in the example in Section 2) we specify the attributes that should not be aggregated (like the `ts` and `s` attributes in the same example). In other words, the operator projects away attributes $\boldsymbol{E}$ and aggregates on all other non-grouping attributes. This method of specification allows data found in newly added attributes to be aggregated (like the `pressure` attribute in Example 2 in Section 2). This definition suggests that Aggregate has operational semantics similar to both Select and Project:

- $[[d, +(X)]]$: Propagate $[[d, +(X)]]$ to $\tilde{S}_O$, adjust internal aggregation to apply the aggregate f to all tuples' X attribute described by d.
- $[[d, -(X)]]$: If $X \in G$, Aggregate fails to support the evolution, and the query aborts execution. If $X \notin G$, propagate $[[d, -(X)]]$ to $\tilde{S}_O$, and eliminate partial aggregates in state for attribute X.

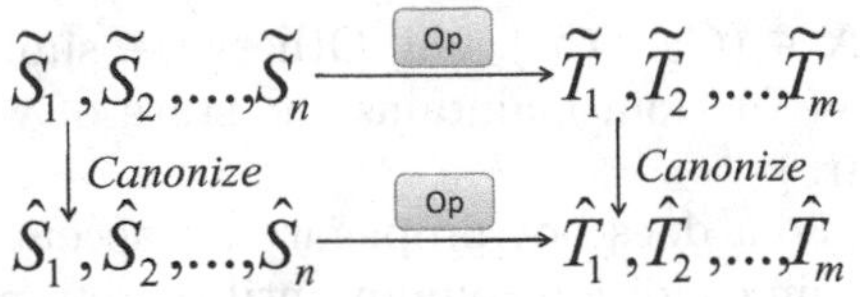

$$\tilde{S}_1, \tilde{S}_2, \ldots, \tilde{S}_n \xrightarrow{\text{Op}} \tilde{T}_1, \tilde{T}_2, \ldots, \tilde{T}_m$$

Canonize Canonize

$$\hat{S}_1, \hat{S}_2, \ldots, \hat{S}_n \xrightarrow{\text{Op}} \hat{T}_1, \hat{T}_2, \ldots, \hat{T}_m$$

Fig. 3. Commutativity diagram for an operator *Op*. Top: Accented streams are processed *Op* on inputs $\tilde{S}_i$ to output accented streams $\tilde{T}_i$. Bottom: Canonical version of the input streams S_i result in canonical output streams T_i. Accented streams can be canonized on both ends to show equivalence.

- $[[d, \wr(X, \alpha, \beta)]]$: If $X \in E$, no accent is output. Fail if d refers to attributes in E. Since partial aggregates are being computed, maintain windows with partial aggregates and output $\beta(X)$ for those windows. Otherwise, propagate a new Alter Data accent that describes groups formed after the alter data evolution and do not apply β.

4.2 Notion of Correctness

Let $s = (s_1, s_2, \ldots, s_n)$ represent a finite substream of adjacent items in an accented stream $\tilde{S}$. A *replacement* for that substream is another finite substream $t = (t_1, t_2, \ldots, t_n)$ such that replacing s with t in-place in $\tilde{S}$ does not affect the information content of the stream. For instance, $([[d, e]], t)$ is a replacement for $(t, [[d, e]])$ for any tuple t that does not match description d. Two finite substreams of accented streams $\tilde{S}$ and $\tilde{S'}$ are *content equivalent* $\tilde{S} \equiv \tilde{S'}$ if one can transform $\tilde{S}$ into $\tilde{S'}$ using replacements. A finite substream of an accented stream is *canonical* when all accents with a $+$ primitive appear at the beginning of the substream, followed by all tuples, followed by all accents with a $-$ primitive, and ending with all accents with a $\wr$ primitive. To *canonize* a substream $\tilde{S}$ is to find a canonical substream (denoted as $\hat{S}$) that is content equivalent to the original substream. We use canonization as part of our formal descriptions; operators do not canonize results as part of operation. For an operator implementation to be *correct*, it must respect the commutativity diagram in Fig. 3 on any input where the query does not abort. The diagram states that for a given set of accented input streams, the operator produces accented output streams whose canonical versions are equivalent to the output the operator produces on canonical input. Accents in output streams must respect accent properties, e.g., that following a drop attribute accent in a stream, all tuples matching the accent's description are undefined for the dropped attribute. Our notion of correctness implies that there is no information loss due to accent-aware stream processing. We have a proof sketch of the Union operator using these notions, and are working on creating proofs for the remaining operators.

5 Related Work

To our knowledge, there is no existing work modeling or implementing evolution in DSMSs, although schema evolution has been amply addressed in DBMSs [10].

Schema evolution research often focuses on mitigating the effect of evolutions on artifacts that rely on the schema, such as embedded SQL [5] or adjusting schema mappings to address new schemas [16].

Extract-transform-load (ETL) workflows are similar to streaming queries; they are a composition of atomic data transformations (called *activities*) that determine data flow through a system. Unlike streaming queries, ETL workflows do not execute continuously and are typically not as resource-constrained as DSMSs. Papastefanatos et al. addressed schema evolution on an ETL workflow by attaching *policies* to each activity. Policies semi-automatically adjust the activity parameters based on schema evolution primitives that propagate through the activities [8,9]. Unlike our approach, ETL research does not need to address how to maintain uptime of queries or the intermingling of schema evolution with data in the presence of accumulated state.

A final area of related work is schema mapping maintenance, where one "heals" a mapping between schemas S and T when either schema evolves (say, from T to T'), thus creating a new mapping (say, from S to T') that respects the semantics of the original mapping, or failing if impossible. Both-as-View (BAV) describes a schema mapping as a sequence of incremental steps [7]. Changes made to a schema become a second sequence of steps that is then composed with the mapping sequence [6]. A similar approach is possible if a mapping is specified using source-to-target tuple-generating dependencies (st-tgds), both when the schema evolution can be reduced into discrete transformations [14,15] or itself specified as st-tgds [16]. The Guava framework is another approach to mapping evolution [12]. A Guava mapping is expressed as a set of algebraic transformations, through which schema evolution primitives expressed against the source schema propagate to the target schema and update it as well.

6 Conclusions and Future Work

We have introduced semantics toward supporting schema and data evolution in stream systems by introducing the notion of an accented stream. Our contributions include adding markers to announce evolutions to operators in a standing query. In general, the work reported here for stream query operators suggests we can extend the capability of conceptual models, as well as the systems that implement them, by expanding the scope of ordinary query operators to handle evolution (as well as other things). We detailed an initial set of evolutions and the effect they have on common stream operators. We provide here a framework that allows stream engines to support schema and data evolution without bringing down standing queries when the evolutions do not render a query meaningless. Moreover, our approach supports simultaneous input streams at different stages in their respective evolutions. Future work will more fully specify the action of the six operators presented here and characterize additional operators, as well as complete the proofs of correctness for stream operators. Some operational considerations were left out in this paper, such as the effect of Alter Data primitives on other accents persistent in state. Future work will address these issues in more detail, in particular in out-of-order architectures [4]. The interaction of accents and punctuations will also be explored, as will the impact of accents on query performance and efficient operator implementation. Future work will also characterize evolution in systems whose consistency is achieved over time.

Acknowledgements

This work is supported by CONACyT México (178258), OTREC, and NSF (0534762, 0612311). We thank Jonathan Goldstein, Badrish Chandramouli, and Kristin A. Tufte for their comments.

References

1. Abiteboul, S., Quass, D., McHugh, J., Widom, J., Wiener, J.L.: The Lorel Query Language for Semistructured Data. Int. J. on Digital Libraries 1(1), 68–88 (1997)
2. Barga, R.S., Goldstein, J., Ali, M.H., Hong, M.: Consistent Streaming Through Time: A Vision for Event Stream Processing. In: CIDR 2007, pp. 363–374 (2007)
3. Johnson, T., Muthukrishnan, S., Shkapenyuk, V., Spatscheck, O.: A Heartbeat Mechanism and its Application in Gigascope. In: VLDB 2005, pp. 1079–1088 (2005)
4. Li, J., Tufte, K., Shkapenyuk, V., Papadimos, V., Johnson, T., Maier, D.: Out-of-Order Processing: A New Architecture for High-Performance Stream Systems. Proc. VLDB Endow. 1(1), 274–288 (2008)
5. Maule, A., Emmerich, W., Rosenblum, D.S.: Impact Analysis of Database Schema Changes. In: ICSE 2008, pp. 451–460 (2008)
6. McBrien, P., Poulovassilis, A.: Schema Evolution in Heterogeneous Database Architectures, a Schema Transformation Approach. In: Pidduck, A.B., Mylopoulos, J., Woo, C.C., Ozsu, M.T. (eds.) CAiSE 2002. LNCS, vol. 2348, pp. 484–499. Springer, Heidelberg (2002)
7. McBrien, P., Poulovassilis, A.: Data Integration by Bi-Directional Schema Transformation rules. In: ICDE 2003, pp. 227–238 (2003)
8. Papastefanatos, G., Vassiliadis, P., Simitsis, A., Vassiliou, Y.: What-if analysis for data warehouse evolution. In: Song, I.-Y., Eder, J., Nguyen, T.M. (eds.) DaWaK 2007. LNCS, vol. 4654, pp. 23–33. Springer, Heidelberg (2007)
9. Papastefanatos, G., Vassiliadis, P., Simitsis, A., Vassiliou, Y.: Design metrics for data warehouse evolution. In: Li, Q., Spaccapietra, S., Yu, E., Olivé, A. (eds.) ER 2008. LNCS, vol. 5231, pp. 440–454. Springer, Heidelberg (2008)
10. Rahm, E., Bernstein, P.A.: An Online Bibliography on Schema Evolution. SIGMOD Record 35(4), 30–31 (2006)
11. W3C: Resource Description Framework (RDF), `http://www.w3.org/RDF/`
12. Terwilliger, J.F.: Graphical User Interfaces as Updatable Views. PhD thesis, Portland State University, Portland, Oregon, USA (2009)
13. Tucker, P.A., Maier, D., Sheard, T., Fegoras, L.: Exploiting Punctuation Semantics in Continuous Data Streams. IEEE Transactions on Knowledge and Data Engineering 15(3), 555–568 (2003)
14. Velegrakis, Y., Miller, R.J., Popa, L.: Mapping Adaptation Under Evolving Schemas. In: VLDB 2003, pp. 584–595 (2003)
15. Velegrakis, Y., Miller, R.J., Popa, L.: Preserving Mapping Consistency Under Schema Changes. VLDB J. 13(3), 274–293 (2004)
16. Yu, C., Popa, L.: Semantic Adaptation of Schema Mappings When Schemas Evolve. In: VLDB 2005, pp. 1006–1017 (2005)

XML Machines

Qing Wang[1] and Flavio A. Ferrarotti[2]

[1] University of Otago, Dunedin, New Zealand
qing.wang@otago.ac.nz
[2] University of Santiago of Chile and Yahoo! Research Latin America, Santiago of Chile
flavio@dcc.uchile.cl

Abstract. In order to capture the dynamics of XML databases a general model of tree-based database transformations is required. In this paper such an abstract computational model is presented, which brings together ideas from Abstract State Machines and monadic second-order logic. The model captures all XML database transformations.

1 Introduction

For a long time already database transformations as a unifying umbrella for queries and updates have been the focus of the database research community [11]. The logical foundations of queries have always been a central focus of interest for database theoreticians.

The sequential ASM thesis [6] defines sequential algorithms by a set of intuitive postulates. Gurevich shows that sequential Abstract State Machines capture these classes of algorithms. In our previous work [8] we picked up on this line of thought characterising database transformations in general. Similar to the ASM theses we formulated five postulates that define database transformations, and proved that these are exactly captured by a variant of ASMs called Abstract Database Transformation Machines (ADTMs).

While the characterisation of database transformations is done without any reference to a particular data model, the presence of backgrounds [1] supposedly enables tailoring the characterisation to any data model of interest. That is, the equivalence between the postulates and ADTMs holds with respect to a fixed background. In [9] we defined tree-based backgrounds and in doing so demonstrated how to adapt our general results to XML database transformations.

The disadvantage of the ADTM-based characterisation of XML database transformations is its lack of linkages to other work on theoretical foundations of XML databases. As XML is intrinsically connected with regular languages, a lot of research has been done to link XML with automata and logics [7]. Weak monadic second-order logics (MSO) are linked to regular tree languages [5,10] in the sense that a set of trees is regular iff it is in weak MSO with k successors.

Therefore, in this paper we define an alternative model of computation for XML database transformations, which exploits weak MSO. Pragmatically speaking the use of weak MSO formulae in forall and choice rules permits more flexible

C.A. Heuser and G. Pernul (Eds.): ER 2009 Workshops, LNCS 5833, pp. 95–104, 2009.

access to the database. As weak MSO subsumes first-order logic, it is straightforward to see that the model of XML machines captures all transformations that can be expressed by the ADTM model with tree-based backgrounds. As our main result in [8] states that already ADTMs capture all database transformations as defined by the intuitive postulates, it should also not come as a surprise that XML machines are in fact equivalent to ADTMs with tree-based backgrounds. For the proof we simply have to show that XML machines satisfy the postulates, i.e. in a sense the hard part of the proof is already captured by the main characterisation theorem in [8].

2 XML Trees

It is common to regard an XML document as an unranked tree, in which nodes may have an unbounded but finite number of children nodes.

Definition 1. An *unranked tree* is a structure $(\mathcal{O}, \prec_c, \prec_s)$ consisting of a finite, non-empty set $\mathcal{O}$ of node identifiers called *tree domain*, ordering relations $\prec_c$ and $\prec_s$ over $\mathcal{O}$ called *child relation* and *sibling relation*, respectively, satisfying the following conditions: (i) there exists a unique, distinguished node $o_r \in \mathcal{O}$ (called the *root* of the tree) such that for all $o \in \mathcal{O} - \{o_r\}$ there is exactly one $o' \in \mathcal{O}$ with $o' \prec_c o$, and (ii) whenever $o_1 \prec_s o_2$ holds, then there is some $o \in \mathcal{O}$ with $o \prec_c o_i$ for $i = 1, 2$.

For $x_1 \prec_c x_2$ we say that x_2 is a *child* of x_1; for $x_1 \prec_s x_2$ we say that x_2 is the *next sibling* to the right of x_1. In order to obtain XML trees from this, we require the nodes of an unranked tree to be labelled, and the leaves, i.e. nodes without children, to be associated with values. Therefore, we fix a finite, non-empty set Σ of *labels*, and a finite family $\{\tau_i\}_{i \in I}$ of data types. Each data type τ_i is associated with a *value domain* $dom(\tau_i)$. The corresponding *universe* U contains all possible values of these data types, i.e. $U = \bigcup_{i \in I} dom(\tau_i)$.

Definition 2. An *XML tree* t (over the set of labels Σ with values in the universe U corresponding to the family $\{\tau_i\}_{i \in i}$ of data types) is a triple (t_t, ω_t, υ_t) consisting of an unranked tree $t_t = (\mathcal{O}_t, \prec_c, \prec_s)$, a total *label function* ω_t: $\mathcal{O}_t \to \Sigma$, and a partial *value function* υ_t: $\mathcal{O}_t \to U$ defined for all leaves in t_t.

The set of all XML trees over Σ (neglecting the universe U) is denoted as T_Σ.

Definition 3. An XML tree t_1 is said to be the *subtree* of an XML tree t_2 at node $o \in \mathcal{O}_{t_2}$ iff there exists an embedding h: $\mathcal{O}_{t_1} \hookrightarrow \mathcal{O}_{t_2}$ satisfying the following properties: (i) the root of tree t_1 is o; (ii) whenever $o_1 \prec_c o_2$ holds in t_1, then $h(o_1) \prec_c h(o_2)$ holds in t_2; (iii) whenever $o_1 \prec_c o_2$ holds in t_2 with $o_1 \in \mathcal{O}_{t_1}$, then also $o_1 \prec_c o_2$ holds in t_1; (iv) whenever $o_1 \prec_s o_2$ holds in t_1, then $h(o_1) \prec_s h(o_2)$ holds in t_2; (v) $\omega_{t_1}(o') = \omega_{t_2}(o')$ holds for all $o' \in \mathcal{O}_{t_1}$; (vi) for all $o' \in \mathcal{O}_{t_1}$ either $\upsilon_{t_1}(o') = \upsilon_{t_2}(o')$ holds or otherwise both sides are undefined.

We use the notation $\widehat{o}$ to denote the subtree of an XML tree t rooted at node o for $o \in \mathcal{O}_t$, and $root(t)$ to denote the root node of an XML tree t. A sequence $t_1, ..., t_k$ of XML trees is called an *XML hedge* or simply a *hedge*, and a multiset $\{\!\{t_1, ..., t_k\}\!\}$ of XML trees is called an *XML forest* or simply a *forest*. The notion of forest is indispensable in situations where order is irrelevant, e.g. when representing attributes of a node, but duplicates are desirable, e.g. for computations in parallel on identical subtrees. ε denotes the *empty hedge*.

In order to define flexible operations on XML trees it will be necessary to select tree portions of interest. Such portions can be subtrees, but occasionally we will need more general structures. This will be supported by XML contexts.

Definition 4. An *XML context* over an alphabet Σ ($\xi \notin \Sigma$) is an unranked tree t over $\Sigma \cup \{\xi\}$, i.e., $t \in T_{\Sigma \cup \{\xi\}}$, such that for each tree t exactly one leaf node is labelled with the symbol ξ and has undefined value, and all other nodes in a tree are labelled and valued in the same way as an XML tree defined in Definition 2.

The context with a single node labelled ξ is called the *trivial context* and denoted as ξ. With contexts we can now define substitution operations that replace a subtree of a tree or context by a new XML tree or context. To ensure that the special label ξ occurs at most once in the result, we distinguish four kinds of substitutions, where $[\widehat{o} \mapsto t]$ indicates substituting t for the subtree rooted at o.

Tree-to-tree substitution: For an XML tree $t_1 \in T_{\Sigma_1}$ with a node $o \in \mathcal{O}_{t_1}$ and an XML tree $t_2 \in T_{\Sigma_2}$ the result $t_1[\widehat{o} \mapsto t_2]$ is an XML tree in $T_{\Sigma_1 \cup \Sigma_2}$.

Tree-to-context substitution: For an XML tree $t_1 \in T_{\Sigma_1}$ with a node $o \in \mathcal{O}_{t_1}$, the result $t_1[\widehat{o} \mapsto \xi]$ is an XML context in $T_{\Sigma_1 \cup \{\xi\}}$.

Context-to-context substitution: For an XML context $c_1 \in T_{\Sigma_1 \cup \{\xi\}}$ with a node $o \in \mathcal{O}_{c_1}$ and an XML tree $t_2 \in T_{\Sigma_2}$, the result $c_1[\widehat{o} \mapsto t_2]$ is an XML context in $T_{\Sigma_1 \cup \Sigma_2 \cup \{\xi\}}$.

Context-to-tree substitution: For an XML context $c_1 \in T_{\Sigma_1 \cup \{\xi\}}$ and an XML tree $t_2 \in T_{\Sigma_2}$ the result $c_1[\xi \mapsto t_2]$ is an XML tree in $T_{\Sigma_1 \cup \Sigma_2}$.

The correspondence between an XML document and an XML tree is straightforward. Each element of an XML document corresponds to a node of the XML tree, and the subelements of an element define the children nodes of the node corresponding to the element. The nodes for elements are labeled by element names, and character data of an XML document correspond to values of leaves in an XML tree. As our main focus is on structural properties of an XML document, attributes are handled, as if they were subelements.

3 Tree Algebra

Our objective is to provide manipulation operations on XML trees at a higher level than individual nodes and edges. So we need some tree constructs to extract arbitrary tree portions of interest. For this we provide two selector constructs, which will result in subtrees and contexts, respectively. For an XML tree $t = (t_t, \omega_t, v_t)$ these constructs are defined: (i) *context* is a binary, partial function defined on pairs (o_1, o_2) of nodes with $o_i \in \mathcal{O}_t$ $(i = 1, 2)$ such that o_1 is an ancestor of o_1, i.e. $o_1 \prec_c^* o_2$ holds for the transitive closure $\prec_c^*$ of $\prec_c$. We have $context(o_1, o_2) = \widehat{o}_1[\widehat{o}_2 \mapsto \xi]$. (ii) *subtree* is a unary function defined on $\mathcal{O}_t$. We have $subtree(o) = \widehat{o}$.

We now need some algebra operations to recombine tree portions to form a new XML tree. We define a many-sorted algebra using three sorts: **L** for labels, **H** for hedges, and **C** for contexts, along with a set $\mathcal{F}=\{\iota, \delta, \varsigma, \rho, \kappa, \eta, \sigma\}$ of function symbols with the following signatures: $\iota : \mathbf{L} \times \mathbf{H} \to \mathbf{H}$, $\delta : \mathbf{L} \times \mathbf{C} \to \mathbf{C}$, $\varsigma : \mathbf{H} \times \mathbf{C} \to \mathbf{C}$, $\rho : \mathbf{H} \times \mathbf{C} \to \mathbf{C}$, $\kappa : \mathbf{H} \times \mathbf{H} \to \mathbf{H}$, $\eta : \mathbf{C} \times \mathbf{H} \to \mathbf{H}$, $\sigma : \mathbf{C} \times \mathbf{C} \to \mathbf{C}$. Given that a fixed alphabet Σ and two special symbols ε and ξ, the set $\mathcal{T}$ of terms over $\Sigma \cup \{\varepsilon, \xi\}$ comprises label terms, hedge terms, and context terms. That is, $\mathcal{T} = T_\mathcal{L} \cup T_\mathcal{H} \cup T_\mathcal{C}$, where $T_\mathcal{L}$, $T_\mathcal{H}$ and $T_\mathcal{C}$ stand for the sets of terms over sorts **L**, **H** and **C**, respectively. The set of label terms $T_\mathcal{L}$ is simply the set of labels, i.e. $T_\mathcal{L} = \Sigma$. The set $T_\mathcal{H}$ contains the subset $T_\mathcal{H}^s$ of tree terms, i.e. we identify trees with hedges of length 1, and is defined by $\varepsilon \in T_\mathcal{H}^s$, $t\langle h \rangle \in T_\mathcal{H}^s$ for $t \in \Sigma$ and $h \in T_\mathcal{H}$, and $t_1, ..., t_n \in T_\mathcal{H}^m$ for $t_i \in T_\mathcal{H}^s$ $(i = 1, ..., n)$. The set of context terms $T_\mathcal{C}$ is the smallest set with $\xi \in T_\mathcal{C}$ and $t\langle t_1, ..., t_n \rangle \in T_\mathcal{C}$ for a label $t \in \Sigma$ and terms $t_1, ..., t_n \in T_\mathcal{H}^s \cup T_\mathcal{C}$ such that exactly one t_i $(i = 1, \dots, n)$ is a context term in $T_\mathcal{C}$. Trees and contexts have a root, but hedges do not (unless they can be identified with a tree). For hedges of the form of $t\langle \varepsilon \rangle$ we use t as a notational shortcut. Furthermore, we use $\#t$ to denote the sort of a term t.

Example 1. Let $\Sigma = \{a, b, c, \tau_1, \tau_2\}$, then a, b, τ_1 and τ_2 are terms of sort **L**, $a\langle b\langle \tau_1 \rangle, c\langle \tau_2 \rangle \rangle$ and $b\langle \tau_1 \rangle, a\langle a\langle \tau_1 \rangle, \tau_2 \rangle$ are terms of sort **H**, and $a\langle a\langle \tau_1 \rangle, \xi, \tau_1 \rangle$ is a term of sort **C**.

Intuitively speaking, the functions ι and δ extend hedges and contexts upwards with labels, and ς and ρ incorporate hedges into non-trivial contexts from left or right, respectively, which takes care of the order of subtrees arising in XML as illustrated in Example 2. The function κ denotes hedge juxtaposition, and likewise σ is context composition. The function η denotes context substitution, i.e. substituting the variable ξ in a context with a hedge, which leads to a tree. A further illustration of these functions is provided in Figure 1, which can be formally defined as follows (the case $n = 0$ for $a\langle t_1, \dots, t_n \rangle$ leads to $a\langle \xi \rangle$): (1) $\iota(a, (t_1, \dots, t_n)) = a\langle t_1, \dots, t_n \rangle$, (2) $\delta(a, c) = a\langle c \rangle$, (3) $\varsigma((t_1, \dots, t_n), a\langle t'_1, \dots, t'_m \rangle) = a\langle t_1, \dots, t_n, t'_1, \dots, t'_m \rangle$, (4) $\rho((t_1, \dots, t_n), a\langle t'_1, \dots, t'_m \rangle) = a\langle t'_1, \dots, t'_m, t_1, \dots, t_n \rangle$, (5) $\kappa((t_1, \dots, t_n), (t'_1, \dots, t'_m)) = t_1, \dots, t_n), t'_1, \dots, t'_m$, (6) $\eta(c, (t_1, \dots, t_n)) = c[\xi \mapsto t_1], \dots, c[\xi \mapsto t_n]$ and (7) $\sigma(c_1, c_2) = c_1[\xi \mapsto c_2]$.

Example 2. As shown in Figure 2, given that a context term $c_1 = a\langle b, \xi \rangle$ and a hedge term $t_1 = b\langle e \rangle, b\langle d \rangle$, we obtain $\varsigma(t_1, c_1) = a\langle b\langle e \rangle, b\langle d \rangle, b, \xi \rangle$ and $\eta(t_1, c_1) = a\langle b, \xi, b\langle \rangle, b\langle d \rangle \rangle$.

The proof of the following proposition is a straightforward exercise.

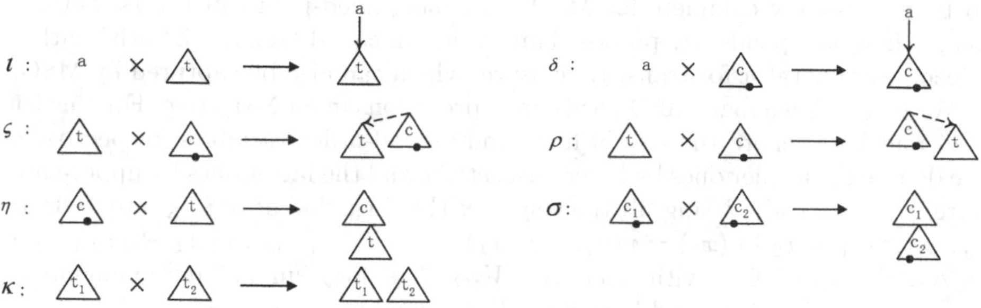

Fig. 1. XML tree algebra

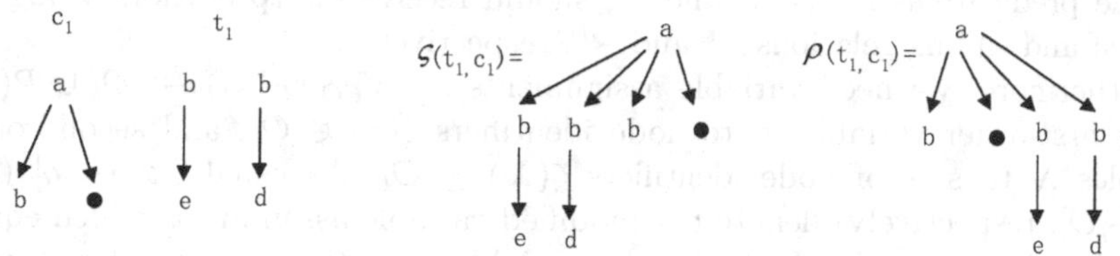

Fig. 2. An illustration of functions $\varsigma(t_1, c_1)$ and $\eta(t_1, c_1)$

Proposition 1. *The algebra defined above satisfying the following equations for $t_1, t_2, t_3 \in \mathcal{T}$ (i.e., whenever one of the terms in the equation is defined, the other one is defined, too, and equality holds): (i) $\eta(\sigma(t_1, t_2), t_3) = \eta(t_1, \sigma(t_2, t_3))$; (ii) $\sigma(\sigma(t_1, t_2), t_3) = \sigma(t_1, \sigma(t_2, t_3))$; (iii) $\kappa(\kappa(t_1, t_2), t_3) = \kappa(t_1, \kappa(t_2, t_3))$; (iv) $\eta(\delta(t_1, t_2), t_3) = \iota(t_1, \eta(t_2, t_3))$; (v) $\varsigma(t_1, \varsigma(t_2, t_3)) = \varsigma(\kappa(t_1, t_2), t_3)$; (vi) $\rho(t_1, \rho(t_2, t_3)) = \rho(\kappa(t_1, t_2), t_3)$; (vii) $\rho(t_3, \varsigma(t_1, t_2)) = \varsigma(t_1, \rho(t_3, t_2))$.*

Example 3. To illustrate equation $\eta(\delta(t_1, t_2), t_3) = \iota(t_1, \eta(t_2, t_3))$ let us take $t_1 = b$, $t_2 = a\langle b, \xi\rangle$ and $t_3 = b\langle e\rangle$. Then the left hand side becomes $\eta(\delta(t_1, t_2), t_3) = \eta(\delta(b, a\langle b, \xi\rangle), b\langle e\rangle) = \eta(b\langle a\langle b, \xi\rangle\rangle, b\langle e\rangle) = b\langle a\langle b, b\langle e\rangle\rangle\rangle$ and the right hand side becomes $\iota(t_1, \eta(t_2, t_3)) = \iota(b, \eta(a\langle b, \xi\rangle, b\langle e\rangle)) = \iota(b, a\langle b, b\langle e\rangle\rangle) = b\langle a\langle b, b\langle e\rangle\rangle\rangle$.

4 Weak Monadic Second-Order Logic

In this section we concentrate on a logic, which permits to to navigate within an XML tree. This can be combined with the use of the tree algebra defined before to manipulate tree structures. We first provide a weak MSO logic over finite and unranked trees adopting the logic from [4] with the restriction that second-order variables can only be quantified over finite sets. The use of MSO logic is motivated by its close correspondence to regular languages, which is known already from early work of Büchi [3]. For XML the navigational part of XPath2.0 can capture first-order logic, and some extensions on XPath have been shown to be expressively complete for MSO, e.g. using fixed-point operators. Several people have independently proposed an extension called "regular XPath" with a Kleene star operator for transitive closure, which can also be captured by MSO.

We first define a logic MSO_X with interpretations in an XML tree. For this let V_{FO} and V_{SO} denote the sets of first- and second-order variables, respectively. We denote the former ones by lower-case letters and the latter ones by upper-case letters, respectively. Using abstract syntax the formulae of MSO_X are defined by: $\varphi \equiv x_1 = x_2 \mid \upsilon(x_1) = \upsilon(x_2) \mid \omega_a(x_1) \mid x \in X \mid x_1 \prec_c x_2 \mid x_1 \prec_s x_2 \mid \neg\varphi \mid \varphi_1 \wedge \varphi_2 \mid \exists x.\varphi \mid \exists X.\varphi$, with $x, x_1, x_2 \in V_{FO}$, $X \in V_{SO}$, unary function symbols υ and ω_a for all $a \in \Sigma$, and binary predicate symbols $\prec_c$ and $\prec_s$.

We interpret formulae of MSO_X for a given XML tree $t = (t_t, \omega_t, \upsilon_t)$ over the set Σ of labels with $t_t = (\mathcal{O}_t, \prec_c^t, \prec_s^t)$. Naturally, the function symbols ω_a and υ should be interpreted by the labelling and value functions ω_t and υ_t, respectively, and the predicate symbols $\prec_c$ and $\prec_s$ should receive interpretations using the children and sibling relations $\prec_c^t$ and $\prec_s^t$, respectively.

Furthermore, we need variable assignments $\zeta : V_{FO} \cup V_{SO} \to \mathcal{O}_t \cup \mathcal{P}(\mathcal{O}_t)$ taking first-order variables x to node identifiers $\zeta(x) \in \mathcal{O}_t$, and second-order variables X to sets of node identifiers $\zeta(X) \subseteq \mathcal{O}_t$. As usual $\zeta[x \mapsto o]$ (and $\zeta[X \mapsto O]$, respectively) denote the modified variable assignment, which equals ζ on all variables except the first-order variable x (or the second-order variable X, respectively), for which we have $\zeta[x \mapsto o](x) = o$ (and $\zeta[X \mapsto O](X) = O$, respectively). For the XML tree t and a variable assignment ζ we obtain the interpretation $val_{t,\zeta}$ on terms and formulae as follows. Terms are either variables x, X or have the form $\upsilon(x)$, thus are interpreted as $val_{t,\zeta}(x) = \zeta(x)$, $val_{t,\zeta}(X) = \zeta(X)$, and $val_{t,\zeta}(\upsilon(x)) = \upsilon_t(\zeta(x))$. For formulae φ we use $[\![\varphi]\!]_{S,\zeta}$ to denote its interpretation by a truth value, and obtain:

- $[\![\tau_1 = \tau_2]\!]_{t,\zeta} = true$ iff $val_{t,\zeta}(\tau_1) = val_{t,\zeta}(\tau_2)$ holds for the terms τ_1 and τ_2,
- $[\![\omega_a(x)]\!]_{t,\zeta} = true$ holds iff $\omega_t(val_{t,\zeta}(x)) = a$,
- $[\![x \in X]\!]_{t,\zeta} = true$ iff $val_{t,\zeta}(x) \in val_{t,\zeta}(X)$,
- $[\![\neg\varphi]\!]_{t,\zeta} = true$ iff $[\![\varphi]\!]_{t,\zeta} = false$,
- $[\![\varphi_1 \wedge \varphi_2]\!]_{t,\zeta} = true$ iff $[\![\varphi_1]\!]_{t,\zeta} = true$ and $[\![\varphi_2]\!]_{t,\zeta} = true$,
- $[\![\exists x.\varphi]\!]_{t,\zeta} = true$ iff $[\![\varphi]\!]_{t,\zeta[x\mapsto o]} = true$ holds for some $o \in \mathcal{O}$,
- $[\![\exists X.\varphi]\!]_{t,\zeta} = true$ iff $[\![\varphi]\!]_{t,\zeta[X\mapsto O]} = true$ for some $O \subseteq \mathcal{O}$,
- $[\![x_1 \prec_c x_2]\!]_{t,\zeta} = true$ iff $val_{t,\zeta}(x_2)$ is a child node of $val_{t,\zeta}(x_1)$ in t, i.e. $val_{t,\zeta}(x_1) \prec_c^t val_{t,\zeta}(x_2)$ holds, and
- $[\![x_1 \prec_s x_2]\!]_{t,\zeta} = true$ iff $val_{t,\zeta}(x_2)$ is the next sibling to the right of $val_{t,\zeta}(x_1)$ in t, i.e. $val_{t,\zeta}(x_1) \prec_s^t val_{t,\zeta}(x_2)$ holds.

The syntax of MSO_X can be enriched by adding $\varphi_1 \vee \varphi_2$, $\forall x.\varphi$, $\forall X.\varphi$, $\varphi_1 \Rightarrow \varphi_2$, $\varphi_1 \Leftrightarrow \varphi_2$ as abbreviations for other MSO_X formulae in the usual way. Likewise, the definition of bound and free variables of MSO_X formulae is also standard. We use the notation $fr(\varphi)$ for the set of free variables of the formula φ. Given that an XML tree t and a MSO_X formula φ with $fr(\varphi) = \{x_1, \ldots, x_n\}$, then φ is said to be *satisfiable* in t with respect to the variable assignment ζ iff $[\![\varphi]\!]_{S,\zeta}(\varphi) = true$.

5 XML Machines

In this section we present *XML machines* (XMLMs), a computational model for XML that adopts Abstract State Machines [2] for the purpose of dealing with XML database transformations. The most important extension of XML machines is the incorporation of MSO_X formulae in forall- and choice-rules and the use of terms from the tree algebra. The other rules used by XML machines are more or less the same except for an added partial update rule that is added for convenience, although it does not add any additional expressive power.

Using an unbounded number of parallel processes, an update operator $\uplus$ merging two hedges into one is needed. With these preliminary remarks we can now define *MSO*-rules in analogy to rules in ASMs. In the following definition the formulae φ always refer to MSO_X formulae as discussed in Section 4.

Definition 5. The set $\mathbf{R}$ of *MSO-rules* over a signature $\Sigma = \Sigma_{db} \cup \Sigma_a \cup \{f_1, \dots, f_\ell\}$ is defined as follows ($var(t)$ is the set of variables occurring in t):

- If t is a term over Σ, and f is a location in Σ such that $\#f = \#t$, then $f := t$ is a rule r in $\mathbf{R}$ called *assignment rule* with $fr(r) = var(t)$.
- If t is a term over Σ, f is a location in Σ and $\uplus$ is a binary operator such that $\#f = \#t$ and $\uplus : \#t^2 \to \#f$, then $f \Leftarrow^{\uplus} t$ is a rule r in $\mathbf{R}$ called *partial assignment rule* with $fr(r) = var(t)$.
- If φ is a formula and $r' \in \mathbf{R}$ is an MSO-rule, then **if** φ **then** r' **endif** is a rule r in $\mathbf{R}$ called *conditional rule* with $fr(r) = fr(\varphi) \cup fr(r')$.
- If φ is a formula with only database variables $fr(\varphi) = \{x_1, \dots, x_k, X_1, \dots, X_m\}$ and $r' \in \mathbf{R}$ is an MSO-rule, then **forall** $x_1, \dots, x_k, X_1, \dots, X_m$ **with** φ **do** r' **enddo** is a rule r in $\mathbf{R}$ called *forall rule* with $fr(r) = fr(r') - fr(\varphi)$.
- If r_1, r_2 are rules in $\mathbf{R}$, then **par** r_1 r_2 **par** is a rule r in $\mathbf{R}$, called *parallel rule* with $fr(r) = fr(r_1) \cup fr(r_2)$.
- If φ is a formula with only database variables $fr(\varphi) = \{x_1, \dots, x_k, X_1, \dots, X_m\}$ and $r' \in \mathbf{R}$ is an MSO-rule, then **choose** $x_1, \dots, x_k, X_1, \dots, X_m$ **with** φ **do** r' **enddo** is an MSO-rule r in $\mathbf{R}$ called *choice rule* with $fr(r) = fr(r') - fr(\varphi)$.
- If r_1, r_2 are rules in $\mathbf{R}$, then **seq** r_1 r_2 **seq** is a rule r in $\mathbf{R}$, called *sequence rule* with $fr(r) = fr(r_1) \cup fr(r_2)$.
- If r' is a rule in $\mathbf{R}$ and ϑ is a location function that assigns location operators ϱ to terms t with $var(t) \subseteq fr(r')$, then **let** $\vartheta(t) = \varrho$ **in** r' **endlet** is a rule r in $\mathbf{R}$ called *let rule* with $fr(r) = fr(r')$.

The definition of associated sets of update sets $\Delta(r, S)$ for a closed MSO-rule r with respect to a state S is again straightforward [8]. We only explain the non-standard case of the partial assignment rule. For this let r denote partial assignment rule $f \Leftarrow^{\uplus} t$, and let S be a state over Σ and ζ a variable assignment for $fr(r)$. We then obtain $\Delta(r, S, \zeta) = \{\{(\ell, a, \uplus)\}\}$ with $\ell = val_{S,\zeta}(f)$ and $a = val_{S,\zeta}(t)$, i.e. we obtain a single update set with a single partial assignment to the location ℓ. As the rule r will appear as part of a complex MSO-rule without free variables, the variable assignment ζ will be determined by the context, and the partial undate will become an element of larger update sets Δ. Then, for a state S, the value of location ℓ in the successor state $S + \Delta$ becomes $val_{S+\Delta}(\ell) = val_S(\ell) \cup \bigcup_{(\ell,v,\cup)\in\Delta} v$, if the value on the right hand side is defined unambiguously, otherwise $val_{S+\Delta}(\ell)$ will be undefined.

Example 4. Assume that the XML tree in Figure 3 is assigned to the variable (tree name) t_{exa}. The following MSO-rule will construct the XML tree in (i) from subtrees of the given XML tree using operators of the tree algebra:

$t_1 := \epsilon$;
forall x,y,z **with** $\prec_c (t_{exa}, root(t_{exa}), x) \wedge \prec_c (t_{exa}, x, y) \wedge \prec_c (t_{exa}, x, z)$
$\wedge \omega(t_{exa}, x) = b \wedge \omega(t_{exa}, y) = c \wedge \omega(t_{exa}, z) = a$
do
$t_1 \leftleftarrows^{\cup} \iota(d, \kappa(subtree(t_{exa}, y), subtree(t_{exa}, z)))$;
enddo ;
$output := \iota(r, t_1)$

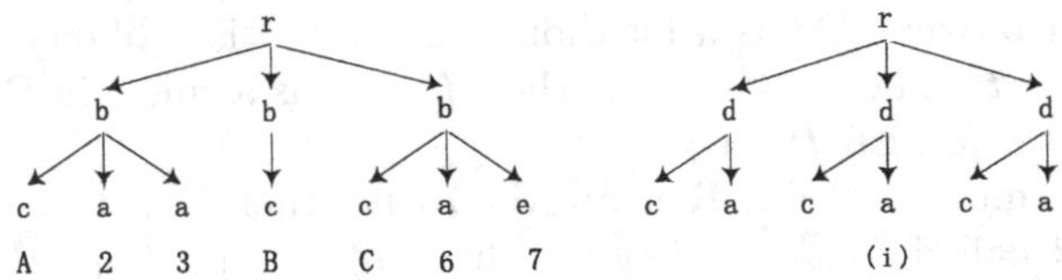

Fig. 3. An XML tree and the result of tree operations

Definition 6. An *XML Machine* (XMLM) $\mathcal{M}$ consists of a set $\mathcal{S}_{\mathcal{M}}$ of states over Σ closed under isomorphisms, non-empty subsets $\mathcal{I}_{\mathcal{M}} \subseteq \mathcal{S}_{\mathcal{M}}$ of initial states, and $\mathcal{F}_{\mathcal{M}} \subseteq \mathcal{S}_{\mathcal{M}}$ of final states, both also closed under isomorphisms, a program $\pi_{\mathcal{M}}$ defined by a closed MSO-rule r over Σ, and a binary relation $\tau_{\mathcal{M}}$ over $\mathcal{S}_{\mathcal{M}}$ determined by $\pi_{\mathcal{M}}$ such that the following holds:

$$\{S_{i+1} \mid (S_i, S_{i+1}) \in \tau_{\mathcal{M}}\} = \{S_i + \Delta \mid \Delta \in \Delta(\pi_{\mathcal{M}}, S_i)\}.$$

Theorem 1. *The XMLMs capture exactly all XML database transformations.*

Proof. According to [8,9] each XML database transformation can be represented by a behaviourally equivalent ADTM with the same tree-based background, and vice verse. As ADTMs differ from XMLMs only by the fact that ADTM-rules are more restrictive than MSO-rules (they do not permit MSO_X formulae in forall- and choice-rules), such an ADTM is in fact also an XMLM.

Thus, it suffices to show that XMLMs satisfy the postulates for XML database transformations in [8]. The first three of these postulates are already captured by the definitions of XMLMs and tree-based background, so we have to consider only the bounded exploration and genericity postulates.

Regarding exploration boundary we note that the assignment rules within the MSO-rule r that defines $\pi_{\mathcal{M}}$ are decisive for the set of update set $\Delta(r, S)$ for any state S. Hence, if $f(t_1, \dots, t_n) := t_0$ is an assignment occurring within r, and $val_{S,\zeta}(t_i) = val_{S',\zeta}(t_i)$ holds for all $i = 0, \dots, n$ and all variable assignments ζ that have to be considered, then we obtain $\Delta(r, S) = \Delta(r, S')$.

We use this to define an exploration boundary witness T. If t_i is ground, we add the access term $(-, t_i)$ to T. If t_i is not ground, then the corresponding assignment rule must appear within the scope of forall and choice rules introducing the database variables in t_i, as r is closed. Thus, variables in t_i are bound by a formula φ, i.e. for $fr(t_i) = \{x_1, \dots, x_k\}$ the relevant variable assignments are $\zeta = \{x_1 \mapsto b_1, \dots, x_k \mapsto b_k\}$ with $val_{S,\zeta}(\varphi) = true$. Bringing φ into a form that only uses conjunction, negation and existential quantification, we can extract a set of access terms $\{(\beta_1, \alpha_1), \dots, (\beta_\ell, \alpha_\ell)\}$ such that if S and S' coincide on these access terms, they will also coincide on the formula φ. This is possible, as we evaluate access terms by sets, so conjunction corresponds to union, existential quantification to projection, and negation to building the (finite) complement. We add all the access terms $(\beta_1, \alpha_1), \dots, (\beta_\ell, \alpha_\ell)$ to T.

More precisely, if φ is a conjunction $\varphi_1 \wedge \varphi_2$, then $\Delta(r, S_1) = \Delta(r, S_2)$ will hold, if $\{(b_1, \dots, b_k) \mid val_{S_1,\zeta}(\varphi) = true\} = \{(b_1, \dots, b_k) \mid val_{S_2,\zeta}(\varphi) = true\}$ holds (with $\zeta = \{x_1 \mapsto b_1, \dots, x_k \mapsto b_k\}$). If T_i is a set of access terms such that whenever S_1 and S_2 coincide on T_i, then $\{(b_1, \dots, b_k) \mid val_{S_1,\zeta}(\varphi_i) = true\} = \{(b_1, \dots, b_k) \mid val_{S_2,\zeta}(\varphi_i) = true\}$ will hold ($i = 1, 2$), then $T_1 \cup T_2$ is a set of access terms such that whenever S_1 and S_2 coincide on $T_1 \cup T_2$, then $\{(b_1, \dots, b_k) \mid val_{S_1,\zeta}(\varphi) = true\} = \{(b_1, \dots, b_k) \mid val_{S_2,\zeta}(\varphi) = true\}$ will hold.

Similarly, a set of access terms for ψ with the desired property will also be a witness for $\varphi = \neg\psi$, and $\bigcup_{b_{k+1} \in B_{db}} T_{b_{k+1}}$ with sets of access terms $T_{b_{k+1}}$ for $\psi[x_{k+1}/t_{k+1}]$ with $val_S(t_{k+1}) = b_{k+1}$ defines a finite set of access terms for $\varphi = \exists x_{k+1}\psi$. In this way, we can restrict ourselves to atomic formulae, which are equations and thus give rise to canonical access terms.

Then by construction, if S and S' coincide on T, we obtain $\Delta(r, S) = \Delta(r, S')$. As there are only finitely many assignments rules within r and only finitely many choice and forall rules defining the variables in such assignments, the set T of access terms must be finite, i.e. r satisfies the exploration boundary postulate.

Regarding genericity assume that $\mathcal{M}$ does not satisfy the genericity postulate. Then there must be a state S and equivalent substructures $S_1, S_2 \preceq S$ such that S_1 is preserved by $\Delta(r, S)$, i.e. $S_1 \preceq S + \Delta_1$ for some $\Delta_1 \in \Delta(r, S)$, but S_2 is not, i.e. $S_2 \not\preceq S + \Delta_2$ for all $\Delta_2 \in \Delta(r, S)$. According to our remark above r must contain a choice rule **choose** $x_1, \dots, x_k$ **with** φ **do** r' **enddo**. For a state S' let $\mathcal{B}_{S'} = \{(b_1, \dots, b_k) \mid val_{S',[x_1 \mapsto b_1, \dots, x_k \mapsto b_k]}(\varphi) = true\}$. Then the automorphism $\sigma : S \to S$ induced by $S_1 \equiv S_2$ is defined by a permutation on $\mathcal{B}$. From this we obtain $\sigma(\Delta(r, S, S + \Delta_1) = \Delta(r, S, \sigma(S + \Delta_1)) = \Delta(r, S, S + \Delta_2)$ for some $\Delta_2 \in \Delta(r, S)$. Thus, $S_2 \preceq S + \Delta_2$ for this Δ_2, which contradicts our assumption. □

6 Conclusion

In this paper we continued our research on foundations of database transformations exploiting the theory of Abstract State Machines. In [8] we developed a theoretical framework for database transformations in general, which are defined by five intuitive postulates and exactly charactericterised by ADTMs, a variant of Abstract State Machines. We argued that specific data model requirements are captured by background classes, while in general only minimum requirements for such backgrounds are postulated.

We now defined an alternative and more elegant computational model for XML database transformations, which directly incorporates weak MSO formulae in forall and choice rules. This leads to so-called XML machines. Due to the intuition behind the postulates it should come as no surprise that the two computation models are in fact equivalent.

This research is part of a larger research agenda devoted to studying logical foundations of database transformations, in particular in connection with tree-based databases. The next obvious step is to define a logic that permits reasoning about database transformations that are specified by XML machines. First steps in this direction have been made in [12].

References

1. Blass, A., Gurevich, Y.: Background, reserve, and gandy machines. In: Clote, P.G., Schwichtenberg, H. (eds.) CSL 2000. LNCS, vol. 1862, pp. 1–17. Springer, Heidelberg (2000)
2. Börger, E., Stärk, R.: Abstract State Machines. Springer, Heidelberg (2003)
3. Büchi, J.R.: Weak second-order arithmetic and finite automata. Zeitschrift für Mathematische Logik und Grundlagen der Mathematik 6(1-6), 66–92 (1960)
4. Comon, H., Dauchet, M., Gilleron, R., Löding, C., Jacquemard, F., Lugiez, D., Tison, S., Tommasi, M.: Tree automata techniques and applications (2007), `http://www.grappa.univ-lille3.fr/tata` (release, October 2007)
5. Doner, J.: Tree acceptors and some of their applications. Journal of Computer and Systems Science 4(5), 406–451 (1970)
6. Gurevich, J.: Sequential abstract state machines capture sequential algorithms. ACM Transactions on Computational Logic 1(1), 77–111 (2000)
7. Kumar, V., Madhusudan, P., Viswanathan, M.: Visibly pushdown automata for streaming XML. In: Proceedings of WW 2007, pp. 1053–1062. ACM, New York (2007)
8. Schewe, K.-D., Wang, Q.: A customised ASM thesis for database transformations (2009) (submitted for publication)
9. Schewe, K.-D., Wang, Q.: XML database transformations (2009) (submitted for publication)
10. Thatcher, J.W., Wright, J.B.: Generalized finite automata theory with an application to a decision problem of second-order logic. MST 2(1), 57–81 (1968)
11. Van Den Bussche, J., Van Gucht, D., Andries, M., Gyssens, M.: On the completeness of object-creating database transformation languages. J. ACM 44(2), 272–319 (1997)
12. Wang, Q., Schewe, K.-D.: Towards a logic for abstract metafinite state machines. In: Hartmann, S., Kern-Isberner, G. (eds.) FoIKS 2008. LNCS, vol. 4932, pp. 365–380. Springer, Heidelberg (2008)

Preface to FP-UML 2009

Juan Trujillo[1] and Dae-Kyoo Kim[2]

[1] University of Alicante, Spain
[2] Oakland University, USA

The Unified Modeling Language (UML) has been widely accepted as the standard object-oriented (OO) modeling language for modeling various aspects of software and information systems. The UML is an extensible language, in the sense that it provides mechanisms to introduce new elements for specific domains if necessary, such as web applications, database applications, business modeling, software development processes, data warehouses. Furthermore, the latest version of UML 2.0 got even bigger and more complicated with more diagrams for some good reasons. Although UML provides different diagrams for modeling different aspects of a software system, not all of them need to be applied in most cases. Therefore, heuristics, design guidelines, lessons learned from experiences are extremely important for the effective use of UML 2.0 and to avoid unnecessary complication. Also, approaches are needed to better manage UML 2.0 and its extensions so they do not become too complex too manage in the end.

The Fifth International Workshop on Foundations and Practices of UML (FP-UML'09) intends to be a sequel to the successful BP-UML'05, BP-UML'06, FP-UML'07, and FP-UML'08 workshops held in conjunction with the ER'05, ER'06, ER'07, and ER'08, respectively. FP-UML'09 intends to be an international forum for exchanging ideas on the best and new practices of the UML in modeling and system developments. Papers focused on the application on the UML in new domains and new experiences with UML 2.0, and foundations, theory, and UML 2.0 extensions are also highly encouraged. As UML 2.0 is oriented towards the software design driven by models, papers applying the Model Driven Architecture (MDA) or the Model Driven Engineering (MDE) to specific domains are also highly encouraged.

The workshop attracted papers from 8 different countries distributed all over the world: Brazil, Spain, USA, Belgium, Netherlands, Cuba, Chile and Canada. We received 14 abstracts and 12 papers were finally submitted. The Program Committee only selected 5 papers, making an acceptance rate of 41.6%. The accepted papers were organized in two sessions. The first one will be focused on Dependability and Agent Modeling, where the first two papers focus on aspect and agent modeling, while the latter is focused on building use case diagrams for secure mobile Grid applications. In the second session, two papers focus on Semantics Representation and Tools will be presented.

We would like to express our gratitude to the program committee members and the external referees for their hard work in reviewing papers, the authors for submitting their papers, and the ER 2009 organizing committee for all their support.

C.A. Heuser and G. Pernul (Eds.): ER 2009 Workshops, LNCS 5833, p. 105, 2009.

Applying AUML and UML 2 in the Multi-agent Systems Project

Gilleanes Thorwald Araujo Guedes and Rosa Maria Vicari

Instituto de Informática
Programa de Pós-Graduação em Computação (PPGC)
Universidade Federal do Rio Grande do Sul (UFRGS) - Porto Alegre - RS - Brasil
gtaguedes@inf.ufrgs.br, rosa@inf.ufrgs.br
http://www.inf.ufrgs.br/pos

Abstract. This article discusses the viability of the AUML and UML languages employment, from the latter's version 2.0 on, in the multi-agent systems project. In this article some works that have used UML for the project of systems that involved agents, as well as some AOSE (Agent Oriented Software Engineering) methodologies that use in some way UML or AUML (or both), are presented. Immediately afterwards the article approaches the AUML language, highlighting the innovations proposed by same and how it can be applied to the multi-agent systems project, identifying its advantages and disadvantages. After that, the paper passes on to describe how UML, from its version 2.0 on, has bypassed AUML and how the former can be applied to the multiagent systems project, pinpointing its positive aspects and its deficiencies.

Keywords: AUML, UML 2, Agents, Use Case Diagram, Actors, Internal Use Cases, Sequence Diagram, Combined Fragments, State Machine Diagram, Composite States, Activity Diagram. Activity Partition.

1 Introduction

Along the years, researchers in the area of software engineering have endeavored to set up methods, techniques, and modeling languages/notations with the goal of creating formal software project patterns while establishing well-defined steps for software building in such a way as to make their development more robust, faster, organized, coherent, trustworthy, easier to maintain and re-use, and presenting better quality. At the same time, in the area of Artificial Intelligence, the employment of intelligent agents as auxiliary aids to software applied to the most diverse dominions is being spread out. This practice has shown to be a good alternative for the development of complex systems, fostering a great increase of agent-supported software development in the several areas.

However, the development of this kind of system has presented new challenges to the software engineering area and this led to the surfacing of a new sub-area, blending together concepts brought over from both the software engineering and artificial intelligence areas, which is known as the AOSE - Agent Oriented Software Engineering, whose goal is that of proposing methods and languages for

C.A. Heuser and G. Pernul (Eds.): ER 2009 Workshops, LNCS 5833, pp. 106–115, 2009.

projecting and modeling agent-supported software. Among the several AOSE methods nowadays extant MaSE, MessageUML, Tropos, and Prometheus can be named. For an example of modeling language we can mention AUML - Agent Unified Modeling Language - that was derived from the UML, though it is possible to find jobs that employ UML proper, that is becoming particularly attractive for multi-agent system projects from its 2.0 version.

In this paper a discussion about the employment of the modeling languages, AUML and UML from its version 2.0, will be presented as an aid for the multi-agent systems project. Firstly, some works that have used UML for the project of systems that involved agents, as well as some AOSE methodologies that use in some way UML or AUML (or both), are presented. The article approaches the AUML language, highlighting the innovations proposed by same and how it can be applied to the multiagent systems project, identifying its advantages and disadvantages. After that, we describe how UML, from its version 2.0 on, has bypassed AUML and how the former can be applied to the multiagent systems project, pinpointing its positive aspects and its deficiencies.

2 UML, AUML and UML 2

UML has become a standard for software modeling and is broadly accepted and used throughout software engineering industry. Therefore, needless to say, many agent-supported software projects use directly the UML language for the designing of this kind of software, for instance [7], [8], [9] and [10]. In addition, some AOSE methods, like MaSE [3] or Tropos [5] employ UML in a partial fashion. For more details about UML see [11] or [12].

2.1 AUML - Agent Unified Modeling Language

Since software agents present specific features in comparison with more traditional software methods, some attempts to adapt UML to these characteristics were made, which brought forth the surfacing of AUML - Agent UML - whose main documentation can be seen in [13].

All the same, the AUML language currently supports only weak notions of agency, representing agents as objects, while employing state-machine diagrams to model their behavior and extended Interaction diagrams to model their communicative acts [1], not support cognitive or social abstractions. Caire in [4] comments that, although this notation is useful, it bears no agent concept in its core, stating also that specifying the behavior of an object in terms of interaction protocols will not change such an object into an agent.

One of the main AUML contributions is the document on interaction diagrams to be found in [13], where it is attempted to extend UML diagrams toward the supporting of communicative acts and the modeling of inter-agent communication. In its original format, this document also proposed some notation for multiple choice representation and parallelism.

However, as can be seen in [14] and in [13] itself, whose document was updated, this symbology was set aside in favor of an alternative proposed by UML 2 itself, as can be seen in [11] and [12], where resources such as the use of combined fragments of both types 'par' (as in <parallel>) and 'alt' (as in <alternative>) was employed in the sequence diagram. Realizing the superiority of the new UML version, [14] has then proposed the adaptation of this new notation to AUML, adding some small innovations to the interaction diagrams, as the lifeline notation for agent representation (as objects). Nevertheless, it is important to highlight that AUML interaction diagrams seem to be the most employed by AUML-based AOSE methods.

Peres in [15] accentuates the existence of scarce documentation on AUML and an even smaller number of updating steps, while also stating that those papers to be found on the language employment always repeat the very same examples presented in the original documentation. The paper also highlights that the diagram definitions are not very accurate into their establishing the degree into which the UML paradigm is to be followed or broken in relation to the multi-agent system particulars. Still, [15] appends that a UML extension will not be sufficient to create an agent-oriented modeling language, because agents need more abstractions and a semantic treatment to focus on their own particular features.

The AUML is currently inactivated, according to [16], where a note informs us that this occurs for three reasons, to wit, first, the launching of UML 2.1, containing many of such agent-related features that AUML required. Secondly, the release of the language, SysML - System Modeling Language - prepared to customize UML, another item to supply many of the characteristics searched by AUML. Thirdly, the UML Profile and Metamodel for Services (UPMS) RFP, that is under elaboration requires a services metamodel and profile for extend UML by means of capacities applicable to service modeling with the employment of SOA (Service Oriented Architecture). All these emerging standards have the stated goal of including agent-related concepts.

Even so, as mentioned above, several methods are trying to apply AUML/UML somehow into their life cycles, especially as related to interaction diagrams, as in MESSAGE/UML [4], Tropos [5] and Prometheus [6]. However, according to [1], many of the AOSE methods present weak notions for agency, never focusing in a satisfactory fashion such cognitive abstractions as beliefs, desires, and intentions, as well as social abstractions like cooperation, competition, and negotiation. In addition, these methods possess strong associations with object-oriented software engineering methods, oftentimes dealing with agents as if they were objects.

The representation of agents as objects conflicts with the definition of agents for, according to [1], 'an agent is a computational process located within an environment and designed to achieve a purpose within said environment by means of autonomous, flexible behavior'.

To adhere to this concept, for cognitive applications an agent should not be represented by classes alone, as happens in AUML, but as an actor able to interact with the system. Probably, along the processes in which it was supposed to partake, an agent would be expected to interact with classes, whose objects store their own knowledge, among other things, but the agent in itself should not be an object instantiated from a class.

The AUML proposal might be valid, however, whenever active objects were represented as agents. Active objects could be compared to reactive agents that, as can be seen in [2] can only react to events and don't own any model of the world in which they are inserted, all the while cognitive agents are endowed with some knowledge about their environment, own mental states, like beliefs, desires, and intentions, and can set up communications with other agents to negotiate help toward achieving their goals. This kind of agents cannot be deal with simply as if they were objects.

2.2 UML 2

As quoted in the prior section, agents can be represented as actors, like happens in [7] and [9]. The latter states that agents can also be represented as actors in the UML use cases diagram and that associations among actors and use cases can be employed to represent perceptions and actions of an agent represented by an actor. However, [9] does not maintain the agent's representation as an actor in the sequence diagram, in which agents are represented as objects, according to AUML proposal in [13]. Agents are already represented by [10] as a bland of actor and object, employing objects with the stereotype control, that modifies the object standard design into a circle shaped by an arrow and inserting within this circle the actor's symbol. Therefore, [10] represents agents as a kind of control-type object, only with features particular to agents. There still is a notation proposed by [17] who suggests modifying the standard design the component, actor, to represent agents, presenting same as square-headed actors. [17] even goes as far as suggesting the creation of cloud-like component to represent the agents' goals. All the same, as [17] identifies active objects as agents, these are modeled like objects in the sequence diagram, as happens in AUML.

In UML, actors are usually modeled, within the use case diagram, as external to the system, for they interact with the system, but cannot be a part of it. Therefore, actors are placed outside the system border, as can be seen in [12].

Anyway, most of the times, the software agents are not external to the software, rather they customarily are inserted in the system environment and, as they are independent, proactive, and able to interact with the software according to their goals, therefore these should be represented as actors. As can be seen in [11], an actor models a type of role played by an entity that interacts with the subject but which is external to the subject In that way, for a multiagent systems project, as states [9], it is necessary to adapt this concept, considering that agents can be internal to the system, that is, an agent can be a part of the subject and, therefore, if we are to represent agents (their roles) as actors, said actors should be internally represented within the system's borders, for they belong in the software, as shows Figure 1. The system boundary is represented by a square the involves the system's functionalities. In the example above, the borderline represents a system of research through Internet. We can notice that there is an external actor who represents a human Internet user, to wit, a real

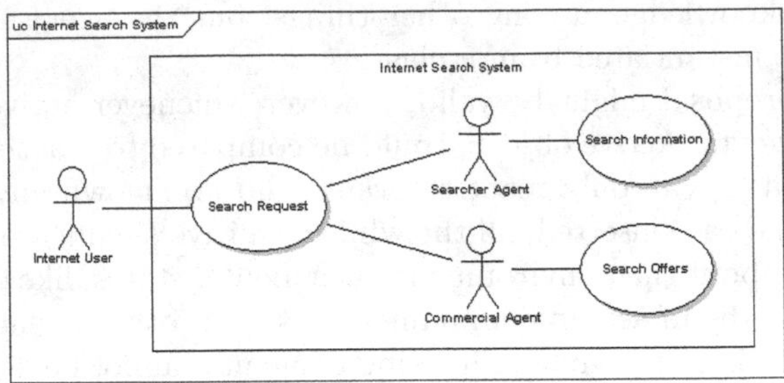

Fig. 1. Example of actors representing agents - based on [12]

person who interacts with the system by means of the use case "Search Request". We can also notice that there are two more actors named "Searcher Agent" and "Commercial Agent" and these are within the system's border because they represent software agents, as occurs in [9] and [17]. "Searcher Agent" is responsible for performing the search required by "Internet User", all the while "Commercial Agent" is responsible for seeking offers associated to the performed research. We can also observe that there are two use cases internal to the system, which represent the processes of search for information and search for offers and these use cases can only be employed by agents internal to the system and are not accessible by the external agents.

The process "Search Request" can be detailed by means of a sequence diagram, as demonstrates Figure 2 (based on [12]). Other than in the AUML approach [13], instead of representing agents as objects, we chose to represent them as actors, in the same way they were represented in the use case diagram. We believe this is more correct in relation to the definition for agent herein adopted and, besides, this keeps coherence with the use case diagram and allows to better differentiate the agents from the real objects instantiated from the class diagram, for in the AUML sequence diagram those objects that represent agents only differ from normal objects by textually identifying the agent's name and the role represented by it in the format "agent/role", instead of being simply shown by the object's name.

The justification for the employment of actors instead of objects, besides that of the concept of agent itself described in [1], is based upon the concept of an actor as described in [11], where is stated that an actor represents a role played by some entity and that an entity may play the role of several different actors and an actor may be played by multiple different entities. Although the concept of role used by UML might be different than the concept of role used in multi-agent systems, they seem to be, at first sight, similar to each other. Thus, the actors seem able to represent the roles of an agent when this interprets more than one. The employment of objects to represent agents could be valid in those projects that contain only reactive agents, that could be taken as active objects; anyway, in such situations that contain cognitive agents, we believe that the representation of same as actors would be more correct.

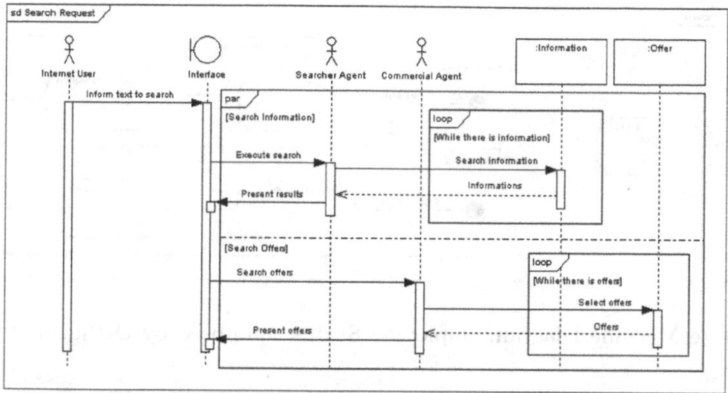

Fig. 2. Detailing of the "Search Request" Process by means of a Sequence Diagram

The diagram presented in Figure 2 also spans those internal use cases that were presented in the Figure 2. They could be detailed in separate diagrams, but the example seems to be more complete and understandable the way it is. In this process, the actor, "Internet User", supplies the text to be researched to the interface and this re-passes same to the actors/agents "Searcher Agent" and "Commercial Agent", that, as modeled in the use cases diagram, are internal to the system and the event, "Inform text for search" is forwarded to them through the system's interface. "Searcher Agent" is responsible for the search of information related to the required research, while "Commercial Agent" is responsible for seeking offers related to said research, like products, books, or courses, for example. Observe that both tasks are performed in parallel and the results are presented simultaneously. This is possible to represent by means of the employment of a type "par" (as in "parallel") combined fragment that demonstrates that concurrence situation within which two agents execute their tasks at the same time. Notice that an interrupted line separates those operations performed by each agent. Each parallel process's acting area is called, interaction operand. Further observe that each interaction operand has a guardian condition to establish its function. Obviously, in a real system there would be included many more classes to represent those information pools that are needed for this kind of research.

The same process can be represented in the state machine diagram below, obviously under a different focus, by means of composite states with orthogonal regions, as shown in the Figure 3 (based on [12]). In the example shown in Figure 3, the process begins by a static state that waits up the user until he types a text for researching. The text for research insertion event generates a transition into a composite state which contains two orthogonal regions. In the first region, like informs its title in the shape of a guard condition, is done the search for information relevant to the research, while in the second region the search for offers related to the looked-for text is performed. The states in each region occur in parallel. The same process can be even more detailed by

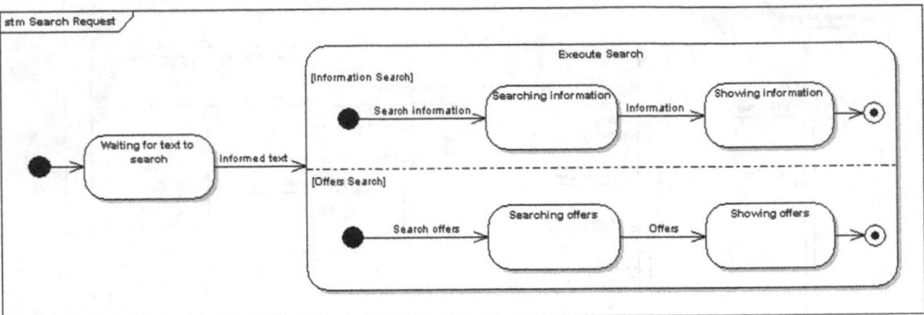

Fig. 3. State Machine Diagram employing States composed by Orthogonal Regions

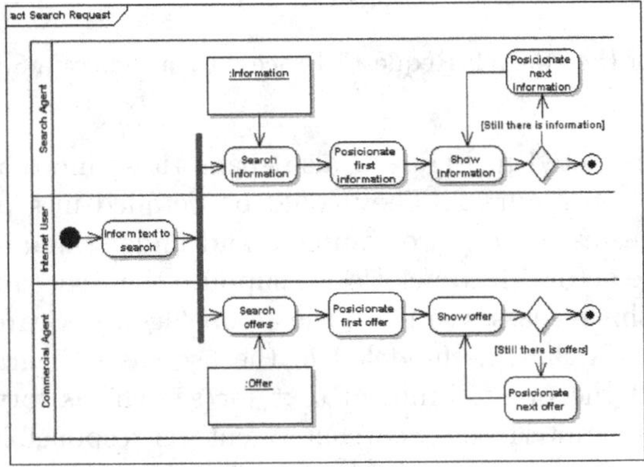

Fig. 4. Activity Diagram employing Activity Partitions

means of the activity diagram, which is an extremely detailed diagram which can be used for the modeling of plans, as state [9] and which is already employed in the Tropos methodology [5]. This parallelism can also be represented through the use of activity partitions and fork/join nodes, as demonstrates Figure 4.

In the example shown by that Figure each actor/agent was represented as an activity partition, something that can also be seen in [8]. The process is initiated when "Internet User" supplies the text to be researched. From that point, the flow is cut by a fork node and divides into two parallel flows, one to each activity partition, where information and offers related to the research performed by the actor will be looked for. For an obvious matter of physical space, this diagram was simplified.

3 Conclusions

AUML, in its present state of development, is still to be found within a rather tentative phase, therefore showing no great innovations in relation to the language from which it was derived. The biggest AUML contribution is an initial proposal for interaction diagrams, whose main original contribution was that of making some changes into the sequence diagram for the representation of parallelism by means of threads, also used in situations of choosing an option over others. This representation used a symbol to multiplex processing into several threads or to represent flow alternatives. However, this notation was abandoned face the notation newly adopted by UML 2, that came to employ combined fragments, a system that looks much more practical and supple.

Considering it is hard to pinpoint palpable differences between AUML and UML and noting how the latter presented great innovations from its version 2.0 on, besides the fact the AUML Project itself acknowledged that many of its goals were achieved by the last few innovations within UML and SYSML, it would seem that UML 2 is the most adequate for multi-agent system projects. However, some authors believe that both UML and AUML are too concerned with object orientation, a feature that would not allow it to be wholly used for multi-agent systems modeling. Maybe UML is adequated to the project of systems in which agents present more reactive characteristics than cognitive ones, but it is possibly insufficient for multi-agent system projects working with BDI architectures, which will require strong agency notions and, if this be proved, there will come up the need to adapt the language for said purpose.

The representation of agents/roles as actors, both in the cases of use diagrams and the sequence diagrams seem to be more correct than their representation as objects, as suggested by AUML. The concept of actor presented by UML is much closer to the concept of agent as the interpreter of a role than the concept of object. The representation of agents as objects would become valid when these will be active objects, but will show itself insufficient when dealing with cognitive agents. Besides, for a matter of coherence, if agents/roles were to be represented as actors in a diagram, like that of use cases, the former should remain so represented in other diagrams, whenever this becomes possible.

There is also the issue of goal representation, much necessary in multi-agent system projects. Goals can be defined as an agent's wishes, that is, whatever he wants to achieve. UML owns no specific constructions for this type of representation, to wit, there is no component to identify a goal to be reached by one or more agents. The closest to this would be the use case component, but any use case would identify a system functionality, that is, a function, service, or task the system is expected to supply, something that is not exactly a goal the way this concept is understood within the area of multi-agent systems. A functionality can represent a goal, but can contain more than one, depending on the situation. Anyway, it might be possible to create a stereotype dubbed, for example, <<goal>>, to identify such use cases that might be considered as goals, other than normal use case and, in that situation in which the goal had sub-goals, another stereotype could be created for this, like <<sub-goal>> and identify these by means of specialization/generalization associations with the main goal. The employment of stereotypes is perfectly valid in UML and its function is precisely that of allowing flexibility to the language, making it possible to attribute new characteristics and functions to already extant components.

The use of stereotypes could also be useful to establish differences between normal actors and agents, by the employment of a stereotype dubbed, for example, <<agent>>. As there is the happenstance that an agent being able to interpret many roles, another stereotype, named <<role>> or <<agent role>> could be created to determine when an actor represents a role interpreted by an agent. In the eventuality of being needed to identify which roles are identified by an agent it would be feasible, within a separate diagram, to create a role hierarchy, where the agent would be identified at the top of the hierarchy and be represented as an actor containing the stereotype <<agent>>, while its roles would be sub-actors, associated to the agent by means of generalization/specialization associations and spanning the stereotype <<agent role>>. This would not stop more than an agent to interpret the same role. Anyway, both the question of stereotype usage for the identification of goals and that for the identification of roles are still to be verified on depth so as to determine whether this alone would be sufficient for this purpose.

Another issue would be that of how to represent the agents' beliefs. Beliefs represent the information agents have about the environment and about themselves. Perhaps it would be feasible to store such information into classes, but it is not still clear how this could be achieved, for beliefs cannot be defined either as attributes or as methods, which are the two kinds of information a class usually contains. To represent this kind of information, we could try to create a profile, like the data-modeling profile, used to map classes on tables, where the <<table>> stereotype is used to represent a class the same way as it were a table and the <<column>> stereotype to represent attributes as colums.

Anyway, it might be possible to create one or more classes that spanned a single attribute (a string-like one, for example) that would simply textually store the agent's beliefs with a single instance to store each of his beliefs. Besides, some languages which implement beliefs usually define same directly in the software codes, without their being stored within a repository. If we were to employ a class diagram to identify the beliefs presented by an agent, it would be necessary to create some sort of mapping to connect their representation in the class diagram and the way they were expected to be implemented.

It seems to be possible to represent communication between agents by means of a sequence diagram, where messages would stand for communicative acts, as is suggested by AUML, but representing the agents/roles as actors and not as objects. Also negotiations, that are the way that agents reach an agreement to help each other so as to further a goal, can be represented by means of messages and combined fragments. Agents collaboration and competition features can be equally represented by state machine diagrams and activity diagrams, employing respectively composite states with orthogonal regions and activity partitions. However, deeper studies would be necessary into the application of said diagram, so as to demonstrate whether it is actually sufficient for this kind of representation.

Finally, we conclude that UML, for all that this is methodology-independent, can be useful in a multi-agent system project when used and possibly adapted to an AOSE methodology specifically oriented toward the development of agent-supported software, in which there are to be defined those UML diagrams that can be shown useful to the methodology and in which moment they are supposed to be applied, as well as which possible adaptations should be performed.

References

1. Vicari, R.M., Gluz, J.C.: An Intelligent Tutoring System (ITS) View on AOSE. International Journal of Agent-Oriented Software Engineering (2007)
2. Alvares, L.O., Sichman, J.S.: Introdução aos sistemas multiagentes. In: Jornada de Atualização em Informática (JAI 1997). cap. 1. Medeiros, C. M. B., Brasília (1997)
3. Deloach, S.A.: Analysis and Design using MaSE and agentTool. In: 2nd Midwest Artificial Intelligence and Cognitive Science Conference, Oxford (2001)
4. Caire, G.: Agent Oriented Analysis Using Message/UML. In: Wooldridge, M.J., Weiß, G., Ciancarini, P. (eds.) AOSE 2001. LNCS, vol. 2222, p. 119. Springer, Heidelberg (2002)
5. Bresciani, P., Perini, A., Giorgini, P., Giunchiglia, F., Mylopoulos, J.: Tropos: An Agent-Oriented Software Development Methodology. Autonomous Agents and Multi-Agent Systems 8, 203–236 (2004)
6. Padgham, L., Winikoff, M.: Prometheus: A Methodology for Developing Intelligent Agents. In: Giunchiglia, F., Odell, J.J., Weiss, G. (eds.) AOSE 2002. LNCS, vol. 2585, pp. 174–185. Springer, Heidelberg (2003)
7. Coelho, N.P., Guedes, G.T.A.: Tutorial Hipermídia Aequator utilizando um Agente Estacionário Reativo Simples: Apresentação da Modelagem. In: 11° Congresso de Informática e Telecomunicações SUCESU-MT. Cuiabá (2006)
8. Kang, M., Wang, L., Taguchi, K.: Modelling Mobile Agent Applications in UML 2.0 Activity Diagrams. In: 3rd International Workshop on Software Engineering for Large-Scale Multi-Agent Systems, pp. 104–111. IEEE Press, Edinburgh (2004)
9. Bauer, B., Odell, J.: UML 2.0 and Agents: How to Build Agent-based Systems with the new UML Standard. Journal of Engineering Applications of AI (2005)
10. Dinsoreanu, M., Salomie, I., Pusztai, K.: On the Design of Agent-Based Systems using UML and Extensions. In: 24th International Conference on Information Technology Interfaces (ITI 2002), Cavtat, Croatia, pp. 205–210 (2002)
11. OMG - Object Management Group. Unified Modeling Language: Superstructure Specification - Version 2.1.1. OMG (2007), `http://www.omg.com`
12. Guedes, G.: UML 2 - Uma Abordagem Prática. Novatec Editora, São Paulo (2009)
13. Huget, M.P., et al.: Interaction Diagrams. Working Documents. AUML Official Website (2003), `http://www.auml.org/`
14. Huget, M.P., Odell, J.: Representing Agent Interaction Protocols with Agent UML. In: Third International Joint Conference on Autonomous Agents and Multi-Agent Systems (AAMAS 2004), New York City (2004)
15. Peres, J., Bergmann, U.: Experiencing AUML for MAS Modeling: A Critical View. In: Software Engineering for Agent-Oriented Systems, SEAS (2005)
16. AUML Official Website, `http://www.auml.org`
17. Depke, R., Heckel, R., Kuster, J.M.: Formal Agent-Oriented Modeling with UML and Graph Transformation. Sci. Comput. Program. 44, 229–252 (2002)

A Collaborative Support Approach on UML Sequence Diagrams for Aspect-Oriented Software

Rafael de Almeida Naufal[1], Fábio F. Silveira[1,2], and Eduardo M. Guerra[1]

[1] Aeronautics Institute of Technology (ITA)
São José dos Campos, SP - Brazil,
Software Engineering Research Group (GPES)
rafael.naufal@gmail.com,
guerraem@gmail.com

[2] Department of Science and Technology,
Federal University of São Paulo (UNIFESP),
São José dos Campos, SP - Brazil
fsilveira@unifesp.br

Abstract. AOP and its broader application on software projects brings the importance to provide the separation between aspects and OO components at design time, to leverage the understanding of AO systems, promote aspects' reuse and obtain the benefits of AO modularization. Since the UML is a standard for modeling OO systems, it can be applied to model the decoupling between aspects and OO components. The application of UML to this area is the subject of constant study and is the focus of this paper. In this paper it is presented an extension based on the default UML meta-model, named MIMECORA-DS, to show object-object, object-aspect and aspect-aspect interactions applying the UML's sequence diagram. This research also presents the application of MIMECORA-DS in a case example, to assess its applicability.

1 Introduction

The progress of Aspect-Oriented (AO) technologies and its mainstream use on software projects had created the importance of distinguish between AO elements and Object-Oriented (OO) components, as, for example, with the application of the Unified Modeling Language (UML) [1] on systems analysis and design. More precisely, it's necessary to represent graphically the behavioral modifications realized by aspects on OO elements. This representation can be achieved with the Sequence Diagram (SD) support, whose responsibility is to highlight the interaction between objects [2].

This paper is intended to extend the UML default meta-model through its default extension mechanism to represent the collaborative modeling of the relationship between aspects and objects in the SD. In this research context, the

C.A. Heuser and G. Pernul (Eds.): ER 2009 Workshops, LNCS 5833, pp. 116–125, 2009.

term collaborative describes how the objects collaborate among themselves and with the aspects to produce some behavior during aspects' static crosscutting. The proposed UML extension will provide: (i) Model the type of aspect execution - before, after or around the execution of Join Points (JP); (ii) An independent technology graphical notation; (iii) Model the object-object, object-aspect, aspect-aspect interactions and unify the representation of JP; (iv) Model the composition of aspects on the SD; and (v) The separation of crosscutting concerns from business domain rules during the design life cycle.

This work addresses the problem related to the limitations of existing approaches to represent concisely and consistently the behavioral modifications applied by aspects on OO elements through the SD.

This paper is structured in the following way: Section 2 shows the related work involving UML proposals to incorporate aspects on the SD. Section 3 presents the proposed UML extension mechanism for the SD, which is named MIMECORA-DS and means Aspect-Oriented Collaborative Interaction Model on SD. Section 4 reports a case example to show the notation's applicability. Concluding remarks and future work are given in Section 5.

2 Related Work

Some solutions were already proposed to create a graphical notation to model existing concepts in AO languages, particularly focused on the SD.

[3], [4], [5] presented an aspects design graphical notation based on the default UML extension mechanism. The authors propose the JP are represented through links (instances of associations in UML) and messages, highlighted on the SD. The proposal submitted by the authors approaches the AspectJ semantics [6], but has some problems concerning aspects modeling on the SD. Their proposal is relevant and well reasoned, but is focused on representing crosscutting behavior resulted only from constructs of AspectJ, thus leaving as a future work the reuse investigation of this technique in other AO modeling environments. Moreover, it does not provide a notation to represent compositions between aspects. Furthermore, the authors do not model whether aspects crosscutting is before, after or during the JP execution.

[7] proposed the accommodation of new elements in UML 2.0 to model the JP, aspects and their relationships with the OO components. Thus they conceived a graphical representation for the JP, which were specified through circles with a cross inside them. Aspects are represented by boxes with crossed lines. The Pointcuts are defined in brackets. They applied the concept of UML packages to separate the functionality of the target application and aspects. While composing modeling elements to represent aspects on the SD, this approach does not describe a notation to show the types of advices that act on OO components. Moreover, the authors do not specify how to model composition of aspects and if the aspects crosscutting is before, after or around the JP. However, this paper describes the modeling of both the types of advices as interactions between aspects.

[8] proposes a graphical notation for AO modeling also based on UMLs existing models. The authors try to model crosscutting concepts of AOP language implementations. The authors divided the occurrence of a particular set of a run-time system scenarios on SD called Join Point Diagram. The notation described by them presents some constraints. There are not elements that represent the dependence and composition of aspects. Further, the authors notation also did not model the advice execution type before, after or around the execution of JPs.

The SD main intention is to document and model the messages exchanging in the object-object interactions. The execution order of messages is an important information that can be extracted from this kind of diagram. Without the representation of the advice execution type (before, after and around) on the object-aspect and aspect-aspect interactions, the execution order is not fully represented because the time that the advices are applied affect the behavior of the instance being crosscutted. MIMECORA-DS highlights new elements to show the time aspects add behavior to the components, as can be seen on the Section 3.

Essentially, the solutions proposed present limitations that are covered by MIMECORA-DS. MIMECORA-DS focuses on solving the problem of modeling the composition and precedence of aspects, thus showing some notations to model these kind of aspect-aspect interactions. It is important to model the composition of aspects because aspects which encapsulate new concerns can be created by these relationships. None of the related alternative solutions showed a proposal to represent the composition and precedence of aspects.

3 MIMECORA-DS and Its Major Artifacts

The UML SD extension to represent the relationship between OO components and aspects is named MIMECORA-DS, which means Aspect-Oriented Collaborative Interaction Model on SD. The MIMECORA-DS is a SD specific notation. The MIMECORA refers to a model more complete and is being applied to other UML diagrams. This research work is focused on its development and its application on the UML SD.

Methods that are candidates for JPs are crosscutted by a new modeling element in order to facilitate the understanding of aspects' action on them. Table 1 presents the elements of the MIMECORA-DS that represents advices' types:

Table 1. Advice notations on MIMECORA-DS [9]

Notation	Description	Stereotype
○	Advice executed before method invocation	<<*before*>>
⊗	Advice executed after method invocation	<<*after*>>
●	Advice executed around method invocation	<<*around*>>

Table 2. Notation of aspects' interactions on MIMECORA-DS

Notation	Description	Stereotype
⊕	Models the composition of aspects	<<*aspc*>>
⦶	Models the precedence of aspects	<<*aspp*>>

Table 3. Aspect instance on MIMECORA-DS

Notation	Description	Stereotype
<<aspect>> *:Aspect	Aspect instance	<<*aspect*>>

Table 2 summarizes the MIMECORA-DS elements which indicate the representation of the composition and precedence of aspects, their symbols and stereotypes.

The element graphically represented by a circle with the sign of sum in denotes the symbol for composition of aspects, which in the context of this research represents the sequential and joined execution of the aspects' advices for the creation of a new concern. It is described by the stereotype <<aspc>>.

The element indicated by a circle with a vertical line within represents the aspects precedence, where the advices are executed on the same or concurrent JP. This element, also referenced by the stereotype <<aspp>>, is responsible for determining the order of precedence in the execution of the advices.

MIMECORA-DS models an instance of an aspect on a SD as it would be an instance of a common class, represented by the stereotype <<aspect>>. The only difference is that its representation has a "X" that describes its crosscutting feature. Table 3 shows its modeling element.

Table 4 describes the various types of messages in the MIMECORA-DS, as their meanings and graphical representation. Such messages characterize object-object,

Table 4. Message's Types on MIMECORA-DS (Adapted from [9])

Notation	Description
——→	Between class instances
←- - - - -	Indicates the return from class instances
— — —↠	Between classes and aspects instances
—·—·—·↠	Between instances from various aspects or from the composition of aspects
—··—··↠	Between instances from various aspects or from precedence of aspects
··········↠	Between instance from the same aspect
Ⓡ←- - - - -	Indicates the return from a crosscutted JP in a crosscutting region

object-aspect and aspect-aspect interactions in a particular scenario represented in the SD. These messages represent the invocation or return from a method in a common OO class or from an aspect interacting with another aspect.

In this section it was described the MIMECORA-DS notation. Next section shows the aspects modeling on a case example, which demonstrates MIMECORA-DS applicability.

4 MIMECORA-DS Application on a Case Example

A Bank Management System (SGB), within the monetary domain, was developed as a case example to show the applicability of the conceived extension and the exemplification of the concepts introduced on this research work. The number of aspects included in this case example highlights the interception of aspects on OO elements as well as the interaction and composition of the aspects included in the target system. Table 5 describes the Functional Requirements (FRs) and Table 6 the Non-Functional Requirements (NFRs) of SGB:

Clearly, if these NFRs were implemented in the OO paradigm, it will end up with code tangling and code scattering. Those interests involve concerns that are not FRs of the monetary domain, because they present crosscutting features [10]. Thus, they are candidate to be modeled by aspects, either by static or dynamic type. This paper presents the dynamic type representation for them on the SD.

Table 5. FRs of SGB

FR 1. Account Requirements	FR 2. Requirements for Special Account (extending the requirements stated in FR 1)
FR 1.1 Register Account	FR 2.1 Register special account
FR 1.2 Conduct financial transactions on accounts, such as withdrawing, deposit and transfer of funds.	FR 2.2 Make possible the existence of customers negative balance, within a threshold established by the bank.

Table 6. NFRs of SGB

NFR 1. Security Requirements	NFR 2. Persistence Requirements
NFR 1.1 Enable access only to customers authenticated through login and password to the financial system.	NFR 2.1 Store in an history all transactions made on the Relational Database Management System (RDBMS), containing the user login, the date, the time and the operation was carried out by him.
NFR 1.2 Persist the history of processed transactions on a log such as date, time, operation and the Automated Teller Machine (ATM) Internet Protocol (IP) address used to make the operation.	

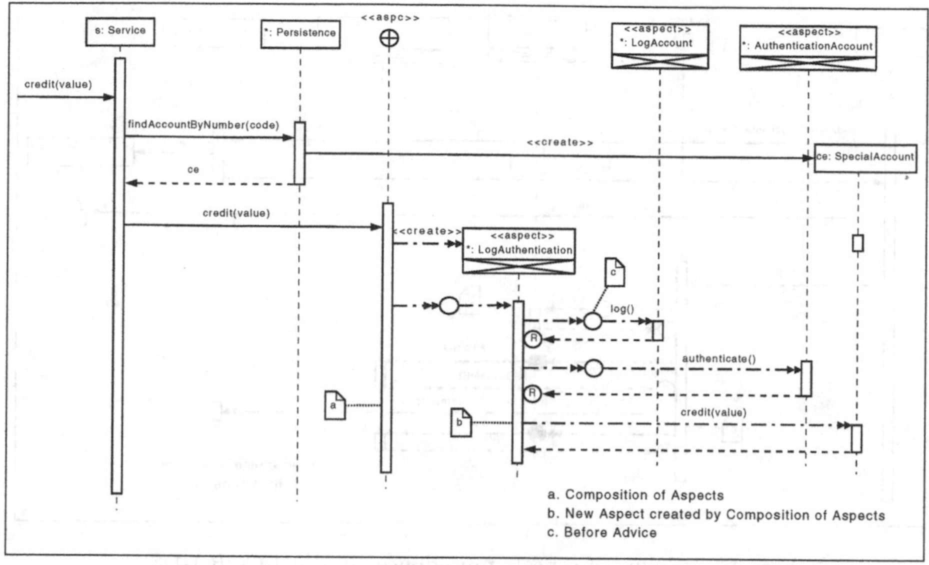

Fig. 1. Modeling of aspects composition on MIMECORA-DS

The aspect-aspect composition relationship on MIMECORA-DS provides the creation of a new concern and is indicated by the merge of the aspects advices which intercept the OO model elements. An example of composition of aspects can be seen in Figure 1. It shows *LogAccount* and *AccountAuthentication* aspects crosscutting together (for the same JPs) the invocation of the JP *credit(value)* before its execution by the *ce* instance of the *SpecialAccount* class. The modeling element "⊕" represented by the <<aspc>> stereotype indicates the merge of the *LogAccount* and *AccountAuthentication* aspects, which creates a new concern, as depicted by the *LogAccount* aspect. The before advice of the *LogAccount* aspect calls the *LogAccount* and *AccountAuthentication* aspects advices. These advices have the before crosscutting type, so they are applied before the execution of the JP (note the presence of the "○" element). This diagram presents a new kind of message provided by MIMECORA-DS to indicate the composition of aspects. This Figure also shows the execution context representation of the message *credit(value)* triggered by the *Service* class instance. This context indicates the moment when the crosscutting behavior is added by aspects.

The aspects *Authentication* and *AuthenticationAccount* are responsible for validating the required credentials of a customer who wants to perform the deposit in a certain special account. These aspects encapsulate the security concern, in order to enable the quantity credit only if the logged customer is actually holding the current account in question.

The precedence of the aspect-aspect relationship at MIMECORA-SD is specified by labels and numeric sequences, which is applied to messages that indicate advices invocation of the corresponding aspects, as depicted in Figure 2. The element "⏀" with the <<aspp>> stereotype indicates the precedence

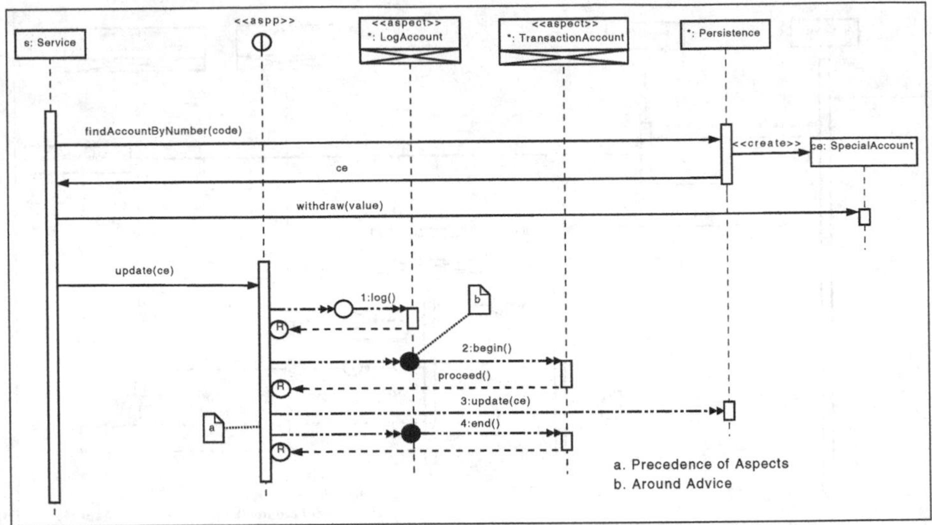

Fig. 2. Modeling of aspects precedence on MIMECORA-DS

aspect-aspect relationship element, from which the *LogAccount* and *Account-Transaction* aspects' advices must be executed. The precedence of the aspect-aspect relationship determines the order of execution of the aspects advices which crosscut the same JPs (their Pointcut select the same Join Points for aspects interception) or concurrent JPs. This Figure shows the *ce* instance update process from the *SpecialAccount* class on the RDBMS, the participation of the *AccountTransaction* aspect to ensure the proper persistence of modifications on the special account in the RDBMS. The interception made by the aspects occurs around the invocation of the JP *update(ce)* from the *Persistence* class (note the presence of the "●" element).

The participation of these aspects encapsulates the persistences concern, responsible for ensuring the ACID properties (Atomicity, Consistency, Isolation and Durability) of transactions created against the RDBMS [11]. The *Account-Transaction* aspect is responsible to create transactional contexts around the *update(ce)* method from the *Persistence* class.

Considering that it is possible to model nested invocations of methods on the DS, the Figure 3 illustrates the notation of MIMECORA-DS for that scenario. Figure 3 shows two before advices to the *balance(code)* from the *Service* class method and also an after advice (specified by the "⊗" element) to the *findAccountByNumber(method)* from the *Persistence* class. Note the presence of the element "Ⓘ" to specify the order of execution of the *LogAccount* and *AuthenticationAccount* aspects advices that crosscut the *balance(code)* method. The ordering of messages of the DS ensures that the *log()* after advice of the *LogAccount* aspect crosscuts the *findAccountByNumber(code)* method instead of the *balance(code)* method.

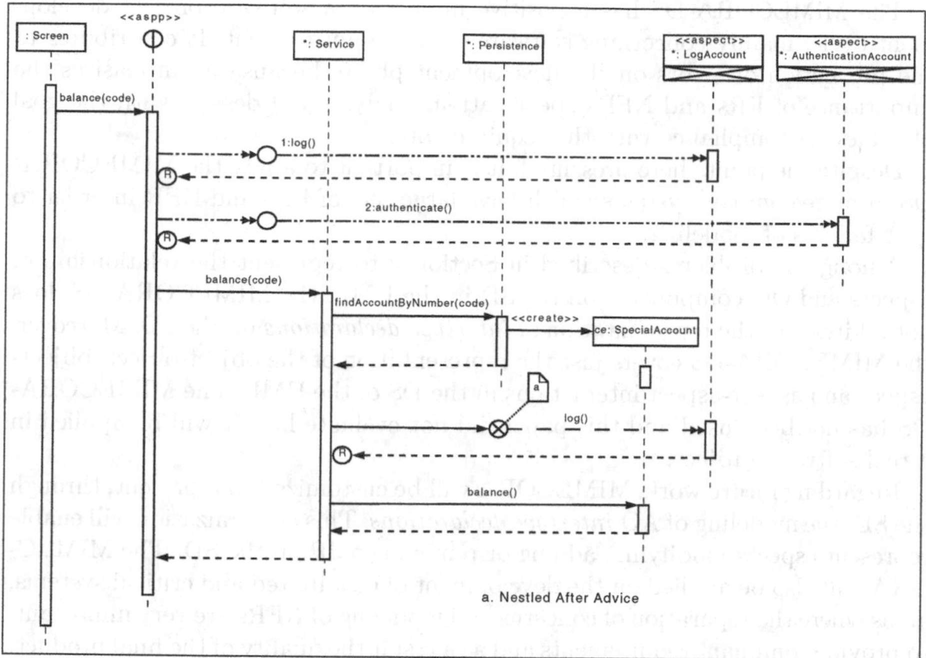

Fig. 3. Example of Nestled After Advice on MIMECORA-DS

5 Conclusion and Future Work

This paper addresses the problem related on the shortcomings of existing work to represent concisely and consistently the behavioral modifications applied by aspects on OO elements through the SD.

To solve this problem it was designed and developed in this paper an UML extension, which was named Aspect-Oriented Collaborative Interaction Model on SD (MIMECORA-DS). The solution provides support for modeling the object-object, object-aspect and aspect-aspect relationships in the SD. This extension allows AO paradigm concepts to be modeled regardless of applied technologies to AO implementation.

The model of the case example has cleared the decoupling between aspects and OO classes in the sense that the classes does not know they are being crosscutted by aspects on design time. Despite the extra effort required because of the introduction of new modeling elements, gains are achieved with this new approach. These gains are due to: (i) Clarity in the representation and modeling of aspects in the target application, (ii) AO modularity in the modeling phase with the representation of composition and precedence of aspects, (iii) Decoupling between classes and aspects, as evidenced in the development of the case example, and (iv) Reduced number of additional elements to represent aspects in the SD.

The MIMECORA-DS has a positive impact on a software project development cycle, mainly concerning software quality improvement. It contributes to the prevention of errors on the development phase, because it emphasizes the importance of FRs and NFRs specification, analysis and design, with the goal of achieving compliance with the requirements.

Despite the points here presented, it is important to apply the MIMECORA-DS in more complex systems which have large sets of FRs and NFR in order to spot failures of modeling.

Among the problems described in Section 2 to represent the relationship of aspects and OO components on the SD in the UML, the MIMECORA-DS does not address yet the representation of *intertype declarations* on the SD. Moreover, the MIMECORA-DS covers just the representation of the object-object, object-aspect and aspect-aspect interactions in the DS of the UML. The MIMECORA-DS has not been used and this paper did not evaluate how it will be applied in a real software project.

Regarding future work, MIMECORA will be customized to represent, through the SD, the modeling of AO *intertype declarations*. This customization will enable represent aspects modifying, adding or removing a JP on the SD. The MIMECORA will also be applied on the development of distributed and critical systems, areas where the separation of concerns and modeling of NFRs are very important, to provide compliant requirements and as a result the quality of the final product. Additionally, a tool will be developed to support MIMECORA-DS notation. This tool will dynamically add and remove aspects from the SD perspective to allow the easy mapping of candidate JPs to be crosscutted by aspects. Thus, the tool might contribute to highlighting in a graphical way MIMECORA-DS elements crosscutting OO components. Thus, this tool can exemplify a practical case of application of MIMECORA-DS on a real project.

References

1. OMG. UML 2.0 Infrastructure Specification (2008), `http://www.omg.org` (accessed on: October 20, 2008)
2. Fowler, M.: UML Distilled: A Brief Guide to the Standard Object Modeling Language, 3rd edn., pp. 53–63. Addison-Wesley, Boston (2004)
3. Stein, D., Hanenberg, S., Unland, R.: Designing aspect-oriented crosscutting i UML. In: Workshop on Aspect-Oriented Modeling with UML, in conjunction with the 1st International Conference on Aspect-Oriented Software Development. Enschede, The Netherlands, p. 6 (2002) (accessed on: September 8, 2008)
4. Stein, D., Hanenberg, S., Unland, R.: On representing join points in the UML. In: Second International Workshop on Aspect-Oriented Modeling with UML In Conjunction with the Fifth International Conference on the United Modeling Language - the Language and its Applications, Enschede, The Netherlands, p. 6 (2002), `http://lglwww.ep.ch/workshops/uml2002/papers/stein.pdf` (accessed on: September 17, 2008)

5. Stein, D., Hanenberg, S., Unland, R.: An UML-based aspect-oriented design notation for aspectj. In: 1st International Conference on Aspect-Oriented Software Development, Enschede, The Netherlands (2002), http://www.dawis.wiwi.uni-due.de/uploads/tx_chairt3-/publications/StHaUn_AspectOrientedDesignNotation_AOSD_2002.pdf
6. Kiczales, G., Hilsdale, E., Hugunin, J., Kersten, M., Palm, J., Griswold, W.G.: An overview of aspectJ. In: Knudsen, J.L. (ed.) ECOOP 2001. LNCS, vol. 2072, p. 327. Springer, Heidelberg (2001), http://hugunin.net/papers/2001-ecoop-overviewOfAspectj.pdf
7. Basch, M., Sanchez, A.: Incorporating aspects into the UML. In: International Workshop on Aspect-Oriented Modeling, p. 5 (2003), http://lglwww.ep.ch/workshops/aosd2003/papers-/BaschIcorporatingAspectsIntotheUML.pdf (accessed on: October 18, 2007)
8. Grassi, V., Sindico, A.: Uml modeling of static and dynamic aspects. In: International Workshop on Aspect-Oriented Modeling, Bonn, Germany, p. 6 (2006) (accessed on: October 16, 2007)
9. Silveira, F.F.: METEORA: Um Método de Testes Baseado Em Estados Para Software de Aplicação Orientado A Aspectos. Doctorate Thesis, Aeronautics Institute of Technology (ITA), Brazil (2007) (accessed on: September 01, 2008)
10. Laddad, R.: AspectJ In Action: Practical Aspect-Oriented Programming, 513 p. Manning Publications Co., Greenwich (2003)
11. Silberschatz, A., Korth, H.F., Sudarshan, S.: Database Systems Concepts, 3rd edn., 821 p. McGraw-Hill, New York (1997)

Applying a UML Extension to Build Use Cases Diagrams in a Secure Mobile Grid Application

David G. Rosado[1], Eduardo Fernández-Medina[1], and Javier López[2]

[1] UCLM. Alarcos Research Group-Information Systems and Technologies Institute.
Information Systems and Technologies Department, ESI, 13071 Ciudad Real, Spain
{David.GRosado, Eduardo.FdezMedina, Mario.Piattini}@uclm.es
[2] University of Málaga, Computer Science Department, Málaga, Spain
jlm@lcc.uma.es

Abstract. Systems based on Grid computing have not traditionally been developed through suitable methodologies and have not taken into account security requirements throughout their development, offering technical security solutions only during the implementation stages. We are creating a development methodology for the construction of information systems based on Grid Computing, which is highly dependent on mobile devices, in which security plays a highly important role. One of the activities in this methodology is the requirements analysis which is use-case driven. In this paper, we build use case diagrams for a real mobile Grid application by using a UML-extension, called GridUCSec-Profile, through which it is possible to represent specific mobile Grid features and security aspects for use case diagrams, thus obtaining diagrams for secure mobile Grid environments.

Keywords: UML extension, Security, Use Cases, secure mobile Grid, secure development.

1 Introduction

With regard to the overall lack of software security in industry, many efforts are currently being made to integrate security into software and software development [1-5]. Systems based on Grid Computing are a type of systems that have clear differentiating features of which security is an extremely important aspect. Grids are centred on sharing resources between dynamic collections of individuals, institutions and resources in a flexible, secure and coordinated manner [6]. Grid environments have special features that make them different from other systems and which must be considered throughout the entire development lifecycle.

The lack of adequate development methods for this kind of systems has encouraged us to build a methodology with which to develop them [7, 8], offering a detailed guide to their analysis, design and implementation. The analysis activity of this methodology is centred on use cases (hereafter UCs) in which we define the behaviour, actions and interactions with those implied in the system (actors), thus obtaining a first approach towards the needs and requirements (functional and non-functional) of the system to be constructed.

C.A. Heuser and G. Pernul (Eds.): ER 2009 Workshops, LNCS 5833, pp. 126–136, 2009.

UML use cases [9] have become a widely used technique for the elicitation of functional requirements [10] when designing software systems. One of the main advantages of UCs is that they are easy to understand with only a limited introduction to their notation, and are therefore very well-suited to the communication and discussion of requirements with system stakeholders. Misuse cases, i.e. negative scenarios or UCs with a hostile intent, have recently been proposed as a new avenue through which to elicit non-functional requirements, particularly security requirements [11-15]. UCs have proved helpful in the elicitation of, communication about, and documentation of functional requirements. The integral development of use and misuse cases provides a systematic way in which to elicit both functional and non-functional requirements [13].

Security requirements exist because certain people and the negative agents that they create (such as computer viruses) pose real threats to systems. Security differs from all other specification areas in that someone is deliberately threatening to break the system. Employing use and misuse cases to model and analyse scenarios in systems under design can improve security by helping to mitigate threats [13].

In the analysis activity of the methodology we use security UCs and misuse cases together with UCs as essential elements of the requirements analysis. These elements must be defined for the context of mobile Grid, and we have therefore extended UML in order to define new UCs, security UCs and misuse cases for mobile Grid systems as a single package (called GridUCSec) of UCs for the identification and elicitation of both functional and non-functional requirements for mobile Grid environments.

A preliminary publication of the methodology has been presented in [8] in which we describe our general approach. [7] provides an informal presentation of the first steps of our methodology which consists of analyzing the security requirements of mobile grid systems directed by misuse cases and security UCs, and which is applied in an actual case study in [16] from which we obtain the security requirements for a specific application by following the steps described in our methodology. We have then gone on to elicit some common requirements of these kinds of systems, and these have been specified to be reused through a UML extension of UCs [17-19]. This paper shows how to apply the UML extension, called GridUCSec-profile, to a real mobile Grid system in order to build UC diagrams, with the help of the reusable UCs available in the repository, using the stereotypes and relationships defined in this profile. One task of the analysis activity of our methodology builds UC diagrams. In this paper we explain how this is achieved.

The remainder of the paper is organized as follows: In section 2, we present the UML extension for secure mobile Grid UCs. In section 3, we apply this UML extension to build UCs diagrams in a mobile Grid application. Finally, we propose our conclusions and future work.

2 UML Extension for Secure Mobile Grid Use Cases

We use the Unified Modeling Language (UML) as the foundation of our work for several reasons: UML is the de-facto standard for object-oriented modelling. Many modelling tools support UML and a great number of developers are familiar with the language. Hence, our work enables these users to develop access control policies

Table 1. Detailed description of Stereotypes for the GridUCSec package

«GridUC»		**Notation**
Description	Specifies requirements of the Grid system and represent the common behaviour and relationships for this kind of systems. It specializes the UseCase within the Grid context defining the behaviour and functions for the Grid system.	G <<GridUC>>
Tagged Values	GridRequirement, ProtectionLevel, SecurityDependence, InvolvedAsset	
«SecurityUC»		**Notation**
Description	Specifies security requirements of the system, describing security tasks that the users will be able to perform with the system.	<<SecurityUC>>
Tagged Values	SecurityRequirement, InvolvedAsset, SecurityDegree, SecurityDomain	
«GridSecurityUC»		**Notation**
Description	This represents specific security features of Grid systems. It adds specific special security features which are covered by this stereotype, and specializes to common security UCs of other applications.	G <<GridSecurityUC>>
Tagged Values	InvolvedAsset, SecurityRequirement, SecurityDegree, SecurityDependence, SecurityDomain	
«MisuseCase»		**Notation**
Description	A sequence of actions, including variants, that a system or other entity can perform, interacting with misusers of the entity and causing harm to certain stakeholders if the sequence is allowed to be completed [12, 21].	<<Misuse>>
Tagged Values	InvolvedAsset, ImpactLevel, RiskLevel, ThreatLikelihood, KindAttack	
«MobileUC»		**Notation**
Description	This represents mobile features of the mobile devices within Grid systems. It defines the mobile behaviour of the system and specializes UseCase within the Grid context and mobile computing defining the behaviour and functions for the mobile Grid system.	M <<MobileUC>>
Tagged Values	MobileRequirement, ProtectionLevel, SecurityDependence, InvolvedAsset, NetworkProtocol	
«Permit»		**Notation**
Description	This relationship specifies that the behaviour of a UC may be permitted by the behaviour of a security UC.	<<permit>>
Tagged Values	PermissionCondition, KindPermission	
«Protect»		**Notation**
Description	This relationship specifies that the behaviour of a UC may be protected by the behaviour of a security UC.	<<protect>>
Tagged Values	InvolvedAsset, ProtectionLevel, KindAttack	
«Mitigate»		**Notation**
Description	This relationship specifies that the behaviour of a misuse case may be mitigated by the behaviour of a security UC.	<<mitigate>>
Tagged Values	SuccessPercentage, KindCountermeasure	
«Threaten»		**Notation**
Description	This relationship specifies that the behaviour of a UC may be threatened by the behaviour of a misuse case.	<<threaten>>
Tagged Values	SuccessPercentage, KindVulnerability, KindAttack	
«GridActor»		**Notation**
Description	This actor specifies a role played by a Grid user or any other Grid system that interacts with the subject.	G «GridActor»
Tagged Values	KindGridCredential, KindGridActor, KindRole, DomainName, Site-Credential	
«MisActor»		**Notation**
Description	This actor specifies a role played by a attacker or misuser or any other attack that interacts with the subject	M «MisActor»
Tagged Values	KindMisActor, HarmDegree	

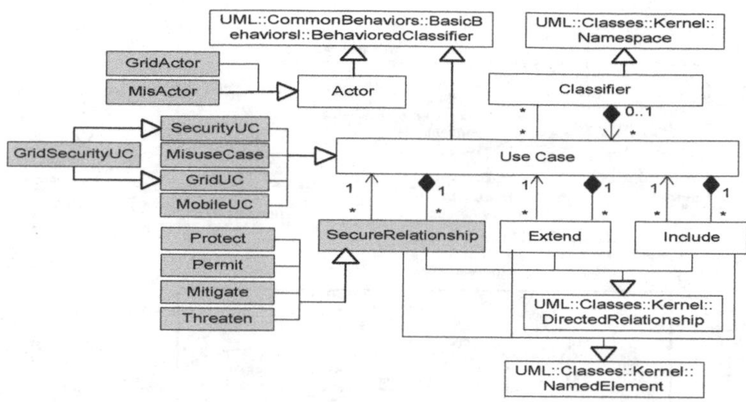

Fig. 1. The concepts used for modeling secure mobile Grid UCs in UML 2.0

using an intuitive, graphical notation. UML offers the possibility of extending the modeling language using well-defined extensibility constructs that are packaged in a so-called UML Profile. In our work, we use *stereotypes* to define new types of model elements and *tagged values* to introduce additional attributes into metamodel types.

In order to define reusable UC diagrams, which are specific to mobile Grid systems, it is necessary to extend the UML 2.0 metamodel and define stereotypes. A stereotype is an extension of the UML vocabulary that allows us to create new building blocks derived from the existing ones but which are specific to a concrete domain, in our case, the Grid computing domain. In this section we present the GridUCSec-Profile extension through which it is possible to represent specific mobile Grid features and security aspects for UC diagrams, thus obtaining UC diagrams for secure mobile Grid environments. This extension has been built as a UML profile which is an extensibility mechanism that allows us to adapt the metaclasses of a model thus making the incorporation of new elements into a domain possible. Fig. 1 shows a UC diagram metamodel in UML 2.0 extended with the new stereotypes of GridUCSec-profile.

In Table 1, we briefly define the stereotypes for the GridUCSec-profile based on the UML 2.0 specification [20]. Three elements are shown in the definition: 1) *Description*: This indicates the purpose and significance for the different users of stereotypes. 2) *Notation*: This corresponds with an icon that it is associated with the stereotype for its graphic notation. 3) *Tagged Values*: This identifies the attributes associated with the stereotype.

3 Applying GridUCSec-Profile to a Real Case

GridUCSec-profile is being validated through a real case application, a business application in the Media domain, defined within the GREDIA European project (www.gredia.eu). This profile will help us to build UC diagrams for a Mobile Grid

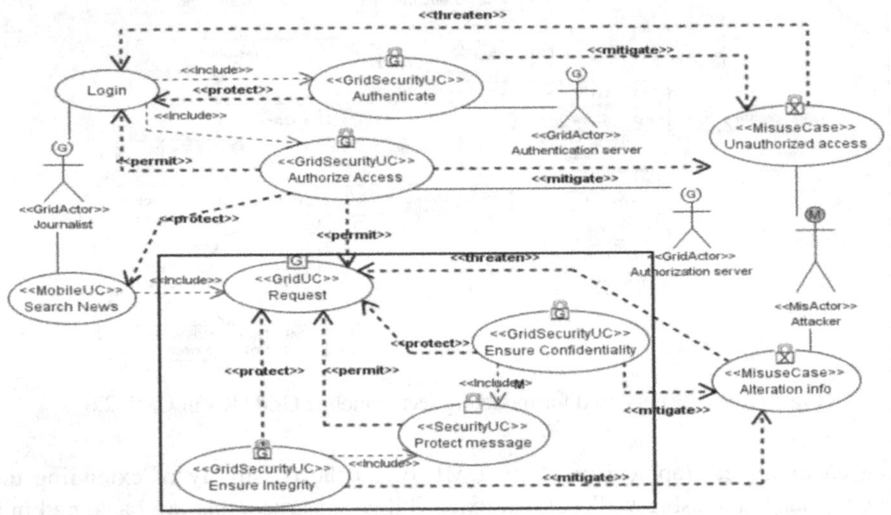

Fig. 2. Main diagram of the application with reusable UCs and reusable sub-diagram

application, which will allow journalists and photographers (actors in the media domain) to make their work available to a trusted network of peers at the same instant as it is produced, either from desktop or mobile devices. We wish to build a system that will cater for the reporter who is on the move with lightweight equipment and wishes to capture and transmit news content.

First, we must identify the functional UCs of the application, but due to space constraints only consider two of them (Login and Search news) are considered here. Second, we must define the possible security needs for these functional UCs (authentication, authorization, confidentiality and integrity). Third, we must identify the possible threats that may attack the system and represent them as misuse cases (unauthorized access and alteration info). Finally, we use the GridUCSec-profile to relate the UCs between them and describe the relevant security aspects that will be necessary in the next activities of the methodology. The resulting diagram is shown in Fig. 2.

The "*«GridSecurityUC» Authenticate*" models the authentication service of the application and is responsible for protecting the "*Login*" UC and for mitigating the "*«MisuseCase» Unauthorized access*" misuse case which threatens the "*Login*" UC. The "*«GridSecurityUC» Authorize access*" models the authorization service and is responsible for protecting the "*«MobileUC» Search news*" UC, for mitigating the "*«MisuseCase» Unauthorized access*" misuse case and for permitting the execution of "*Login*" and "*«GridUC» Request*". We also have the "*«MisuseCase» Alteration info*" misuse case that threatens the modification or alteration of the information exchanged in the messages every time that a request is sent to the system. This threat is mitigated by the "*«GridSecurityUC» Ensure Confidentiality*" and "*«GridSecurityUC» Ensure Integrity*" UCs which are part of the reusable sub-diagram stored in the repository. Finally, the "*«MobileUC» Search News*" UC is identified as a mobile UC due to the possible mobility of the user who requests information from the system from the mobile devices. This mobile UC includes the "*«GridUC» Request*" UC which is responsible for making the request in a secure manner.

In order to build the resulting diagram, we have used a reusable UCs diagram (sub-diagram shown in Fig. 2) which is availability in the repository and is defined by using our UML profile, to model a common scenario that ensures confidentiality and integrity of a request in Grid environments, which is required of our application. This sub-diagram shows how the "*«GridUC» Request*" UC is protected, through «protect» relationships, by the "*«GridSecurityUC» Ensure Confidentiality*" and "*«GridSecurityUC» Ensure Integrity*" security UCs which mitigate the "*«MisuseCase» Alteration info*" misuse case that threatens "*«GridUC» Request*". It also establishes a «permit» relationship from the "*«SecurityUC» Protect message*" security UC, meaning that once the message is protected, the request can be carried out.

Table 2 shows the detailed information of the reusable sub-diagram stored in the repository according to GridUCSec-profile. In this table we can see the different values for the tagged values of the stereotypes used in the sub-diagram. So, for example, we assign the following values to the "*«GridSecurityUC» Ensure Confidentiality*" UC:

- *SecurityRequirement: {Confidentiality}*. This indicates that this UC establishes confidentiality in the diagram, incorporating this security requirement in the application.
- *InvolvedAsset: {Message, Data}*. This indicates that the important assets in this UC are message and data, thus establishing confidentiality in both messages and data.
- *SecurityDomain: SecNews*. This identifies the security domain of the application in which security controls are carried out. This application contains SecNews.
- *SecurityDegree: {High}*. This is used to establish confidentiality in messages. It adds a high degree of security to the message exchanges and communication in the system.
- *SecurityDependence: {VLow}*. This value indicates that this UC has a very low risk level and does not, therefore, need to be protected by others.

This security UC protects the "*«GridUC» Request*" UC and mitigates the "*«MisuseCase» Alteration info*" misuse case. Many values of the tagged values of these stereotypes must therefore coincide, indicating the relationships between them to fulfil their purposes. The "*InvolvedAsset*" tagged value for the "*«GridUC» Request*" UC is therefore "Message", indicating that messages are the asset to be protected from threats and attacks which may damage them. This protection is carried out by both "*«GridSecurityUC» Ensure Confidentiality*" and "*«GridSecurityUC» Ensure Integrity*". The value for the "*InvolvedAsset*" tagged value of the *«protect»* stereotypes must also coincide and are assigned the "Message" value. The message is also one of the assets that may be threatened by the "*«MisuseCase» Alteration info*" misuse case, which we shall deal with next. The values in the other stereotypes shown in Table 2 are assigned by following the same criteria.

Table 2. Detailed definition for the reusable subdiagram using GridUCSec-profile

<table>
<tr><th>Stereotype</th><th colspan="2">Tagged Values</th></tr>
<tr><td rowspan="3">«GridSecurityUC»
Ensure Confidentiality
(EC)</td><td colspan="2">SecurityRequirement: {Confidentiality}</td></tr>
<tr><td>InvolvedAsset: {Message, Data}</td><td>SecurityDomain: SecNews</td></tr>
<tr><td>SecurityDegree: {High}</td><td>SecurityDependence: {VLow}</td></tr>
<tr><td rowspan="3">«GridSecurityUC»
Ensure Integrity (EI)</td><td colspan="2">SecurityRequirement: {Integrity}</td></tr>
<tr><td>InvolvedAsset: {Message, Data}</td><td>SecurityDomain: SecNews</td></tr>
<tr><td>SecurityDegree: {High}</td><td>SecurityDependence: {VLow}</td></tr>
<tr><td rowspan="3">«SecurityUC»
Protect Message (PM)</td><td colspan="2">SecurityRequirement:{Confidentiality, Integrity, Privacy}</td></tr>
<tr><td>InvolvedAsset: {Message}</td><td>SecurityDomain: SecNews</td></tr>
<tr><td colspan="2">SecurityDegree: {High}</td></tr>
<tr><td rowspan="2">«GridUC»
Request (R)</td><td>GridRequirement: {Interoperatibility}</td><td>SecurityDependence: {Medium}</td></tr>
<tr><td>ProtectionLevel: {Medium}</td><td>InvolvedAsset: {Message}</td></tr>
<tr><td rowspan="2">«Protect»
EC – R</td><td>InvolvedAsset: {Message, Data}</td><td>ProtectionLevel: {High}</td></tr>
<tr><td colspan="2">KindAttack: {MasqueradingAtt}</td></tr>
<tr><td rowspan="2">«Protect»
EI – R</td><td>InvolvedAsset: {Message, Data}</td><td>ProtectionLevel: {High}</td></tr>
<tr><td colspan="2">KindAttack: {EavesdroppingAtt, MasqueradingAtt}</td></tr>
<tr><td rowspan="2">«Permit»
PM - R</td><td colspan="2">PermissionCondition: messages encrypted and signed</td></tr>
<tr><td colspan="2">KindPermission: {Execute, Include, Protect}</td></tr>
</table>

It is next necessary to define the relationships between all the UCs that are part of the main diagram (reusable or not) and their relationships with the UCs from the sub-diagram to be integrated into the main diagram. In Table 3, we define these relationships and any relevant information that it is necessary to obtain for the following activities or tasks of the methodology. In the reusable sub-diagram, we have defined security UCs which permit us to establish *«mitigate»* relationships with misuse cases. So, for example, the confidentiality of messages can mitigate and prevent the modification or alteration of the messages that are exchanged in the system, and this is represented with the *«mitigate»* relationship between the "*«GridSecurityUC» Ensure Confidentiality*" UC and the "*«MisuseCase» Alteration info*" misuse case. The values defined for this relationship are the following:

- *SuccessPercentage: {High}*. This indicates a high percentage of attack mitigation with message confidentiality.
- *KindCountermeasure: encrypt message*. This indicates the countermeasure that it is recommendable to take to protect the security against this attack.

For the "*«MisuseCase» Alteration info*" misuse case it is necessary to define the values which detail the main features of the attack, and which assist us towards a better knowledge of this type of attacks in order to make decisions regarding how to protect to our system from them. The values assigned to this misuse case are:

- *InvolvedAsset: {Message, Identity, Data}*. This indicates the assets that may be attacked by this UC. In this case, the alteration of information affects messages, data and identity stored in the mobile device. The message is the asset to be protected by the security UCs and which is threatened by the misuse cases in this application.
- *ImpactLevel: {High}*. This threat produces a high impact level in the system if the alteration of the messages is carried out successfully.

Table 3. Detailed description of the elements of the main diagram using GridUCSec-profile

Stereotype	Tagged Values	
«MobileUC» Search News (SN)	MobileRequirement: {Integrity, Delegation}	
	SecurityDependence: {High}	InvolvedAsset: {Message}
	NetworkProtocol: {WAP}	ProtectionLevel: {VHigh}
«MisuseCase» Alteration info (AI)	InvolvedAsset: {Message, Identity, Data}	
	ImpactLevel: {High}	RiskLevel: {High}
	ThreatLikelihood: {Frequent}	KindAttack: {MasqueradingAtt}
«MisuseCase» Unauthorized access (UA)	InvolvedAsset: {Message, Identity, Data}	
	ImpactLevel: {High}	RiskLevel: {High}
	ThreatLikelihood: {Frequent}	KindAttack: {MasqueradingAtt}
«GridSecurityUC» Authorize Access (AA)	Securityrequirement: {Confidentiality}	
	SecurityDegree: {High}	InvolvedAsset: {Message}
	SecurityDependence: {VLow}	SecurityDomain: SecNews
«GridSecurityUC» Authenticate (Auth)	Securityrequirement: {Confidentiality}	
	SecurityDependence: {VLow}	InvolvedAsset: {Message}
	SecurityDegree: {High}	SecurityDomain: SecNews
«Threaten» AI – R	KindVulnerability: messages by wireless network	SuccessPercentage: {High}
	KindAttack: {MasqueradingAtt, EavesdroppingAtt}	
«Threaten» UA – Login	KindVulnerability: identity and credential stored	SuccessPercentage:{VHigh}
	KindAttack: {AccessControlAtt, MaliciousAtt}	
«Mitigate» EC – AI	SuccessPercentage: {High}	KindCountermeasure: encrypt message
«Mitigate» EI – AI	SuccessPercentage: {High}	KindCountermeasure: digital sign
«Mitigate» AA–UA	SuccessPercentage: {VHigh}	KindCountermeasure: check privilegies
«Mitigate» Auth – UA	SuccessPercentage: {VHigh}	KindCountermeasure: check identity
«Protect» Auth - Login	InvolvedAsset: {Credential, Identity}	ProtectionLevel: {High}
	KindAttack: {AccessCOntrolAtt, IntruderAtt}	
«Protect» AA – SN	InvolvedAsset: {Identity, Resource}	ProtectionLevel: {VHigh}
	KindAttack: {MaliciousAtt, AccessControlAtt}	
«Permit» AA – R	PermissionCondition: check privilegies	KindPermission: {CheckExecute}
«Permit» AA - Login	PermissionCondition: check access rights	
	KindPermission: {CheckExecute, Protect}	
«GridActor» Journalist	KindGridActor: {Mobile User}	DomainName: News
	KindRole: journalist	KindGridCredential: {UserPass, X509}
	Site-Credential: {(News, UserPass),(SecNews,X509)}	
«GridActor» Authentication server	KindGridActor:{Service}	KindRole: security server
	KindGridCredential:{X509}	DomainName: SecNews
«GridActor» Authorization server	KindGridActor:{Service}	KindRole:security server
	KindGridCredential:{X509}	DomainName: SecNews
«MisActor» Attacker	KindMisActor: hacker	HarmDegree: {Medium}

- *RiskLevel: {High}*. With regard to the assets involved in this misuse case, this attack produces a high risk level of damage to the assets.
- *ThreatLikelihood: {Frequent}*. This specifies a frequent (monthly) likelihood that this threat will occur in the system to alter information in the messages.
- *KindAttack: {MasqueraddingAtt}*. The masquerading attack could permit the disclosure or modification of information.

The UC that has most relationships with the other UCs is the *"«GridSecurtyUC» Authorize access"* which protects *"«MobileUC» Search News"*, grants permission for the realization of *"«GridUC» Request"* and *"Login"* UCs, and mitigates the

"*«MisuseCase» Unauthorized access*" misuse case. This UC therefore defines 4 types of relationships, which are shown in Table 3. For example, for the *«Protect»* relationship, we have defined the following values:

- *«Protect» Authorize Access – Search News (AA – SN).* This relationship defines values for the tagged values:
  - *InvolvedAsset: {Identity, Resource}.* This indicates that the assets which should be protected by authorization rules are the identity of the user and the resource owned by this identity.
  - *ProtectionLevel: {VHigh}.* This relationship specifies a very high protection level that the origin UC offers to the destination UC.
  - *KindAttack: {MaliciousAtt, AccessControlAtt}.* This relationship can protect UCs from malicious and access control attacks.

Table 3 shows the remaining values for the tagged values of the stereotypes of the diagram in Fig. 2. Each value is obtained as we have shown previously.

4 Conclusions and Future Work

In order to study the needs and particularities of mobile Grid systems, it was necessary to define an extension of UML UCs that would capture the performance, functions, properties and needs that arise in this kind of systems. The UML extension for UCs makes it possible to analyse the system's security requirements from the early stages of development, to enrich UC diagrams with security aspects and to define values that are essential if we are to interpret and capture what will be required in the following activities of our development process.

This UML profile permits us to identify features, aspects and properties that are important in the first stages of the life cycle and will be very useful when making decisions about which security mechanisms, services, etc. to use in the design activity. The application of this profile to a real case has helped us to refine and improve the definition of the profile by adding or changing new values, properties or constraints that were not initially considered. For example, we have defined mobile UCs because it is necessary to capture the mobile behaviour, and we have also defined new tagged values because we found aspects that must be included in our analysis and which were not initially included. Furthermore, this extension will permit us to build more detailed, complete and richer UC diagrams in terms of semantics.

As future work, we aim to complete the details of this methodology (activities, tasks, etc.) through the research-action method by integrating security requirements engineering techniques (UMLSec, etc.) and defining the traceability of artifacts. We will complete the real case by describing all of the application's functional UCs with GridUCSec-profile.

Acknowledgments. This research is part of the following projects: QUASIMODO (PAC08-0157-0668) financed by the "Viceconsejería de Ciencia y Tecnología de la

Junta de Comunidades de Castilla-La Mancha" (Spain), ESFINGE (TIN2006-15175-C05-05) granted by the "Dirección General de Investigación del Ministerio de Educación y Ciencia" (Spain), y SISTEMAS (PII2I09-0150-3135) financed by the "Consejería de Educación y Ciencia de la Junta de Comunidades de Castilla-La Mancha".

References

1. Bass, L., Bachmann, F., Ellison, R.J., Moore, A.P., Klein, M.: Security and survivability reasoning frameworks and architectural design tactics. SEI (2004)
2. Breu, R., Burger, K., Hafner, M., Jürjens, J., Popp, G., Lotz, V., Wimmel, G.: Key issues of a formally based process model for security engineering. In: International Conference on Software and Systems Engineering and their Applications (2003)
3. Haley, C.B., Moffet, J.D., Laney, R., Nuseibeh, B.: A framework for security requirements engineering. In: Software Engineering for Secure Systems Workshop, Shangai, China, pp. 35–42 (2006)
4. Jürjens, J.: Secure Systems Development with UML. Springer, Heidelberg (2005)
5. Mouratidis, H., Giorgini, P.: Integrating Security and Software Engineering: Advances and Future Vision. IGI Global (2006)
6. Foster, I., Kesselman, C.: The Grid2: Blueprint for a Future Computing Infrastructure, 2nd edn. Morgan Kaufmann Publishers, San Francisco (2004)
7. Rosado, D.G., Fernández-Medina, E., López, J., Piattini, M.: Engineering Process Based On Grid Use Cases For Mobile Grid Systems. In: The Third International Conference on Software and Data Technologies- ICSOFT 2008, Porto, Portugal, pp. 146–151 (2008)
8. Rosado, D.G., Fernández-Medina, E., López, J., Piattini, M.: PSecGCM: Process for the development of Secure Grid Computing based Systems with Mobile devices. In: International Conference on Availability, Reliability and Security (ARES 2008). IEEE Computer Society, Barcelona (2008)
9. The Object Management Group (OMG): OMG Unified Modeling Language (OMG UML), Version 2.2 (2007), http://www.omg.org/spec/UML/2.1.2/Infrastructure/PDF/
10. Alexander, I., Maiden, N.: Scenarios, Stories, Use Cases: Through the Systems Development Life-Cycle. John Wiley & Sons, Chichester (2004)
11. Sindre, G., Opdahl, A.L.: Templates for misuse case description. In: 7th International Workshop on Requirements Engineering: Foundation for Software Quality, Austria (2001)
12. Sindre, G., Opdahl, A.L.: Capturing Security Requirements by Misuse Cases. In: 14th Norwegian Informatics Conference (NIK 2001), Tromsø, Norway (2001)
13. Alexander, I.: Misuse Cases: Use Cases with Hostile Intent. IEEE Software, 58–66 (2003)
14. Firesmith, D.G.: Security Use Cases. Journal of Object Technology, 53–64 (2003)
15. Sindre, G., Opdahl, A.L.: Eliciting security requirements with misuse cases. Requirements Engineering Journal 10, 34–44 (2005)
16. Rosado, D.G., Fernández-Medina, E., López, J.: Obtaining Security Requirements for a Mobile Grid System. International Journal of Grid and High Performance Computing (2009) (to be published in April 1, 2009)
17. Rosado, D.G., Fernández-Medina, E., López, J.: Extensión UML para Casos de Uso Reutilizables en entornos Grid Móviles Seguros. XIV Jornadas de Ingeniería del Software y Bases de Datos - JISBD 2009, San Sebastián (2009)

18. Rosado, D.G., Fernández-Medina, E., López, J., Piattini, M.: Towards an UML Extension of Reusable Secure Use Cases for Mobile Grid systems. IEICE Transactions on Information and Systems (2009) (submitted)
19. Rosado, D.G., Fernández-Medina, E., López, J.: Reusable Security Use Cases for Mobile Grid environments. In: Workshop on Software Engineering for Secure Systems, in conjunction with the 31st International Conference on Software Engineering, Vancouver, Canada, pp. 1–8 (2009)
20. OMG: OMG Unified Modeling Language (OMG UML), Superstructure, V2.1.2 (2007), `http://www.omg.org/spec/UML/2.1.2/Infrastructure/PDF/`
21. Røstad, L.: An extended misuse case notation: Including vulnerabilities and the insider threat. In: XII Working Conference on Requirements Engineering: Foundation for Software Quality, Luxembourg (2006)

The MP (Materialization Pattern) Model for Representing Math Educational Standards

Namyoun Choi, Il-Yeol Song, and Yuan An

College of Information Science and Technology, Drexel University
Philadelphia, PA 19104, USA
{namyoun.choi, songiy}@drexel.edu, yuan.an@ischool.drexel.edu

Abstract. Representing natural languages with UML has been an important research issue for various reasons. Little work has been done for modeling imperative mood sentences which are the sentence structure of math educational standard statements. In this paper, we propose the MP (Materialization Pattern) model that captures the semantics of English sentences used in math educational standards. The MP model is based on the Reed-Kellogg sentence diagrams and creates MP schemas with the UML notation. The MP model explicitly represents the semantics of the sentences by extracting math concepts and the cognitive process of math concepts from math educational standard statements, and simplifies modeling. This MP model is also developed to be used for aligning math educational standard statements via schema matching.

Keywords: Unified Modeling Language, P (Materialization Pattern) model.

1 Introduction

Representing natural languages with UML (United Modeling Language) [1], for instance English, has been an important research issue for various reasons – including the transition of natural language software requirements into modeling ([6], [7], [9]), natural language query sentence processing [2], and representation of knowledge [4] which is extracted from the text by an automatic tool.

In this paper, we present the MP (Materialization Pattern) model for modeling imperative mood sentences used in math educational standard statements. Math educational standards (math standards) state the mathematical understanding, knowledge, and skills that students should obtain from pre-kindergarten through grade 12. Two examples of math standard statements are as follows: 1) *Write fractions with numerals and number words.* 2) *Model, sketch, and label fractions with denominator to 10.* The purpose of the MP model is to represent the semantics of such imperative mood sentences using the UML notation so that the semantics of statements from different math standards can be compared. Our MP model captures math concepts and the cognitive process of math concepts from a math standard statement. Hence, the MP model enables us to compare the level of similarity of two statements from different math standards in terms of math concepts and the cognitive process of the

C.A. Heuser and G. Pernul (Eds.): ER 2009 Workshops, LNCS 5833, pp. 137–146, 2009.

math concepts. In this paper, we focus on how to represent math standard statements using the MP model.

The MP model is developed at a sentence level for each statement from typical math standards. We call these sentences *MP statements*. We classify MP statements based on different MP schemas. Our sentence analysis is based on the Reed-Kellogg sentence diagram [10]. We identify a math concept as the MP class or a noun class, and the cognitive process [11] of a math concept as a verb stereo type class.

The contributions of our paper are as follows: (a) We propose the MP model that can explicitly model the semantics of imperative mood sentence structures used in math standards. (b) We identify eleven types of MP schemas that classify the types of statements used in math standards and then develop heuristics to convert the statements to the MP schemas. (c) A distinct feature of the MP model is to extend the granularity of modeling with a verb stereo type class, in which a verb is reified as a class, and thus simplifies modeling of sentences by a Materialization Pattern in a domain class diagram.

The paper is organized as follows: We examine related work in Section 2. In Section 3, we present terminologies used in the MP model, the components of the MP model, heuristics and examples in the MP modeling, and different types of MP statements. In Section 4, a conclusion and future work are presented.

2 Related Work

Illieva (2007) and Illieva and Boley (2008) divide English sentences into three basic groups such as the subject, the predicate, and the object in a tabular presentation of sentences and build a graphical natural language for UML diagram generation. If the sentences lack a subject, the position of the subject is kept empty in a table and it will be filled by the analyst in an interactive mode. Math standard statements have only one subject "student," and the subject is omitted because all the statements are imperative mood sentences. Tseng and Chen (2008) briefly mention how to model an imperative mood sentence of English sentences in UML for transforming natural language queries into relational algebra through the UML class diagram notation. Their approach for modeling an imperative mood sentence of English sentences is summarized as follows: 1) Find out hidden associations between classes., or 2) If the verb does not transfer an action, there is no association at all and the English sentence is modeled as a class hierarchy only without including a verb as an association or a class. In math standards, it is not easy to find out hidden associations on a sentence level, and MP verbs are reified as classes in the MP model. Bryant et al. (2003) describe the method of translating requirements in natural language into UML models and/or executable models of software components. Their method depends on whole requirements in natural language rather than a sentence level. The requirements are refined and processed for creating a knowledge base using natural language processing techniques. And then the Knowledge base is converted into TLG (Two-Level Grammar) [5] which is used as an intermediate representation between the informal knowledge base and the formal specification language representation. TLG can be converted into UML at the final step.

Our survey shows that most research in this area has not been focused on an imperative mood sentence. Our MP model is focused on capturing semantics of imperative mood sentences for modeling math standards.

3 Developing the MP Model

In this section, we discuss terminologies, the components, heuristics, and examples of the MP model, and different types of MP statements.

3.1 Terminology

An English sentence has the subject and the predicate. The sentence structure of math standard statements is an imperative mood sentence. It only has a predicate which consists of verbs with verb modifiers and nouns with noun modifiers. See Fig. 1 for the sentence structure of a math standard statement in the Reed-Kellogg system [10].

- *Reed-Kellogg system*: It is a graphic representation of a sentence structure and represents relationships between the elements of sentences and their modifiers. The horizontal main line is for elements such as the subject, the verb, the direct object, and the complement. Modifiers are placed under elements they modify.

We now explain these following terminologies using a math standard statement such as "Model, sketch, and label fraction with denominator to 10."

- *MP statement*: It is a math standard statement in Fig. 1. For example, "Model, sketch, and label fraction with denominator to 10." is an MP statement.
- *MP verb*: It is a verb in Fig. 1 at the beginning of an MP statement and can be more than one. For example, "model, sketch, and label" are MP verbs.
- *MP noun*: It is a noun in Fig. 1. For example, "fraction" is an MP noun.
- *MP verb modifier*: A verb modifier which modifies an MP verb.
- *MP noun modifier*: A noun modifier which modifies an MP noun. For example, "with denominator to 10" is an MP noun modifier.
- *MP nouns with MP noun modifiers* are *math concepts* and their *properties*, and imply *what students are learning*.
- *Cognitive process of a math concept*: It describes *how students are learning regarding a math concept by MP verbs.*
- *MP schema:* It is created when we model an MP statement using the UML notation.
- *Materialization Pattern (MP)*: It represents an *MP class* and *its (verb) materialization hierarchy that realizes the behaviors of the MP class*. A *MP class* represents a *concept* represented by a noun. A *materialization hierarchy* is a verb *hierarchy that models the behaviors of the MP class*. The *relationship* between the MP class and the materialization hierarchy is represented as *a realization relationship of UML.*

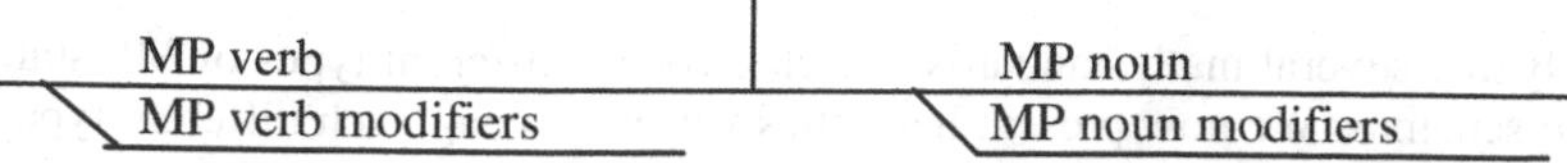

Fig. 1. An MP statement in the Reed-Kellogg system

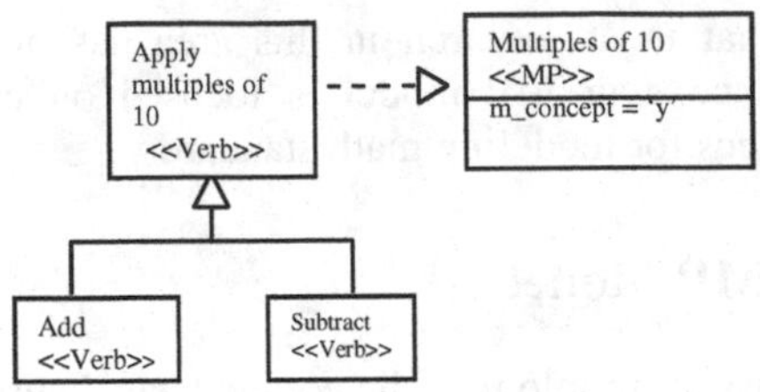

Fig. 2. An MP model for a math standard statement "Add and subtract multiples of 10." A class "*Multiples of 10*" is an MP class. The "m_concept = 'y' " implies that the class "Multiples of 10" is a math concept. "*Add*" and "*Subtract*" are verb stereo type classes which are the cognitive process of the math concept "Multiples of 10". The relationship between the class "Multiples of 10" and the (verb) materialization hierarchy which includes "Apply multiples of 10", "Add", and "Subtract" is represented by a realization relationship.

3.2 The Components of the MP Model

Table 1. Classes and relationships in the MP model

			semantics	notation
class	MP class		Abstraction class	<<MP>>
	Verb stereo type class		A verb is reified as a class.	<<Verb>>
	Noun class		Regular UML class	Regular UML notation
relationship	Predicative relationship	Association	It connects two concepts using a verb [4].	
		Aggregation	It is more specific than association and represents a whole-part relationship.	
	Dependency relationship		One class depends on another.	<<use>>
	Prepositional relationship		It connects two concepts using a preposition [4].	<<preposition>>
	Transitive verb relationship		It relates a noun to another noun or a verb.	<<preposition>>
	Realization relationship		The verb stereo type class realizes the behavior that the MP class specifies.	
	Generalization relationship		" is a " relationship	

3.3 The Components of the MP Model

By analyzing several math standards, we classify 11 different types of MP statements and MP schemas. We now present heuristics for converting each different type of MP statements to the MP schema.

3.3.1 Heuristics in the MP Modeling. We present heuristic to determine classes and relationships.

1. Heuristic to determine classes

   1) All *MP verbs* are converted to *verb stereo type classes*, which represent the *cognitive process of math concepts.*
   2) All *MP nouns* except *Type 2 MP Statement* and *Type 3 MP Statement* are converted into *MP classes*. These *MP classes* are *math concepts.* A superclass is created as an MP class if a superclass exists *when* more than one MP class exist.
   3) All nouns in an MP noun modifier or an MP verb modifier are converted to noun classes. A superclass is created if a superclass exists *when* more than one noun class in an MP modifier or an MP verb modifier exist.
   4) If association, aggregation, or dependency relationships exist between noun classes or between an MP class and noun classes then *noun classes* are also *math concepts.*

2. Heuristic to determine relationships

   Identify relationships as follows:

   - Between an MP class and noun classes in an MP noun modifier,
   - Between noun classes in an MP modifier or an MP verb modifier,
   - Between an MP verb class and noun classes in an MP verb modifier.

   1) There is always a realization relationship between an MP class and a verb stereo type class.
   2) There is a predicative relationship when two concepts are connected using a verb. If it represents a whole-part relationship, then the predicative relationship is an aggregation. In other cases, the predicative relationship is an association.
   3) There is a prepositional relationship when two concepts are connected using a preposition.
   4) There is a transitive verb relationship when a transitive verb relates a concept to another concept [8] or a verb stereo type class.
   5) There is a dependency relationship when an MP verb modifier or an MP noun modifier starts with "using."
   6) There is a generalization relationship if a superclass exists.

3.3.2 Examples of the MP Model. In order to develop an MP schema diagram, we take four different steps for each type of MP statements as follows:

1. Step 1: Write an MP statement which is a math educational standard statement.
2. Step 2: Create the table format of the MP statement.
3. Step 3: Draw a diagram of the MP statement based on the Reed-Kellogg system.
4. Step 4: Develop a MP schema using the UML notation from the Reed-Kellogg diagram of the MP statement.

Due to the lack of space we only present all steps of *Type 1 and Type 2.*

1. Type 1 MP Statement: It only has MP verbs and an MP noun(s) without modifiers.

1. Step 1: An example of MP statements (math standard statements): Recognize, classify, order, and compare whole numbers.
2. See Table 2, Fig. 3, and Fig. 4 for Step 2, Step 3, and Step 4, respectively.

Table 2. A Type 1 MP statement

MP verb	MP noun
Recognize, classify, (and) compare	whole numbers

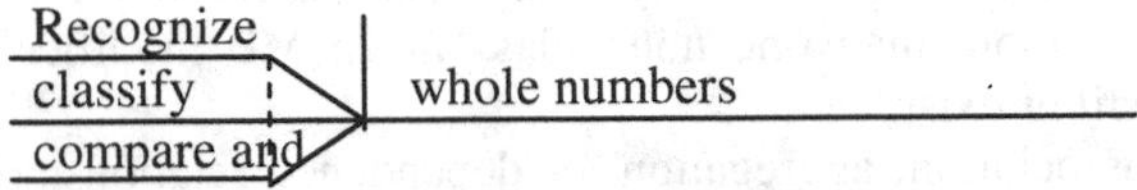

Fig. 3. A Type 1 MP Statement in the Reed-Kellogg system

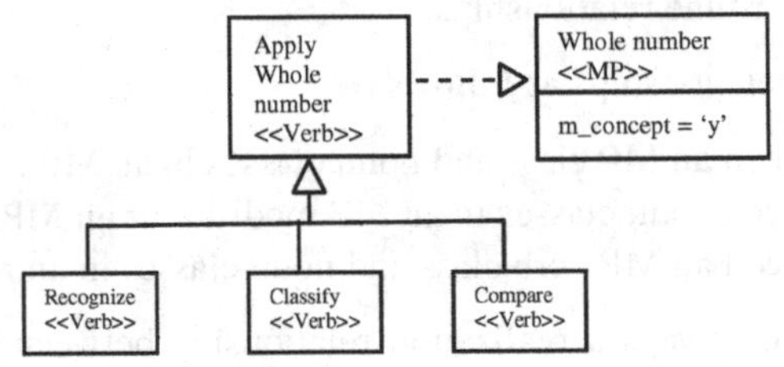

Fig. 4. An MP schema of a Type 1 MP Statement: A class "Whole number" is an MP class. The "m_concept = 'y' implies that the class "Whole number" is a math concept. Verb stereo type classes "Recognize", "Classify", and "Compare" are the cognitive process of the class "Whole number" that is a math concept.

Type 2 MP Statement:

1. Step 1: A math standard statement: Estimate the value of irrational numbers.
2. See Table 3 and Fig. 5 for Step 2 and Step 3, respectively.

Table 3. A Type 2 MP Statement

MP verb	MP noun	MP noun modifier
Estimate	(the) value	of irrational numbers

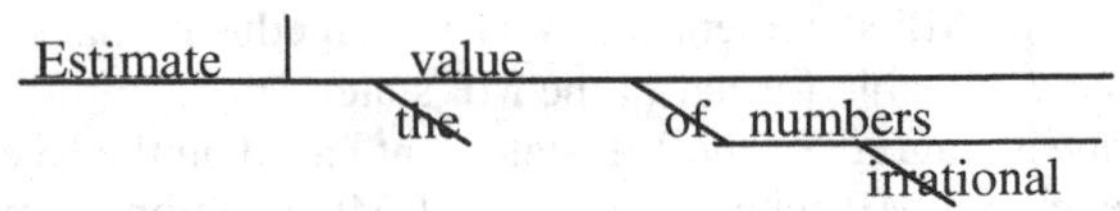

Fig. 5. A Type 2 MP Statement in the Reed-Kellogg system

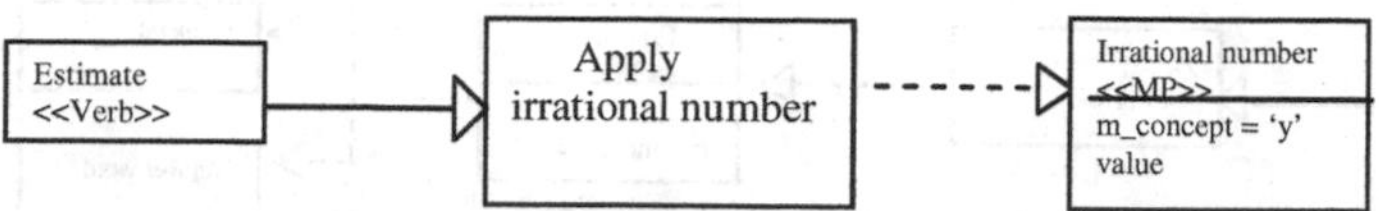

Fig. 6. An MP schema of a Type 2 MP Statement: A class "Irrational number" is an MP class which has an attribute "value". The "m_concept = 'y'" implies that the class "Irrational number" is a math concept. A verb stereo type class "Estimate" implies the cognitive process of the MP class "Irrational number".

3. Step 4: An MP schema using the UML notation is as follows:

- An MP noun modifier is a complement in the form of a prepositional phrase such as "of irrational numbers".
- Model a noun in the MP noun modifier as an MP class, which is a *math concept* and an MP noun(s) as an *attribute* of the MP class.

3. *Type 3 MP Statement:*
   - A math standard statement: Demonstrate an understanding of concepts of time.
   - An MP noun modifier is a prepositional phrase which has more than one "of (or for)" such as "*of concepts of time*".
   - Model the second noun (for example, "Time") in the MP noun modifier as a noun class which is *a math concept* and the first noun(s) (for example, "concept") in the MP noun modifier as an *attribute* of this noun class
   - A prepositional relationship (for example, <<of>>) exists between an MP class (for example, "Understanding") and a noun class (for example, "Time") in the prepositional phrase.
4. *Type 4 MP Statement*
   - An MP statement: Write fractions with numerals and number words.
   - An MP noun modifier is a prepositional phrase which starts with other prepositions except "of" or "for" such as "with numerals and number words".
   - A prepositional relationship (for example <<with>>) exists between an MP class (for example, Fraction) and noun classes (for example, Numeral and Number word) in the prepositional phrase.

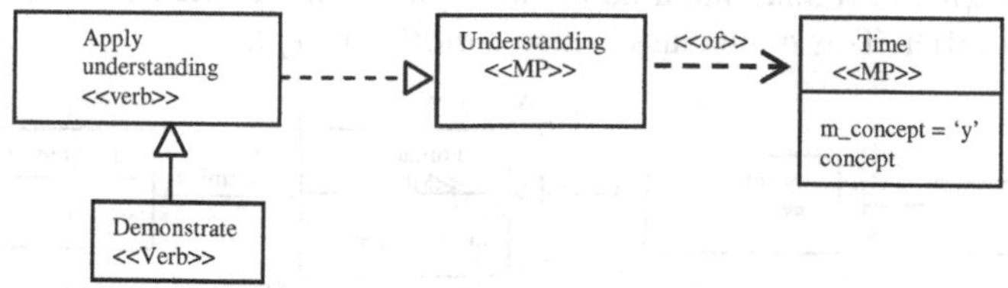

Fig. 7. An MP schema of a Type 3 MP Statement

5. *Type 5 MP Statement:*
   - A math standard statement: Estimate and use measuring device with standard and non-standard unit to measure length.

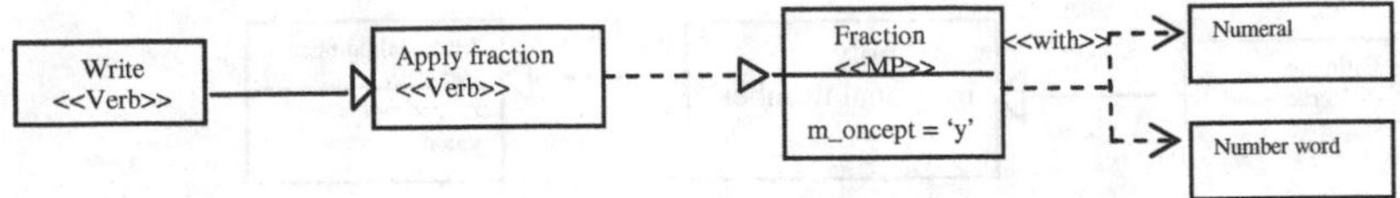

Fig. 8. An MP schema of a Type 4 MP Statement

- An MP noun modifier is a prepositional phrase such as "with standard and non-standard unit to measure length".
- A prepositional relationship (for example, << with>>) exists between an MP class (for example, Measuring device) and noun classes (for example, Unit of measure, Standard unit, Non-standard unit) in the MP noun modifier.
- An association relationship (for example, Measure) exists between noun classes (for example, Unit of measure, and Quantity) in the MP noun modifier. *These noun classes* are also *math concepts*.

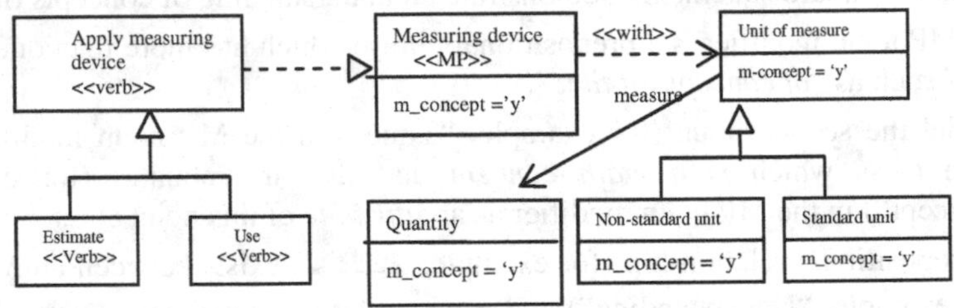

Fig. 9. An MP schema of a Type 5 MP Statement

6. *Type 6 MP Statement:*

- A math standard statement: Develop formulas for determining measurements.
- An MP noun modifier is an infinitive phrase, a prepositional phrase with gerund (for example, for determining measurements), or a pronoun clause.
- An association relationship (for example, determine) exists between an MP class (for example, Formula) and a noun class (for example, Measurement) in the MP noun modifier. This noun class is also a *math concept*.

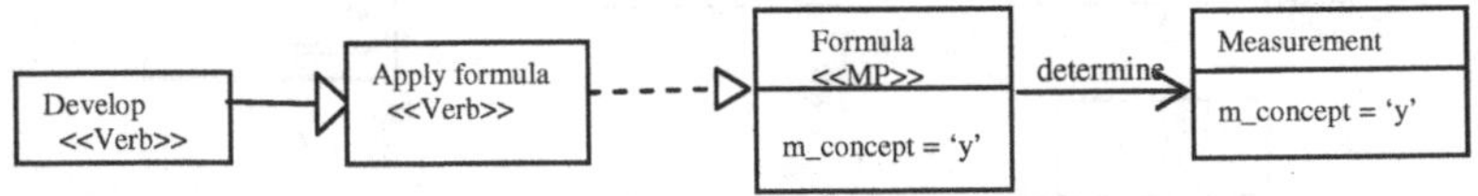

Fig. 10. An MP schema of a Type 6 MP Statement

7. *Type 7 MP Statement*:

- A math standard statement: Create two-dimensional designs that contain a line of symmetry.

- An aggregation relationship exists between an MP class (for example, Two-dimensional design) and a noun class (for example, Line of symmetry) in an MP noun modifier. This noun class is also a *math concept*.

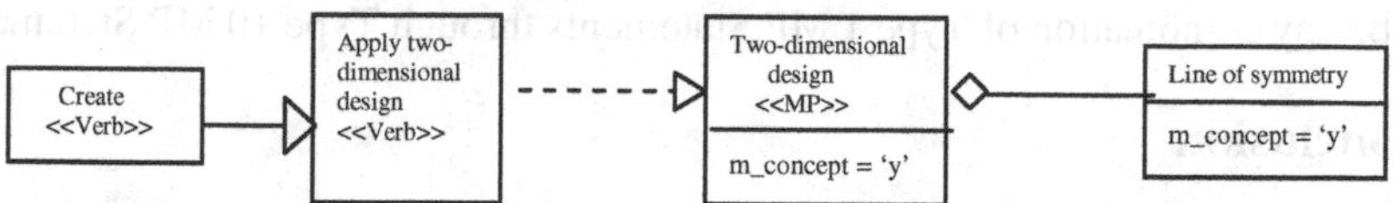

Fig. 11. An MP schema of a Type 7 MP Statement

Type 8 MP Statement:

- A math standard statement: Demonstrate skills for using fraction to verify conjectures, conform computations, and explore complex problem-solving situation.
- A predicative or dependency relationship (for example, <<use>>) can exist between an MP class (for example, Skill) and a noun class (for example, Fraction) in an MP noun Modifier. An association relationship (for example, verify, conform, and explore) exists between noun classes in the MP noun modifier. These noun classes (for example, Conjecture, Computation, and Problem-solving) are also *math concepts*.

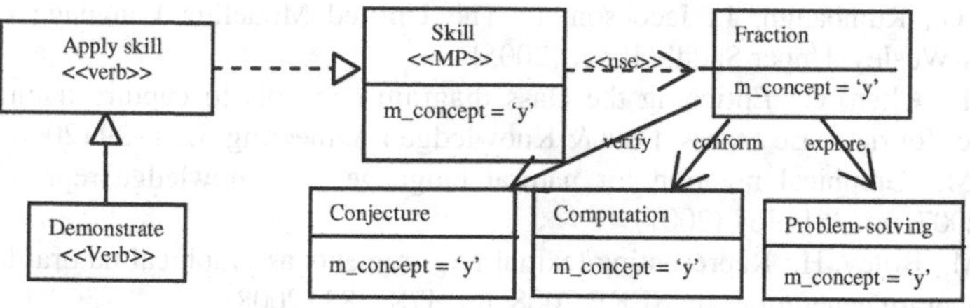

Fig. 12. An MP schema of a Type 8 MP Statement

9. *Type 9 MP Statement*:

- A math standard statement: Apply number theory to rename a number quantity.

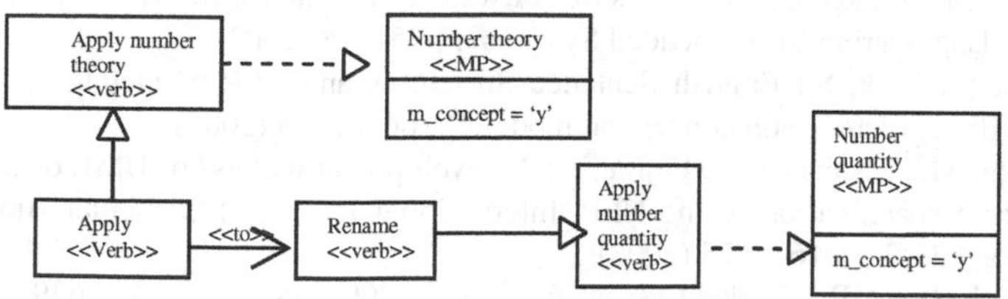

Fig. 13. An MP schema of a Type 9 MP Statement

10. *Type 10 MP Statement:* It has the form of computation tasks in numbers.
    - A math standard statement: Create and solve word problems involving addition, subtraction, multiplication, and division of whole numbers.
11. *Type 11 MP Statement*:

It can be any combination of Type 1 MP Statements through Type 10 MP Statements.

4 Conclusion

In this paper, we have presented the MP (Materialization Pattern) model for capturing the semantics of imperative mood sentences of math educational standard statements. We classified math standard statements into 11 types and provided heuristics for determining classes and relationships for the MP model. We illustrated our method using an example of MP statements for each type. Our method has focused on capturing the semantics of English sentence diagrams based on the Reed-Kellogg system by identifying math concepts, the cognitive process of math concepts, relationships, and classes.

Our MP model is useful for aligning math educational standard statements by schema matching because it extracts math concepts and the cognitive process of math concepts. It can be also useful for modeling other educational standards with minor modification. In the future, we intend to develop a semi-automatic tool for creating MP schemas and will use it for aligning math educational standard statements.

References

1. Booch, G., Rumbaugh, J., Jacobson, I.: The Unified Modeling Language User Guide. Addison-Wesley, Upper Saddle River (2005)
2. Tseng, F., Chen, C.: Enriching the class diagram concepts to capture natural language semantics for database access. Data & Knowledge Engineering 67, 1–29 (2008)
3. Ilieva, M.: Graphical notation for natural language and knowledge representation. In: SEKE 2007, pp. 361–367 (2007)
4. Ilieva, M., Boley, H.: Representing textual requirements as graphical natural language for UML diagram generation. In: SEKE 2008, pp. 478–483 (2008)
5. Bryan, B., Lee, B.: Two-level Grammar as an Object-Oriented Requirements Specification. In: 35th Hawaii Int. Conf. System Sciences (2002)
6. Ilieva, M., Ormandjieva, O.: Automatic Transition of Natural Language Software Requirements Specification into Formal Presentation. In: Montoyo, A., Muńoz, R., Métais, E. (eds.) NLDB 2005. LNCS, vol. 3513, pp. 392–397. Springer, Heidelberg (2005)
7. Bryant, B., Lee, B., Cao, F., Zhao, W., Gray, J., Burt, C.: From Natural Language Requirements to Executable Models of Software Components. In: Monterey Workshop on Software Engineering for Embedded Systems, pp. 51–58 (2003)
8. Hartman, S., Link, S.: English Sentence Structures and EER Modeling. In: The fourth Asia-Pacific conference on conceptual modeling, pp. 27–35 (2007)
9. Takahashi, M., Takahashi, S., Fujita, Y.: A development method of UML documents from requirement specifications using NLP. International Journal of Computer Applications in Technology 33(2-3), 164–175 (2008)
10. Reed, A., Kellogg, B.: Graded Lessons in English (2006) ISBN 1-4142-8639-2
11. Bloom, B., Krathwohl, D.: Taxonomy of educational objectives: the classification of educational goals by a committee of college and university examiners Handbook 1: Cognitive domain, New York, Longmans (1956)

XMI2USE: A Tool for Transforming XMI to USE Specifications

Wuliang Sun[1], Eunjee Song[1], Paul C. Grabow[1], and Devon M. Simmonds[2]

[1] Department of Computer Science
Baylor University
Waco, TX 76798, USA
{wuliang_sun, eunjee_song, paul_grabow}@baylor.edu
[2] Department of Computer Science
University of North Carolina at Wilmington
Wilminton, NC 28403, USA
simmondsd@uncw.edu

Abstract. The UML-based Specification Environment (USE) tool supports syntactic analysis, type checking, consistency checking, and dynamic validation of invariants and pre-/post conditions specified in the Object Constraint Language (OCL). Due to its animation and analysis power, it is useful when checking critical non-functional properties such as security policies. However, the USE tool requires one to specify (i.e., "write") a model using its own textual language and does not allow one to import any model specification files created by other UML modeling tools. Hence, to make the best use of existing UML tools, we often create a model with OCL constraints using a modeling tool such as the IBM Rational Software Architect (RSA) and then use the USE tool for model validation. This approach, however, requires a manual transformation between the specifications of two different tool formats, which is error-prone and diminishes the benefit of automated model-level validations. In this paper, we describe our own implementation of a specification transformation engine that is based on the Model Driven Architecture (MDA) framework and currently supports automatic tool-level transformations from RSA to USE.

Keywords: Model Transformation, MDA, XMI, OCL, Modeling Tool, USE.

1 Introduction

The Object Constraint Language (OCL) is a declarative language for describing rules applied to models and is an important supplement for the Unified Modeling Language (UML), providing expressions that have neither the ambiguities of natural language nor the inherent difficulty of using complex mathematics [14]. OCL is defined as a standard "add-on" to the UML, the Object Management Group (OMG) standard for object-oriented analysis and design [14][18]. During software development, constraints can be written in OCL to supply complementary information at a conceptual level, to achieve higher precision and accuracy

C.A. Heuser and G. Pernul (Eds.): ER 2009 Workshops, LNCS 5833, pp. 147–156, 2009.

within the model and to improve the expressiveness of certain artifacts in the analysis and design phases [17]. In particular, OCL can be used to a) specify invariants on classes and types in the class model, b) specify type invariants for stereotypes, c) describe preconditions and postconditions on operations and methods, d) describe guards, e) specify constraints on operations, f) and serve as a navigation language [14].

Because the OCL is widely used for model validation and verification, most of the current UML tools support the OCL, but are typically limited to storing and presenting constraints. For example, IBM Rational Software Architect (RSA) is a powerful UML tool which integrates comprehensive modeling features with a standard Java/J2EE development IDE. However, for the OCL, RSA only provides syntax highlighting, content assist, and syntax parsing [12]. Therefore, other tools, which are more powerful for validating constraints, have been created, such as the UML-based Specification Environment (USE) [9]. Compared to UML modeling tools such as RSA, USE is more capable of validating a UML class model by evaluating its OCL constraints. However, a significant drawback of the USE tool is that every model that we want to validate must be manually translated into a textual specification written in the USE-specific language. Therefore, to take advantages of both tools, we must first use a UML tool such as IBM RSA to create the model with OCL constraints, and then perform a manual transformation between the specifications of the two different tools. As commonly observed, manual transformations are time-consuming and error-prone. To address these problems, we defined and implemented an automated transformation based on the Model Driven Architecture (MDA) framework [11] from RSA to USE.

The remainder of the paper is organized as follows. In Section 2, we summarize related work in the area of model transformation and modeling/analysis tool support. In Section 3, we describe a class model example with OCL constraints that can be used as an example source model in our transformation. In Section 4, we present our specification transformation approach. In Section 5, we discuss the mapping completeness between RSA and USE, and finally conclude in Section 6.

2 Related Work

XMI is an OMG standard for exchanging metadata information via the Extensible Markup Language (XML). It can be used for any model whose meta-model can be expressed using the Meta-Object Facility (MOF) [11]. The most common use of XMI is as an interchange format for UML models [17]. IBM RSA is the latest generation Rational modeling tool which provides the important features of the previous generation of Rational modeling tools, integrates comprehensive modeling features, and uses a standard Java/J2EE development IDE. RSA is based on the Eclipse Modeling Framework (EMF) technology. RSA diagrams can be used to edit and display models derived from any EMF-based meta-model. EMF provides a generic customizable XML or XMI resource implementation. The combination of RSA and EMF provides a powerful capability for integrating domain-specific languages (DSLs) with UML in a single tool set for design and development. RSA supports the XMI format (version 2.1) specification and allows a user to import and export a UML XMI model specification [12].

There are several OCL tools that support an XMI or XML specification. The Dresden OCL compiler [10] supports code generation by allowing the compilation of the OCL into Java code. For this tool, one can load the UML model from an XMI file (version 1.2) generated by the Argo/UML tool [3]. The ModelRun tool [4] allows interactive verification of OCL properties and can load the UML model from the files created by other tools such as Rose 2000. The OCLE [6] provides model validation against methodological, profile or target implementation language rules expressed in OCL. Also, the OCLE supports UML model exchange using XMI (version 1.0 or version 1.1). However, these above OCL tools can not validate whether the object model of the system conforms to the OCL constraints defined in the class model of the system.

USE [9] [7] is an OCL tool that has been used both in research and in industry for validating models with constraints thanks to its powerful snapshot generation feature. In USE, a snapshot shows a system state of the specified system at a particular point in time. As a system evolves, a sequence of system states is produced. For each snapshot, the OCL constraints are automatically checked and system state information is given as graphical views. The USE can also be employed to animate the model by creating the sequence diagram and thus validate it according to the system's requirement that are represented through OCL constraints (invariants and pre- and postconditions), which can be evaluated during the animation of the model. Additional OCL expressions can be entered and evaluated to query a system state and sequence diagram operations in a model can be visualized and evaluated as well. Karsten *et al.* [16], for example, have used USE to validate authorization constraints in UML models. However, USE uses its own textual specification as the only input and cannot import or export the XMI specification for sharing model information with OCL constraints with other powerful UML tools such as IBM RSA.

The model transformation framework in MDA defines a model transformation by mapping each metamodel element (i.e., each language construct) of the source language into a metamodel element of the target language. [11]. The source and target model can be written in the same language (e.g., for refactoring), but this framework can be applied to a transformation between two different languages as well [11]. For example, Denis *et al.*[8] use Alloy to validate the radiation therapy machine designed by UML with the manual translation from UML to Alloy. Kyriakos [2] and Bordbar [5] propose model-based techniques for the automated transformation of UML class diagrams with OCL constraints to Alloy code. Motivated by the work in [2] [5], we have used an MDA technique to implement a transformation between the XMI specification exported from RSA and USE specification.

3 A Class Model Example with OCL Constraints

The example in Fig. 1 (taken from [7]) represents a Company UML Model with OCL constraints. The model has three classes: Project, Employee and Department. OCL constraints are represented by the four invariants as shown in Fig. 1. *MoreEmployeesThanProjects* specifies that the number of employees working in a department must be greater or equal to the number of projects controlled by the department. *MoreProjectHigherSalary* requires that employees get a higher salary when they work on more projects. *BudgetWithinDepartmentBudget* indicates that the budget of a project must not exceed the budget of the controlling department. *EmployeesInControllingDepartment* specifies that the employees working on a project must also work in the controlling department.

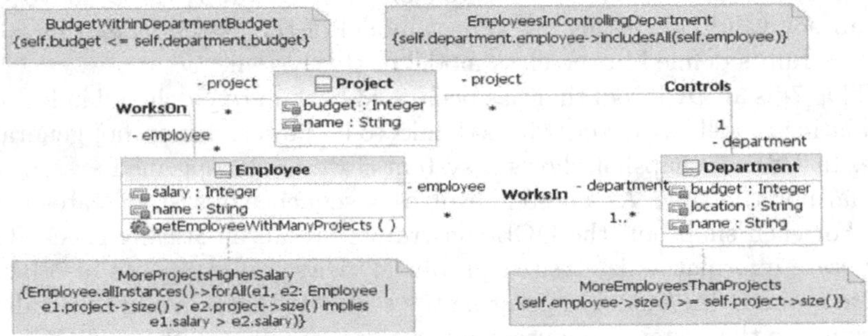

Fig. 1. Company Model Created by IBM RSA

```
<packageElement xmi:type="uml:Class"xmi:id="_cM-2XhbyEd2PjtdFhWRrAg"  name="Employee">
  <ownedRule xmi:type="uml:Constraint" xmi:id="_cM-2XxbyEd2PjtdFhWRrAg"  name="MoreProjectsHigherSalary"
                        constraintElement="_cM-2XhbyEd2PjtdFhWRrAg">
    <specification xmi:type="uml:OpaqueExpression" xmi:id="_cM-2YBbyEd2PjtdFhWRrAg">
      <language>OCL</language>
        <body>Employee.allInstances()->forAll(e1, e2:Employee|e1.project->size() > e2.project->size()
                        implies e1.salary > e2.salary)</body>
    </specification>
  </ownedRule>
  <ownedAttribute xmi:type="uml:Property" xmi:id="_cM-2YRbyEd2PjtdFhWRrAg"  name="project"
          visibility="private" type="_cM-2aRbyEd2PjtdFhWRrAg" association="_cM-2ghbyEd2PjtdFhWRrAg">
    <upperValue xmi:type="uml:LiteralUnlimitedNatural" xmi:id="_cM-2YhbyEd2PjtdFhWRrAg"  value="*"/>
    <lowerValue xmi:type="uml:LiteralInteger" xmi:id="_cM-2YxbyEd2PjtdFhWRrAg"/>
  </ownedAttribute>
  <ownedAttribute xmi:type="uml:Property" xmi:id="_cM-2ZBbyEd2PjtdFhWRrAg"  name="department"
          visibility="private" type="_cM-2dhbyEd2PjtdFhWRrAg" association="_cM-2gxbyEd2PjtdFhWRrAg">
    <upperValue xmi:type="uml:LiteralUnlimitedNatural" xmi:id="_cM-2ZRbyEd2PjtdFhWRrAg"  value="*"/>
    <lowerValue xmi:type="uml:LiteralInteger" xmi:id="_cM-2ZhbyEd2PjtdFhWRrAg"  value="1"/>
  </ownedAttribute>
  <ownedAttribute xmi:type="uml:Property" xmi:id="_cM-2ZxbyEd2PjtdFhWRrAg"  name="name" visibility="private">
    <type xmi:type="uml:PrimitiveType" href="http://schema.omg.org/spec/UML/2.1.1/uml.xmi#String"/>
  </ownedAttribute>
  <ownedAttribute xmi:type="uml:Property" xmi:id="_cM-2aBbyEd2PjtdFhWRrAg"  name="salary" visibility="private">
    <type xmi:type="uml:PrimitiveType" href="http://schema.omg.org/spec/UML/2.1.1/uml.xmi#Integer"/>
  </ownedAttribute>
</packagedElement>
```

Fig. 2. Partial Example of XMI Specification Exported from RSA

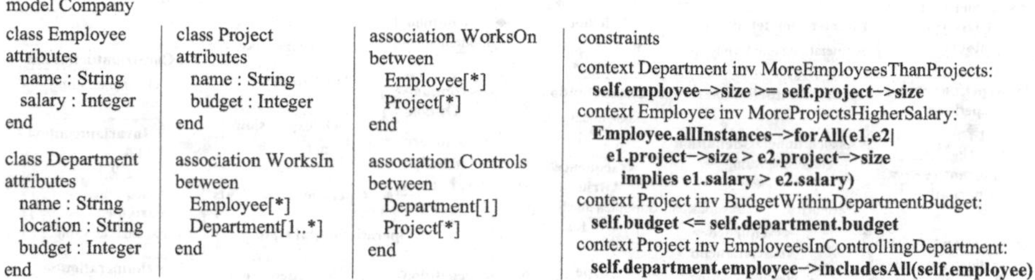

```
model Company

class Employee
attributes
  name : String
  salary : Integer
end

class Department
attributes
  name : String
  location : String
  budget : Integer
end

class Project
attributes
  name : String
  budget : Integer
end

association WorksIn
between
  Employee[*]
  Department[1..*]
end

association WorksOn
between
  Employee[*]
  Project[*]
end

association Controls
between
  Department[1]
  Project[*]
end

constraints

context Department inv MoreEmployeesThanProjects:
  self.employee->size >= self.project->size
context Employee inv MoreProjectsHigherSalary:
  Employee.allInstances->forAll(e1,e2|
    e1.project->size > e2.project->size
      implies e1.salary > e2.salary)
context Project inv BudgetWithinDepartmentBudget:
  self.budget <= self.department.budget
context Project inv EmployeesInControllingDepartment:
  self.department.employee->includesAll(self.employee)
```

Fig. 3. USE Specification of Company Model

Fig. 2 and Fig. 3 show how the Company model given earlier in Fig. 1 is represented as text either as an XMI specification or as a USE specification. While an XMI specification of the model can be exported from the RSA, as shown in Fig. 2, to use the USE for evaluating constraints, we have to rewrite a textual description of a UML model as shown in Fig. 3. This USE-specific textual description is readable only by the USE tool and does not conform to any other exchangeable specification standard [17].

4 XMI to USE Automated Transformation Overview

Fig. 4 gives an overview of our specification transformation approach. RSA can import or export an XMI specification of a UML model that conforms to the UML metamodel. In USE, however, a model must be first described as a textual specification whose syntax is defined in terms of EBNF [1] (refer to [7] for details).

Generating the USE Specification Metamodel: To use the MDA-based transformation framework for an automated tool-level transformation between

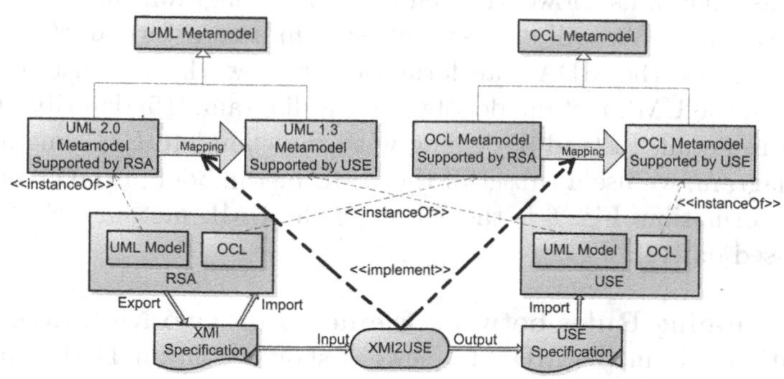

Fig. 4. XMI to USE Transformation Overview

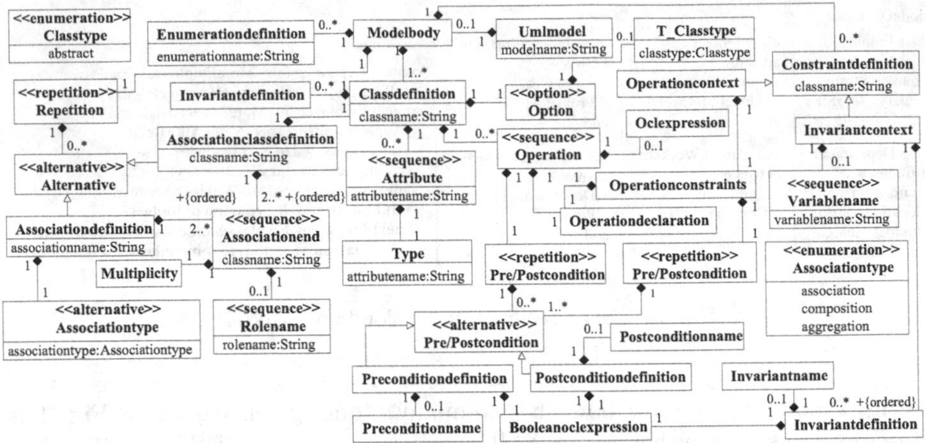

Fig. 5. Partial Metamodel of the USE Specification

RSA and USE, we first have to convert the concrete syntax of the USE specification given in EBNF (i.e., grammarware) to a MOF-compliant syntax representation (i.e., modelware or a metamodel of USE specification). We have used a two-step grammarware-to-modelware conversion approach proposed in [19] to generate a simplified and tailored metamodel of the USE specification language from its EBNF representation. Fig. 5 shows the generated (partial) USE specification metamodel that includes selected major elements such as model definition, association definition, class definition, constraint definition, enumeration definition and association class definition.

Identifying the Scope of Available Mappings: RSA supports the modeling of use case diagrams, class diagrams, sequence diagrams, communication diagrams, activity diagrams, state machine diagrams and so on. USE only supports the modeling of class diagrams. Therefore, the applicable mapping only exists for class diagrams. However, even for a class diagram, the mapping is not completely seamless. We will discuss the issues in a transformation mapping in the Sect. 5. To use the MDA transformation framework, we must refer to the elements from the UML metamodel of the class diagram. [15] describes the UML metamodel in detail. Instead of dealing with the complete UML metamodel of the class diagram, we use a subset of the UML metamodel of the class diagram in the transformation. Fig. 6 is the subset of the UML metamodel of the class diagram based on [15].

Defining Mapping Rules between Elements in Two Metamodels: Each transformation rule in the MDA framework should contain 1) the source language reference, b) the target language reference, c) optional transformation parameters, d) a bidirectional indicator, e) the source language condition, f) the target language condition and g) a set of mapping rules. Every transformation

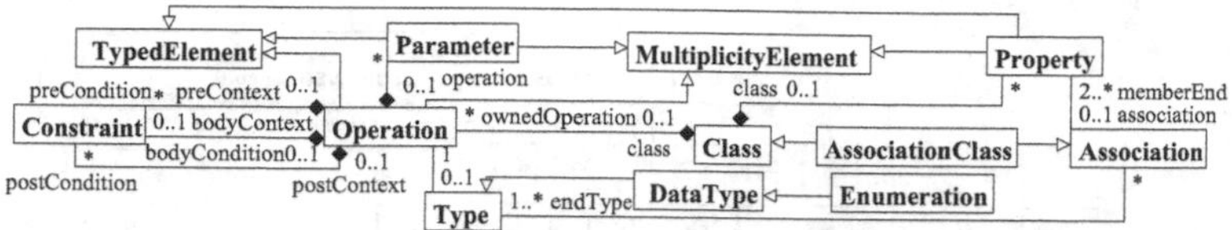

Fig. 6. Subset of the UML Metamodel of the Class Diagram

rule starts with the keyword *Transformation* followed by a name. The source and target languages are identified by listing both language names between parentheses following the transformation name. The parameters to each transformation rule are written as a list of variable declarations following the keyword *params*. The source and target language model elements are written as variable declarations following the keywords *source* and *target*. The directional indicator is given by the keyword *bidirectional* or *unidirectional*. The source and target language conditions are written as OCL boolean expressions after the keywords *source condition* and *target condition*. All mapping rules come after the keyword *mapping*. In the notation for mapping rules, the symbol <~> is used. (refer to [11] for details).

Our transformation engine implements the mapping between the UML and OCL metamodel supported by RSA(i.e., represented by the XMI specifications exported by RSA) and the UML and OCL metamodel supported by USE. Therefore any instance of the UML and OCL metamodel described by the XMI specification, can be a source to our engine and the corresponding USE specification is generated as a target specification of the transformation.

5 Issues in a Transformation Mapping

Here we consider a transformation mapping to be sufficiently complete if all elements in the set of original language constructs are mapped onto language constructs in the set of target language constructs. For our current work, the mapping is only in one direction (i.e., from RSA to USE) and can be either for the UML metamodels or for the OCL metamodels. For the UML metamodel, RSA supports the elements related to the class diagram, sequence diagram, communication diagram, activity diagram, and, state machine diagram, while USE supports the elements related to the class diagram only. Consequently, our mapping is only related to class diagrams. However, there are two complications:

- RSA supports a subset of UML 2.0 [12], while USE supports a subset of UML 1.3 [7].
- USE only supports the basic metamodel elements (e.g., class, association, and operation) of the UML class diagram, while RSA supports additional metamodel elements (e.g., package) of the UML class diagram as well.

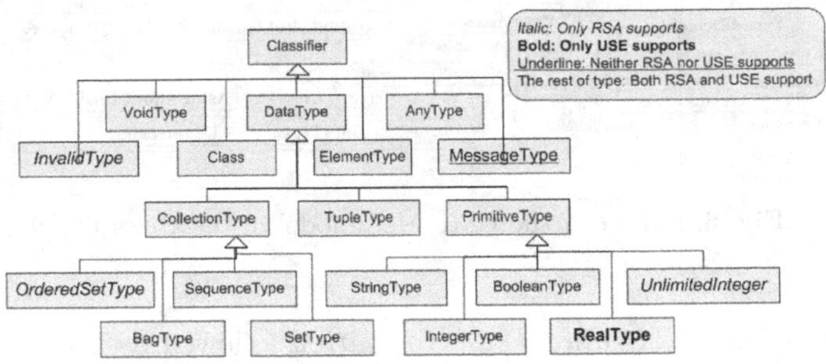

Fig. 7. Metamodel for OCL Types (Simplified)

Therefore, our mapping supports a least common denominator between UML 2.0 and UML 1.3 class constructs. Fig. 7 presents a simplified metamodel for the *Type* element that is common to both UML and OCL metamodels [14]. We use different fonts to distinguish constructs support by RSA or USE, or both. *InvalidType*, *OrderedSetType* and *UnlimitedInteger* (in italics) are supported only by RSA. **RealType** (in bold) is supported only by USE. Neither RSA nor USE supports "MessageType" (underlined). Other types in normal font are supported by both tools. Some types, such as "CollectionType", are supported by both tools, but representing an empty collection is different in each tool. The following summarizes more mapping problems we have discovered during the transformation between XMI and USE:

- Incompatible primitive types: There exists primitive type compatible problem between RSA and USE. The former can identify Integer, Boolean, String and UnlimitedNatural, while the latter can identify Integer, Real, Boolean and String.
- Unsupported collection type in USE, OrderedSet: There is a collection type compatible problem RSA and USE. Standard OCL supports four kinds of collections types: Set, OrderedSet, Bag, and Sequence. The OCL in RSA includes all of them, while the OCL in USE only includes Set, Bag and Sequence.
- Empty collection representation: Standard OCL supports empty collections. To represent the same empty set (Set{} in standard OCL) for the class Employee in Fig. 1, Set{null} is used in RSA, while oclEmpty(Set(Employee)) is used in USE.

For the OCL metamodel, each element supported by USE always has a corresponding element supported by RSA and vice versa when we exclude *Type* element, explained earlier. However, there are still minor differences between RSA and USE when expressing the OCL standard operations although both RSA and USE claim that they support OCL 2.0 [14]. For example, RSA supports several operations that USE does not, such as *oclIsInvalid()*, *product()*. Also, for an OCL standard operation, *isUndefined()* supported by USE, but it has a different name, *oclIsUndefined()*, in RSA. However, due to our page limitation, we do not present solutions to these issues in this paper.

6 Conclusion and Future Work

The USE tool is one of few OCL tools allowing interactive monitoring of OCL invariants and pre- and postconditions and the automatic generation of non-trivial system states. However, USE expects a textual description of a model and its OCL constraints that are not compatible with other UML modeling/analysis tools. In this paper, we have described an MDA-based transformation approach with a transformation tool, called XMI2USE, that currently provides an automatic specification transformation from RSA to USE. We have presented the metamodel of the USE specification, identified the scope of the applicable mappings between RSA and USE, introduced the mapping rules based on the MDA framework and discussed the mapping completeness between the UML and OCL metamodels.

To validate our approach, we have used the XMI2USE in our graduate-level software engineering class that uses advanced OCLs for secure models. Students used the tool for modeling and analyzing access control policies and we have found that the automated transformation increased the productivity and quality of the project. We are currently working on extending our tool to support the transformation from USE to RSA. Our future work includes the elaboration of our transformation engine so that it can support UML modeling tools other than RSA.

Acknowledgments. This work was supported in part by funds from the Vice Provost for Research at Baylor University. We would like to thank all anonymous reviewers for their constructive and valuable feedback.

References

1. Alanen, M., Porres, I.: A Relation Between Context-Free Grammars and Meta Object Facility Metamodels. Technical report, Turku Centre for Computer Science (2003)
2. Anastasakis, K., Bordbar, B., Georg, G., Ray, I.: UML2Alloy: A Challenging Model Transformation. In: ACM/IEEE 10th International Conference on Model Driven Engineering Languages and Systems (2007)
3. ArgoUML, an open source UML modeling tool by Tigris.org, `http://argouml.tigris.org/`
4. Boldsoft, Boldsoft OCL Tool Model Run, Boldsoft, Stockholm (2002)
5. Bordbar, B., Anastasakis, K.: UML2Alloy: A tool for lightweight modelling of Discrete Event Systems. In: Guimaraes, N., Isaias, P. (eds.) IADIS International Conference in Applied Computing (2005)
6. Chiorean, D.: Using OCL Beyond Specifications. In: Workshop of the pUML-Group held together with the UML 2001 on Practical UML-Based Rigorous Development Methods (2001)

7. Database Systems Group, Bremen University, USE: A UML based Specification Environment (Preliminary Version 0.1) (2007), http://www.db.informatik.uni-bremen.de/projects/USE/use-documentation.pdf
8. Dennis, G., Seater, R., Rayside, D., Jackson, D.: Automating commutativity analysis at the design level. In: ISSTA 2004: ACM SIGSOFT international symposium on Software testing and analysis, pp. 165–174. ACM Press, New York (2004)
9. Gogolla, M., Buttner, F., Richters, M.: USE: A UML-Based Specification Environment for Validating UML and OCL. Science of Computer Programming 69, 27–34 (2007)
10. Hussmann, H., Demuth, B., Finger, F.: Modular architecture for a toolset supporting OCL. In: Evans, A., Kent, S., Selic, B. (eds.) UML 2000. LNCS, vol. 1939, pp. 278–293. Springer, Heidelberg (2002)
11. Kleppe, A., Warmer, J., Bast, W.: MDA Explained — The Model Driven Architecture: Practical and Promise. Addison-Wesley Longman Publishing Co., Inc., Boston (2005)
12. Leroux, D., Nally, M., Hussey, K.: Rational Software Architect: A tool for domain-specific modeling. IBM Systems Journal 45(3) (2006)
13. Object Management Group (OMG), MOF 2.0/XMI Mapping Specification, v2.1 (2005)
14. Object Management Group (OMG), Object Constraint Language (OCL) Version 2.0. OMG Document ptc/06-05-01 (2006)
15. Object Management Group (OMG), Unified Modeling Language (UML), Infra-and Superstructure, V2.1.2 (2007)
16. Sohr, K., Ahn, G., Gogolla, M., Migge, L.: Specification and Validation of Authorisation Constraints Using UML and OCL. In: de di Vimercati, S.C., Syverson, P.F., Gollmann, D. (eds.) ESORICS 2005. LNCS, vol. 3679, pp. 64–79. Springer, Heidelberg (2005)
17. Toval, A., Requena, V., Fernandez, J.: Emerging OCL tools. Software and Systems Modeling 2(4), 248–261 (2003), http://www.um.es/giisw/ocltools/ (last updated on, 12/20/2006)
18. Warmer, J., Kleppe, A.: The object constraint language: Getting Your Models Ready for MDA. Addison-Wesley Longman Publishing Co., Inc., Boston (2003)
19. Wimmer, M., Kramler, G.: Bridging grammarware and modelware. In: Bruel, J.-M. (ed.) MoDELS 2005. LNCS, vol. 3844, pp. 159–168. Springer, Heidelberg (2006)

Preface to MOST-ONISW 2009

Martin Doerr[1], Fred Freitas[2], Giancarlo Guizzardi[3], and Hyoil Han[4]

[1] Foundation for Research and Technology – Hellas (FORTH)
[2] Federal University of Pernambuco, Recife, Brazil
[3] Federal University of Espírito Santo, Vitória, Brazil
[4] LeMoyne College, USA

Ontology is a cross-disciplinary field concerned with the study of concepts and theories that can be used for representing shared conceptualizations of specific domains. Ontological Engineering is a discipline in computer and information science concerned with the development of techniques, methods, languages and tools for the systematic construction of concrete artifacts capturing these representations, i.e., models (e.g., domain ontologies) and metamodels (e.g., upper-level ontologies). In recent years, there has been a growing interest in the application of formal ontology and ontological engineering to solve modeling problems in diverse areas in computer science such as software and data engineering, knowledge representation, natural language processing, information science, among many others.

A crucial question is whether ontologies can replace information models. But whereas ontologies work quite well as virtual schemata in mediation systems, they may perform poorly as information models and on the user interface level. On the theoretical side, there is a lack of understanding of the effective relation and interplay of ontological and epistemological features in information models and systems. Furthermore there are still open questions concerning good scientific practice in developing ontologies. On the practical side, there is still a lack of good practice of how to integrate existing information systems into ontology driven applications and few experiences at all with creating good new data structures from ontologies directly for interoperation in complex and diverse application environments.

The objective of MOST-ONISW 2009 is to bring together researchers and practitioners in Information Management interested in the relation between ontology and information models, and theoretical topics such as formal ontology, formal logics, conceptual modelling, computational linguistics, cognitive science, knowledge representation, the Semantic Web, and MDE (Model-Driven Engineering), as well as more practical topics as a result of applications of ontologies in diverse fields, such as knowledge management, informatics for education, ontology-based information and database integration, e-commerce, information processing (retrieval, classification and extraction), to mention just a few.Among the issues are:

What is the *difference and relation* between information models and ontologies? Which criteria must ontologies match in order to provide a sound basis for an information system? How to interact and relate the ways of knowing and what can be known with the form of knowledge in information systems? Are there systematic kinds of information elements associated with information management processes that are not of ontological nature? What is the *epistemological impact* on ontologies?

C.A. Heuser and G. Pernul (Eds.): ER 2009 Workshops, LNCS 5833, pp. 157–158, 2009.

How should we construct *ontologies from information* models for semantic interoperability, and create and manage mapping specifications for mediators, data transformation systems, Web service wrappers via ontologies. What are the characteristic *cases of heterogeneity* and how can they be managed generically. What are the *languages* and *tools* for mapping and transformation algorithm *generators*?

How can we effectively *enable domain experts* to specify the semantics of their information systems in order to exploit Semantic Web technology? How can we visualize the ontology and mapping information in a user-friendly way?

How can we make effective *information models*, i.e. database schemata, data entry forms, Web service interfaces, and simplified query interfaces *from ontologies*? Ontologies can help to objectively describe the loss of information and reasoning capabilities due to necessary simplifications in information structures. What are the problems, mechanisms, and rules in order to preserve semantic interoperability?

How does *argumentation and information system* content relate? Current argumentation models, systems for collaborative work model and Web2.0 applications visualize the flow of arguments or register resulting propositions, but do not model how argumentation operates on information system contents expressed in terms of ontologies, so that a full externalization of multiple arguments and understanding of their integrated effect on information system contents can be achieved.

What is the relation between formal ontologies and natural languages? How can we link knowledge represented in an ontological way to every day language? Can we map layperson communication to domain expert-governed ontologies?

How should we utilize ontologies and conceptual modelling for data management, integration and interoperability in Semantic web applications, particularly in e-science, life sciences, e-business and cultural applications? What are *architectures and models of good practice*? Are there domain-overarching global core ontologies? What are their characteristics?

What is semantics? Are semantics logical formulae? Is ontological commitment a set of formulae or an interpretation function to real world things and phenomena in the user's mind? What role does ontological commitment play in conceptual modelling and database integration?

Researchers and practitioners are invited to submit theoretical, technical and practical research contributions that directly or indirectly address the issues above. Particularly welcome are e-science, life-sciences, e-business and cultural applications. The workshop foresees a technical discussion on the relation of ontologies and conceptual modelling.

Analysis Procedure for Validation of Domain Class Diagrams Based on Ontological Analysis

Deisymar Botega Tavares, Alcione de Paiva Oliveira, José Luís Braga, and Jugurta Lisboa Filho

Departamento de Informática - UFV -Av. P. H. Rolfs s/n - 36.570-000, Viçosa - MG - Brazil
{dbotega,alcione,zeluis,jugurta}@dpi.ufv.com
http://www.dpi.ufv.br

Abstract. A well-conceived conceptual model is essential to obtain systems that are easier to maintain. The UML class diagram is a powerful tool that can be applied at this step, but the developer has to have a clear understanding of the domain concepts in order to yield a diagram that captures the concepts and the relations of the domain. In order to verify the adequacy of the class diagram, an analysis of the object's essence and its permitted relations can be accomplished. This analysis is called ontological analysis, but its execution can be quite difficult because it is necessary to master sophisticated philosophical concepts like identity and rigidity. This article presents a procedure that aims to accomplish the ontological analysis of the UML class diagram without exposing the complexity of the concepts that underlies the procedure.

Keywords: software procedure, Conceptual Model, Class Diagram, Ontological Analysis.

1 Introduction

One of the software development process challenges is to generate reliable domain conceptual models. Ontological analysis techniques have been created with the purpose of validating conceptual schemes expressed by UML class diagrams. Among the techniques we can mention the VERONTO Technique [6] and the OntoUML Profile [2]. Combining features of VERONTO and the OntoUML Profile, there is the OntoCon Technique [5]. This latter technique was created aiming to help improve the class diagrams, focusing mainly on providing the rules and restrictions to validate generalization/specialization relationships. However ontological analyses are difficult to apply because they make use of philosophical concepts, rarely mastered by software modelers. To cope with the complexity of the application of the OntoCon technique, a procedure, entitled PrOntoCon, was developed. The procedure guides the modeler through a sequence of steps with the aim to accomplish the validation of existing UML class diagrams, by checking its hierarchies or detecting the absence of them. The procedure also aims to be a better approach to role modeling through the use of analysis patterns adopted

C.A. Heuser and G. Pernul (Eds.): ER 2009 Workshops, LNCS 5833, pp. 159–168, 2009.

from the OntoUML profile. The procedure was applied to several UML class diagrams and the diagrams that it yielded were analyzed by skilled modelers, who judged they were more scalable, with less redundancy and easier to integrate than the original ones. Besides all these advantages, the procedure PrOntoCon facilitated the application of philosophical concepts that underlies the OntoCon technique.

The paper is organized as follows. Section 2 introduces the OntoCon technique. Section 3 describes the PrOntoCon procedure. Section 4 presents a study case applying the procedure. Section 5 summarizes the contributions.

2 The OntoCon Technique

The OntoCon technique combines features of the VERONTO technique [6] and the OntoUML profile [2]. The VERONTO and the OntoUML profile are based on the meta-properties rigidity, identity, unity and external dependence. Succinctly, as stated by [1], a rigid property (+R) is a property that is essential to all its instances. Therefore, if an element is an instance of a property, it will remain an instance throughout its existence. For example, the class Person is said to be rigid because one instance of that class will always be a person. Conversely, a non-rigid (∼R) property is a property that is not essential to some of its instances. For example, the class Student is said to be non-rigid, because it always possible for one instance to leave this condition. For further details about the meta-properties the reader should refer to [1].

Based on a combination of these meta-properties the authors of [1] proposed a formal ontological property classification. Table 1 shows part of the OntoCon stereotypes associated with the combination of meta-properties. The names of the stereotypes were adopted from the formal ontological property classification. It was established, for every stereotype, the stereotypes that may occur as supertypes. Besides the already mentioned +R (∼R) notation, there is the following notation: +D (-D) is used to indicate that the element in question has (has not) the external dependence meta-property; the notation +I (-I) indicates that the

Table 1. OntoCon technique stereotypes and hierarchical restrictions. Adapted from [5].

Stereotypes	Meta-properties	Stereotypes allowed as Supertypes
`<<type>>`	+O +I +R -D	`<<category>> <<type>> <<quasi-type>>`
`<<quasi-type>>`	-O +I +R -D	`<<type>> <<quasi-type>> <<category>>`
`<<material role>>`	-O +I ∼R +D	`<<type>> <<quasi-type>> <<phased sortal>> <<material role>> <<formal role>> <<category>>`
`<<phased sortal>>`	-O +I ∼R -D	`<<type>> <<quasi-type>> << phased sortal>> << category>>`
`<<formal role>>`	-O -I ∼R +D	`<<category>> <<formal role>>`
`<<category>>`	-O -I +R -D	`<<category>>`

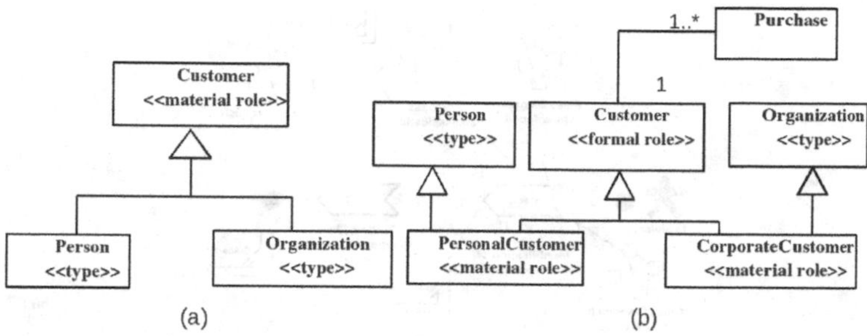

Fig. 1. (a) Role modeling misconception. (b) Solution through the application of an analysis pattern. Source: adapted from [5].

element has (has not) a identity condition; and the notation +O (-O) means that the element provides (does not provide) a condition of identity.

The analysis pattern inherited from OntoUML [2] profile enables the OntoCon technique to correct role modeling misconceptions, as the one shown in Fig. 1a. The Customer class is a anti-rigid class (~R), i.e. its instances will not necessarily be a customer throughout its existence. On the other hand the classes Person and Organization are rigid (+R). Thus, the class Person and Organization can not be subclass of the Customers class.

The application of the analysis pattern solves the problem and imposes the creation of two additional classes (PrivateCustomer and CorporateCustomer), as shown in Fig. 1b. Due to lack of space, other restrictions that are part of the OntoCon are not described here. A more complete description of the technique can be found in [5].

3 The PrOntoCon Procedure

The PrOntoCon main goal is to conduct the modeler through a sequence of steps aiming at the validation of UML Class Diagrams without exposing the modeler to the difficulties inherent to the philosophical concepts underlining the OntoCon technique. The PrOntoCon has four phases: (i) stereotype identification; (ii) hierarchy checking; (iii) application of the analysis pattern to adjust the modeled roles, (iv) UML constructors checking. For a better visualization of these phases, each one of them is represented by an activity diagram modeled using the SPEM (Software Process Engineering Metamodel) UML profile [4], that exhibits the steps the modeler should follow in order to validate the class diagram. Fig. 2 shows a SPEM diagram for the first phase of the procedure. First of all it is necessary to identify the proper stereotype of each class of the diagram. To accomplish this it is necessary to apply a decision tree analysis procedure as a first activity of phase 1. The goal of this phase is to guide the modeler through the decisions necessary to map each class to an OntoCon stereotype. The decision tree starts with a question over the identity meta-property. Depending on

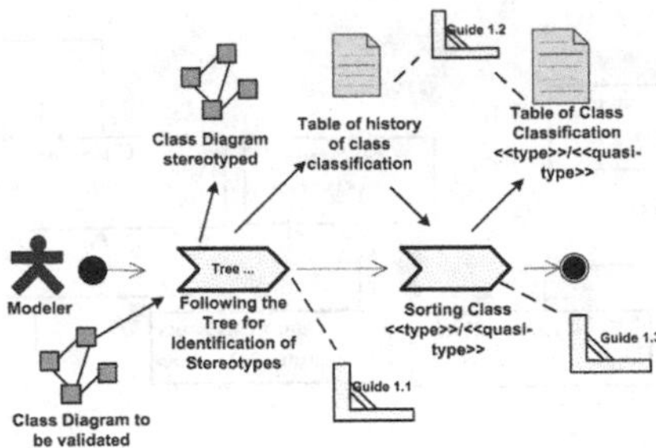

Fig. 2. SPEM Diagram for Phase 1: Stereotype Identification

the answer (yes or no) the user is led to another question until it comes to a suggested stereotype for the target class. For the sake of simplicity, the questions were elaborated without the terms concerning the ontological analysis and related philosophical concepts. For example, the first question is: "*Do all the class instances own a common feature with a unique value for each individual? For instance: for the members of the class PERSON there is, among other features, the finger prints. Thus one can say that the class PERSON has an identity.*[1]" To help even further the modeler, the process also contains a set of examples and counterexamples.

The decision tree does not help to distinguish between the stereotypes `<<type>>` and `<<quasi-type>>`. For this reason some class will be tagged as `<<type>>/ <<quasi-type>>`. It was not possible to create a simple question that could lead the modeler to identify if a class is an identity supplier (+O). This decision was postponed to the second phase of the procedure.

The second activity of the PrOntoCon first phase require the modeler to group the stereotyped classes as `<<type>>/ <<quasi-type>>` in one of the categories *agent*, *object*, *event* or *moment*, all based on the Unified Foundational Ontology [3]. The *agent* category comprises entities that can be considered as physical (e.g. person) or social agents (e.g. organization) capable of actions. The *object* category comprises entities that can be physical (home) or social objects (currency). *Event* is something that causes a transformation (e.g., sale) and finally the *moment* category comprises entities that can only exist in other entities such as: color and symptom.

The artifacts produced by the first phase are the partly stereotyped class diagram and one table containing the grouping of the class into the categories previously described. This artifact will be the input data for the PrOntoCon second phase, entitled *Hierarchy Checking*.

[1] For the sake of correctness it is necessary to state that the fingerprint, iris, DNA and other unique characteristics of a person are all an identity condition and must not be confused with the concept of identity discussed by philosophy. The identity condition stems from the existence of philosophical identity but it is not the same thing. It is a manifestation of the existence of an identity.

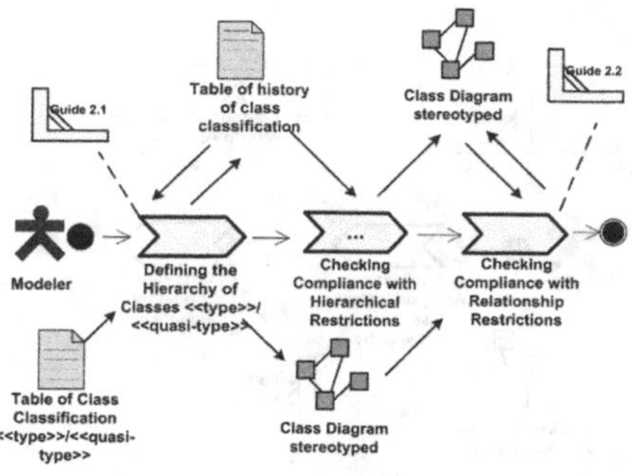

Fig. 3. SPEM Diagram for Phase 2: Hierarchy Checking

Fig. 3 shows the SPEM activity detail diagram for the second phase of the procedure. The first activity of this phase aims to distinguish between `<<type>>/` `<<quasi-type>>` classes and to establish a hierarchy among them. In order to do that the modeler will have to detect if among the classes that belongs to this category, there is or there should be a generalization/specialization relationship. Those classes that standalone and do not participate in a hierarchical relationship will be classified as `<<type>>`. Among the classes that participate in a hierarchical relationship, the ones that belong to the highest level of the hierarchy will be stereotyped as `<<type>>` and the lower level ones, if they inherit the identity of the highest-level class, will be stereotyped as `<<quasi-type>>`.

The next activity of phase 2, entitled *Checking Compliance with Hierarchical Restrictions*, is responsible for checking compliance of the hierarchical relationship with the restrictions imposed by the OntoCon technique. This activity is further divided into two sub activities. First of all, one must check whether the existing generalization/specialization relationships in the diagram are correct. This is done by inspecting each line of inheritance to check whether the existing subclass and superclass are allowed. Whenever an error in the hierarchy is found, like an inversion of `<<type>>/<<material role>>` hierarchy, the procedure leads to an application of the OntoUML analysis pattern. The OntoUML analysis pattern application belongs to the PrOntoCon third phase and will be explained further ahead. If other types of errors are found, the modeler will be conducted to redo the stereotype classification, seeking to identify a possible error.

The second sub activity deals with the absence of hierarchies that could exist. Every class stereotyped as `<<phased sortal>>`, `<<quasi-type>>` or `<<material role>>` must, necessarily, be a subclass of a `<<type>>` class. Thus the sub activity called *Mandatory Hierarchies Checking*, induces the modeler to check those classes. If the absence of a superclass is detected then the modeler should try to identify on the diagram a possible candidate to be the missing superclass. In case that a coherent candidate is not found in the diagram then the modeler must create the required superclass.

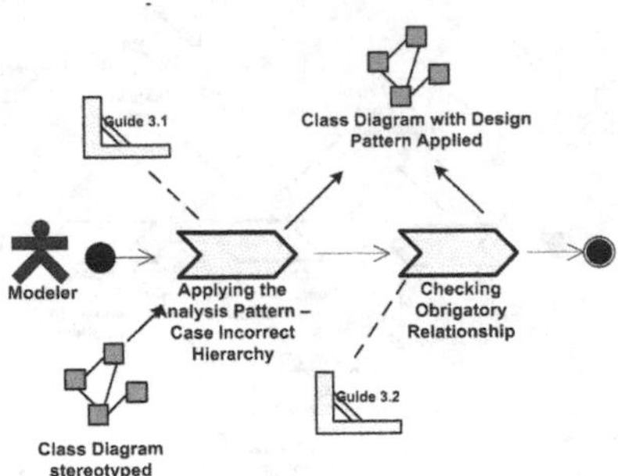

Fig. 4. SPEM Diagram for Phase 3: Analysis Pattern Application

The existence of some mandatory relationships between classes is checked also, constituting the third activity of the second phase of PrOntoCon. Classes stereotyped as `<<material role>>` or `<<formal role>>` must have a relationship with another class, and this relationship should have a cardinality with minimum value of 1, because such classes are dependent of other classes. These restrictions are checked by the activity *Mandatory Relationships Checking*.

The PrOntoCon third phase deals with role modeling misconceptions. It deals with two cases. The first one handles the problem of `<<type>>` class being subclass of `<<material role>>` class. As mentioned before, this fact is detected in the second phase of the procedure. Once the problem is detected the modeler is conducted by the activity diagram of Fig. 4.

Fig. 4 diagram shows two activities: applying the analysis pattern and obligatory relationship checking. The execution of the first activity will lead to the creation of as many `<<material role>>` classes as there are `<<type>>` classes involved in the misconceived hierarchy. For instance, as Fig. 1b shows, the `<<material role>>` class becomes a `<<formal role>>` and the superclass of the created `<<material role>>` classes. The latter ones becomes also subclass of different `<<type>>` classes from which they inherit their identity.

Having completed this activity, it is necessary to check whether there exists the proper dependency relationship between the class `<<formal role>>` and some other class, with cardinality of at least 1 (see example in Fig. 1b). Such relationship is required since `<<formal role>>` classes have a dependence metaproperty (+D), as can be seen in table 1. Therefore, any instance of a `<<formal role>>` class must be related with at least one instance of the class dependent.

The second problem of role modeling misconception handled by the application of the analysis pattern is the one that takes place when one `<<material role>>` class is subclass of more than one `<<type>>` class. For instance, one can imagine a situation where the `<<material role>>` class Client is subclass of the Organization and Person classes. This is an improper case of *Multiple inheritance and the application of the analysis pattern corrects it in a similar fashion to the former case.*

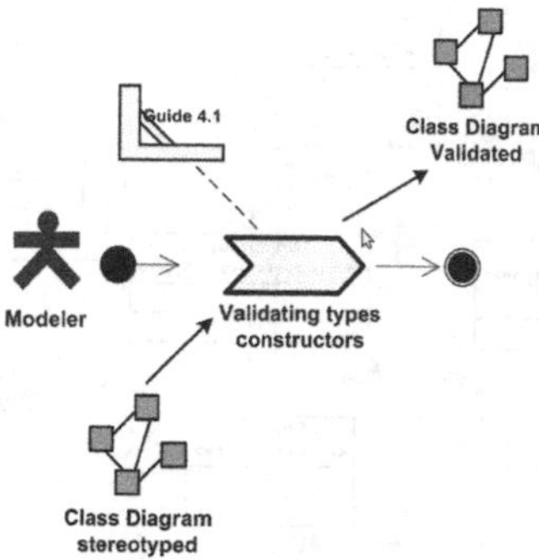

Fig. 5. SPEM Diagram for Phase 4: Constructors Checking

The fourth and last phase of the procedure is responsible for validating the UML constructors of each class with respect to their stereotype. The activity diagram of this phase, shown in Fig. 5, takes the modeler to check the UML constructors according to the following restrictions: classes stereotyped as `<<type>>`, `<<quasi-type>>`, `<<material role>>` or `<<phased sortal>>` must be concrete or abstract classes; and classes stereotyped as `<<category>>` or `<<formal role>>` must not be concrete classes.

4 Case Study

This section presents an example of the use of the PrOntoCon procedure. Fig. 6 shows the domain class diagram of a financial contract management system for postgraduate courses of a specific academic institution. The classes in the diagram of Fig. 6 already have the stereotypes detected in the first activity of the first phase. To become a graduate student in that institution it is necessary to set a contract between the institution and a sponsor (a person or organization). Note that the sponsor does not have to be the student himself. The sponsor becomes a client of the institution. Companies can make a financial contract for a group of students. Each financial contract is linked to a type of payment plan.

As shown in Section 3, the purpose of the first activity of the first phase is to identify the class stereotypes using a decision tree. So for the class Person, for example, one has to answer the first question of the tree. As every person has a unique characteristic fingerprint and it is unique for each person, then the answer to the first question is “yes”. Afterwards the procedure takes the modeler to answer the second question: *Will every object that stands as a class instance be an instance of this class throughout their existence in all possible domains?* The answer to that question is yes, because every instance of a Person will remain as a person throughout its existence. With these answers, the class Person gets the stereotype `<<type>>/ <<quasi-type>>`. After answering all the questions of the tree to all of the classes in the diagram, the modeler gets the stereotyped diagram shown in Fig. 6.

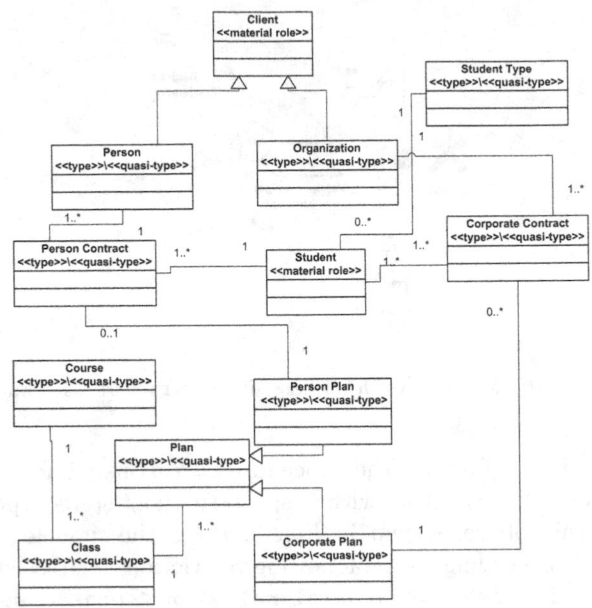

Fig. 6. UML class diagram partially stereotyped by the PrOntoCon procedure

After performing the second activity of the first phase the classes classified as <<type>>/ <<quasi-type>> were grouped as follows: classes Plan, Personal Plan, Corporate Plan, Course and Student Type as *Object*; classes Organization and Person as *Agent*; and classes Personal Contract, Corporate Contract and (Attendant) Class as *Event*. The procedure advances to the second phase. After the first phase of this activity, it was obtained the final stereotypes for the classes classified as <<type>>/ <<quasi-type>>. It should be noted that the class Contract was created as a generalization of the classes Personal Contract and Corporate Contract. Moving on to the phase second activity, one should consider whether the existing hierarchical relationships are correct. The analysis reveals that the relationships between the Client class and the Person and Organization classes are incorrect. As the error involves the modeling of roles, the procedure now leads the modeler to apply the analysis patterns. Applying the analysis patterns renders a new class diagram but due to lack of space, only the final diagram is shown (Fig. 7). It should be Noted that two new classes <<material role>> were created and the Client class turned into a <<formal role>> having a mandatory relationship with the Contract class.

Continuing with the PrOntoCon, the next step is the mandatory hierarchies checking. It is observed that the class Student with the stereotype <<material role>> must necessarily have a superclass with the stereotype <<type>>. The natural candidate is the Person class. After the UML constructor validation in done in the fourth and final stage of the procedure, it is realized that the Client class should be an abstract class.

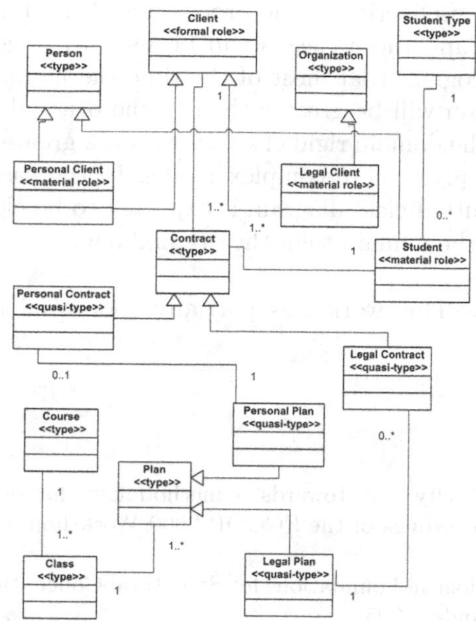

Fig. 7. UML Class Diagram after the application of the procedure

Observing the final class diagram, shown in Fig. 7, it is possible to see some improvements. First there is the data redundancy reduction with the creation of the Contract class and with the detection of the relationship between the Person class and the Student class. Data that in the original diagram, could be duplicated, such as the student name and its SSN, won't be due to the final diagram. Hence the consistency of the data will be preserved. Another improvement was the role modeling rectification, generating a more accurate diagram of the domain. It can also be perceived an increase in the system scalability, since new types of customers that may arise will be easily added to the model without causing major changes. The creation of views is improved as well, since there is a well structured separation between the client roles and their respective rigid classes.

5 Conclusions

The objective of establishing an ontology based procedure to check class diagrams where the modeler does not need to be acquainted with the philosophical concepts that underlines it was reached. With the PrOntoCon, the modeler is equipped with an instrument to guide him through the early stages of development, which is still poorly explored by software engineering techniques. The following benefits can be reached through use of the procedure: (i) increase the systems scalability, i.e., the systems become more prepared for the increasing demands, (ii) reduction of the data redundancy (iii) easier integration of the modules because the application of the procedures elicits rigid classes common across the diagrams, and those classes can be used to integrate the modules. It is important to recognize that most of the time the number of classes in the validated class diagram will be greater than in the original one, because it always makes explicit the domain rigid class. However, a greater number of classes do not necessarily imply a more complex model. In the case of the procedure PrOntoCon the outputted class diagram is expected to be clearer, more general and more faithful to the domain than the original one.

Acknowledgments. This work was partially sponsored by FAPEMIG and CNPq.

References

1. Guarino, N., And Welty, C.: Towards a methodology for ontology based model engineering. In: Proceedings of the ECOOP 2000 Workshop on Model Engineering (2000)
2. Guizzardi, G.: Ontological Foundations for Structural Conceptual Models. Universal Press, The Netherlands (2005)
3. Guizzardi, G., Falbo, R., And Guizzardi, R.: A importância de ontologias de fundamentação para a engenharia de ontologias de domínio: o caso do domínio de processos de software. IEEE Latino-americano (2008)
4. OMG. Software and systems process engineering meta-model specification. Technical report. Object Management Group - OMG (2008)
5. Tavares, D.B., Oliveira, A.P., Braga, J.L., Lisboa Filho, J.: Validação de Diagrama de Classes por meio da Técnica OntoCon. In: CLEI 2008 / XXXIV Conferencia Latinoamericana de Informática, Santa Fe, pp. 330–339 (2008)
6. Villela, M.L.B.: Validação de diagramas de classe por meio de propriedades ontológicas. Master's thesis, Universidade Federal de Minas Gerais, UFMG - Belo Horizonte - MG - Brasil. Dissertação de Mestrado (2004)

Ontology for Imagistic Domains: Combining Textual and Pictorial Primitives

Alexandre Lorenzatti, Mara Abel, Bruno Romeu Nunes, and Claiton M.S. Scherer

Institute of Informatics of Federal University of Rio Grande do Sul
Caixa Postal 15.064 – CEP 91.501-970 – Porto Alegre – RS – Brazil
Institute of Geosciences of Federal University of Rio Grande do Sul
Caixa Postal 15.001 – CEP 91.509-900 – Porto Alegre – RS – Brazil
{alorenzatti,marabel,brnunes}@inf.ufrgs.br,
claiton.scherer@ufrgs.br

Abstract. This paper proposes a knowledge model for representing concepts that requires pictorial as well as conceptual representation to fully capture the ontological meaning. The model was built from the proposition of pictorial primitives to be associated to the original conceptual primitives. The formalized pictorial content is then used to provide an organization to the domain, based on the visual characteristics of the objects as humans are used to do. The combination of both primitives allows the definition of domain ontologies to support visual interpretation activities. The approach was applied to build the Stratigraphy ontology for the definition of Sedimentary Facies and Structures.

Keywords: Visual knowledge, Ontology, Conceptual modeling, Stratigraphy.

1 Introduction

Visual knowledge modeling is an intense area of research. There are many approaches searching for a formal way of representing the visual content of the concepts. Human-interpretation activities are usually strongly based on visual information and images. An efficient formal representation of visual knowledge would allow extracting meaning of pictures, searching for content of images through the Internet, indexing documentation using visual content, and developing expert system to automate the decision process in imagistic domains.

Medicine, stock market analysis, aerial traffic monitoring and Geology are examples of imagistic domains [1], which require from the problem solver the ability of applying visual recognition of objects, and from this initial recognition, to start the search and analytical methods in order to interpret these objects [2].

A formal (computer processable) representation of the visual content is required in order to reach all these goals. Ontologies are being applied to represent visual knowledge because they are formal and allow automatic processing over the represented knowledge. However, ontological representations are mainly

C.A. Heuser and G. Pernul (Eds.): ER 2009 Workshops, LNCS 5833, pp. 169–178, 2009.

conceptual, based on the textual definition of a vocabulary associated with the characterisation of each word and their relationship [3]. However, Abel's thesis [4] proved that, in order to support problem solving in imagistic domain, some concepts are necessary, which are not fully represented through a vocabulary.

This paper proposes a set of primitives that combine textual ontological representation with pictorial primitives to formalise concepts in an imagistic domain. These primitives allow the addition of pictorial content to domain ontologies. This is especially significant to those concepts that the experts are not able to externalise the full meaning without the complement of a pictorial representation. Most of the features related to rocks and field that support Geological interpretation are like that and, even more often, those related to Stratigraphy, where we develop our study.

Stratigraphy is an area of Geology that tries to understand the history of formation of some terrain based on the identified sedimentary structures imprinted on sedimentary rocks. The sedimentary structures are the visual aspect of the spatial organisation of the grains of a rock as a result of the process of deposition of these grains during the rock formation [5].

This paper is organised as follows. Section 2 presents previous work in the representation of visual knowledge. The approach of this work is presented in Section 3. Sections 4 and 5 are reserved respectively for discussions and conclusions.

2 Related Work on Visual Knowledge Representation

As much as images are becoming a common content in information sources, more approaches are being studied and developed for visual knowledge representation. Most of them try to deal with the semantic gap that is found between the image representation (commonly digital files that represent maps, graphs or pictures) and the significant content that is recognised in the image and is used by someone to take decisions. The mapping between the low level representation and the semantic content of some image is related to the *symbol grounding problem* by Harnad [6].

Hudelot [7] presents a visual knowledge modelling approach for the symbol grounding problem in an application for the recognition of greenhouse rose leaf diseases. The approach divides the conceptual representation of visual knowledge in three semantic levels, namely *Image level*, *Visual level* and *Semantic level*, each one treating visual knowledge in distinct levels of abstraction. It uses two ontologies (a visual concept ontology and an image processing ontology) and a knowledge base to represent visual knowledge. The image processing ontology is used in the Image level to describe basic forms extracted from images through algorithmic processes. The visual concept ontology is applied in the Visual level where image concepts are linked with domain concepts from the Semantic level through its visual description. The knowledge base encodes in a declarative manner the symbol grounding knowledge employed to map concepts between the Image and Semantic levels.

Santin [8] and Fiorini [9] present similar approaches for visual knowledge modeling in the Geology domain applied for the interpretation of visual features for petroleum exploration. Both divide the conceptual representation in semantic levels, similar to [7] although these approaches apply distinct strategies to deal with image processing algorithms and visual content. Santin extracts visual knowledge from images by combining manual and automatic segmentation and associates the image content to polygons for further interpretation. Fiorini segments the image applying wavelets that recognise significant features in the image according to previously defined experts criteria.

Silva [10] also formalises visual knowledge applied in the evaluation of the quality of rocks as petroleum reservoirs. The approach differs by representing the knowledge in two levels of expertise. The first level uses a domain ontology to formalise the visual knowledge that is easily recognised by a novice. The knowledge formalised in this level is represented by atomic concepts, attributes and values. The expert level represents abstractions and the tacit knowledge applied by the expert in recognising diagnostic features over the images.

Liu [11] presents a framework to formalise the visual knowledge applied for visual classification of birds. The framework is composed by a domain ontology and a shape ontology. The domain ontology formalises the ornithologists vocabulary when classifying birds. The shape ontology is organised according to the visual features of the birds (body, beak and wing shapes) which captures the visual information that supports the classification. Both ontologies represent different aspects from the domain and the mapping between ontologies establishes the relationship of domain and visual knowledge. The approach proposes the automatic construction of the shape ontology through the clusterization of real images taken from animals.

Bertini [12] presents an approach for visual knowledge modelling applied for video digital libraries annotation in the soccer domain. The approach is composed by a domain ontology and a set of visual concepts. The domain ontology is expressed in linguistic terms and defined by domain experts. The visual concepts are used represent the visual counterpart of abstract linguistic concepts enriching ontologies with pictorial content. These concepts are automatically defined through a visual clustering process of videos and images.

The formalisation of the pictorial content aggregated in ontologies is an important issue in our research. In our work we propose the representation of the pictorial content inspired the idea of *inferential "free-rides"*. Shimojima [13] formally defines inferential free-rides as the capture of semantic information from a visual symbol, i.e. the visual symbol is built in order to express the right semantic information. Fig.1, extracted from [14], depicts this kind of immediate inference. Fig.1-a shows a sentential language, which describes the relationship among the objects A, B and C, which is equivalent to the graphical language presented in Fig.1-b, whose conclusion is reached in a more straightforward way.

Our research also addresses the same general objective of the previous described works: capturing the visual knowledge in formal representations in order to support interpretation tasks in distinct domains. However, our approach differs from the previous presented ones in the following aspects that will be further detailed in this paper:

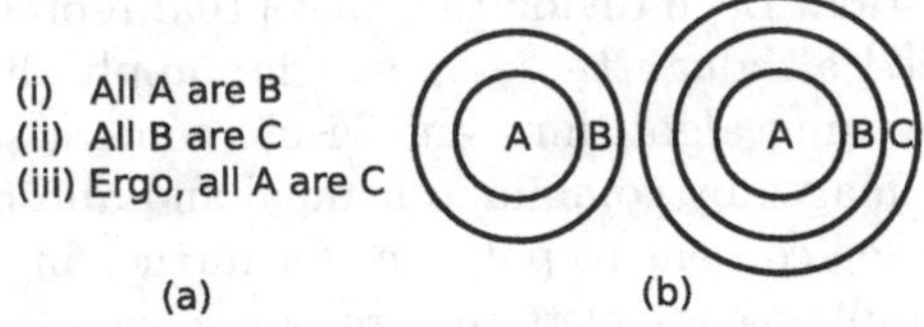
(i) All A are B
(ii) All B are C
(iii) Ergo, all A are C
A
B
A
B
C
(a)
(b)

Fig. 1. Example of immediate inference. Extracted from [14].

- The knowledge representation is not going to be fully conceptual as in Silva [10] and Hudelot [7];
- The visual knowledge is not captured through image processing techniques as in Santin [8], Fiorini [9], Liu [11] and [12];
- We use a dual representation – conceptual and pictorial – to capture the knowledge in representational primitives, differently from Hudelot [7], Silva [10], Santin [8] and Fiorini [9];
- The pictorial content does not constitute only a documentation, but is used to define the structure of the domain;
- Pictorial representation is not automatically created (using clusterization) as in Liu [11] and [12].

3 Representing Visual Knowledge through Primitives and Pictorial Content

The cognitive mechanism applied by geologists, when interpreting sedimentary structures to define the stratigraphic history of a rock, is the same than doctors apply when interpreting X-Ray exams in Medicine.

The Stratigraphic concepts represented in our ontology were captured through recorded interviews with the expert, in which he presented the problem and the main principles of the domain in retrospective protocols. The film recordings were later on transcripted to textual files and the concepts were manually identified in the text and incorporated to a list. This list was refined and organised by the expert in a hierarchical structure presented in Figure 2. The initial structure was further populated by new concepts extracted by the indicated literature. Further details of the knowledge elicitation techniques applied can be found in [2].

The main concepts studied in our approach are the *Sedimentary Facies* and the *Sedimentary Structure*. A sedimentary facies is a particular organisation of a rock in a spatial arrangement, that, along with the preserved fossil content, identifies the depositional environment in which the existent sediment has been deposited and consolidated in that rock. The sedimentary structure is the external visual aspect of that internal spatial arrangement. It is the more striking visual object recognised in the domain and the first one to be used in raising interpretation hypotheses. Both concepts comprise two main challenges for ontology engineering. The first is the incapability of the geologists in defining the instances in a pure verbal way, requiring a drawing or a picture to complete the

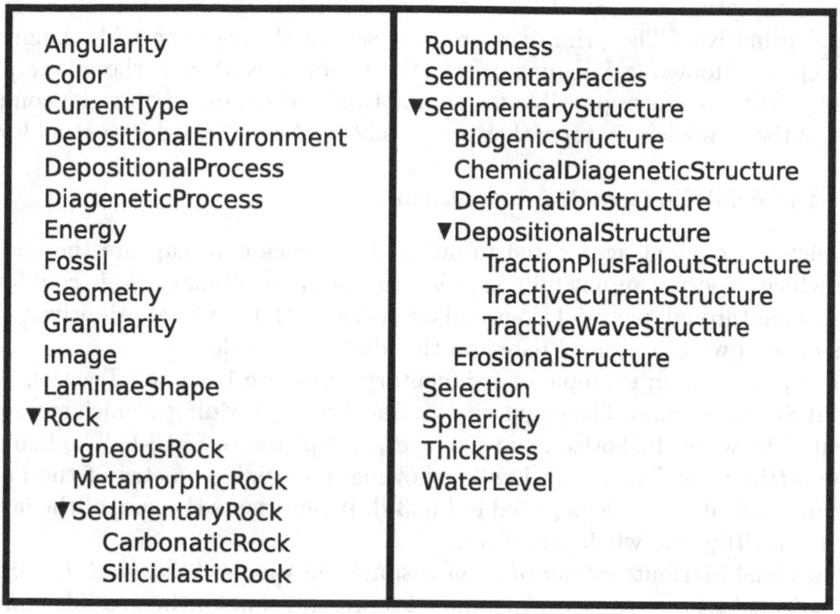

- Angularity
- Color
- CurrentType
- DepositionalEnvironment
- DepositionalProcess
- DiageneticProcess
- Energy
- Fossil
- Geometry
- Granularity
- Image
- LaminaeShape
- ▼Rock
  - IgneousRock
  - MetamorphicRock
  - ▼SedimentaryRock
    - CarbonaticRock
    - SiliciclasticRock
- Roundness
- SedimentaryFacies
- ▼SedimentaryStructure
  - BiogenicStructure
  - ChemicalDiageneticStructure
  - DeformationStructure
  - ▼DepositionalStructure
    - TractionPlusFalloutStructure
    - TractiveCurrentStructure
    - TractiveWaveStructure
  - ErosionalStructure
- Selection
- Sphericity
- Thickness
- WaterLevel

Fig. 2. Concepts hierarchy

idea. The second is the fact that the terminology associated to the concepts is still informally treated in the domain. Even the specialised literature does not present a formal organisation of the vocabulary and the definition of sedimentary structures [15]. The consequence is the existence of many examples of ambiguous terminology, overloading vocabulary and multiple denominations for the same geological feature.

Our intention in developing the ontology for Stratigraphy is providing a defined vocabulary to be shared and used by geologists in the description of exploration well cores and outcrops. Achieving a shared accepted vocabulary formally defined will provide the adequate basis for developing knowledge systems for stratigraphic documentation and interpretation. This is the long term aim of our project.

The knowledge acquisition process has allowed us to identify the geometrical attributes that are used by geologists to visually recognise sedimentary structures. These attributes were the basis to organize, by visual criteria the concepts in a hierarchy. The preliminary hierarchy is shown in Fig. 2 under the concept *Sedimentary Structure*. Besides visually providing an organization for the sedimentary structures, this hierarchy also represents the organization of the depositional processes that create the sedimentary structures, since each process imprints int the rock exactly one kind of structure. The identified attributes come from the Depositional Structure class, namely *angularity*, *laminae shape* and *thickness*.

The conceptual content of the visual knowledge is described using a set of textual primitives. The primitives are represented through the CML language (Conceptual Modelling Language) from the CommonKADS methodology [16]. These primitives are responsible for nominating and characterizing the components of the geological features that are possible to be described in textual form.

3.1 Pictorial Content Representation

The pictorial content aggregated to ontologies is meant to capture the knowledge which experts can not fully express through a vocabulary. This content is represented through pictorial icons and described with the visual primitives. The icons created were conceived based on the idea of free-rides.

Fig.3-a depicts an example of sedimentary structure from the Traction plus Fallout Structure class. The angularity attribute of the beddings, which measures the angle between the horizon and the layer, is depicted in Fig.3-b. The laminae shape attribute is depicted in Fig.3-c showing a special geometry of the layer. The thickness attribute is depicted in Fig.3-d. It represents the sum of the layers which constitute the whole structure.

The visual attribute *angularity* can assume the values: *horizontal*, *low angle* and *high angle*. These values are a nominal representation for the possible numerical values assumed by angle A in Fig.3-b. Horizontal means an angle between $0°$(zero) and $2°$ degrees, low angle means an angle between $2°$ and $10°$ degrees and high angle means angles over than $10°$ degrees. The respective pictorial representation of the angularity attribute values is depicted in Figure 4-{(a), (b) and (c)}.

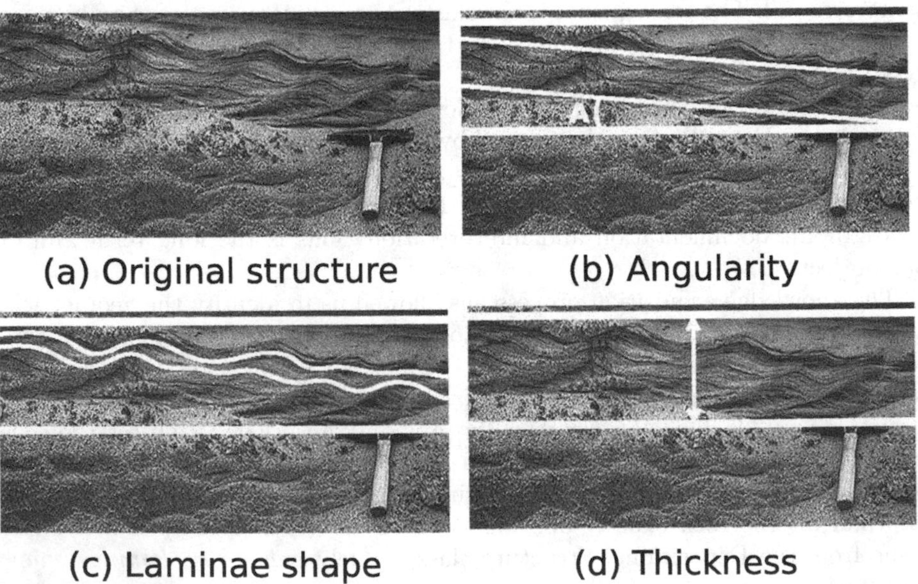

Fig. 3. Example of sedimentary structure with visual attributes emphasized

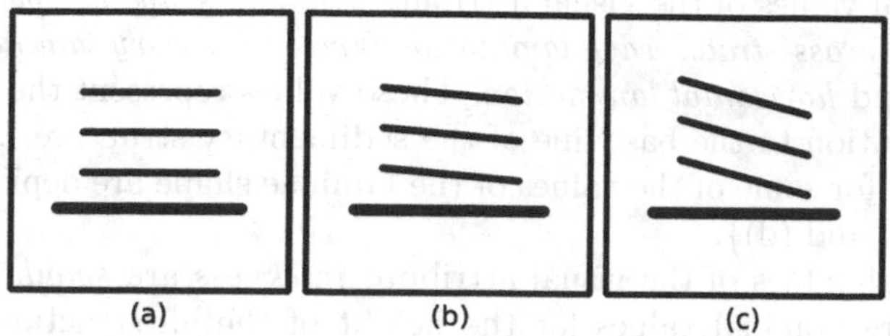

Fig. 4. Icons representing the values for the angularity visual attribute: (a) horizontal; (b) low angle; (c) high angle

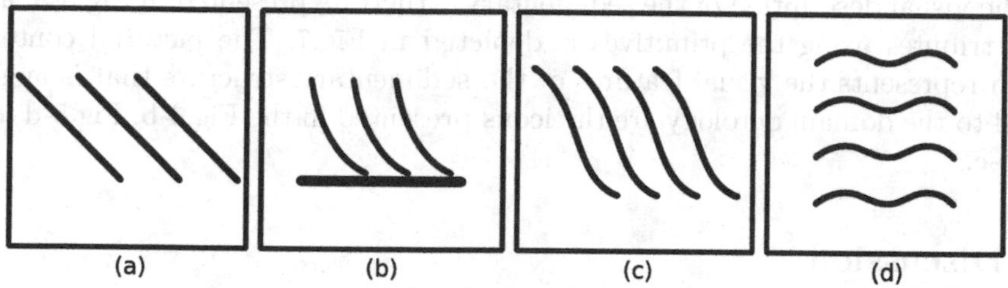

Fig. 5. Icons representing some of the values for the laminae shape visual attribute: (a) planar; (b) tangential; (c) sigmoidal cross-strata; (d) wavy lamination

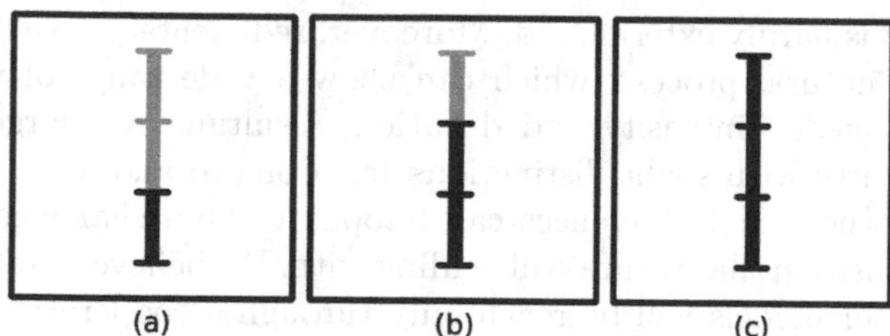

Fig. 6. Icons representing the values for the thickness visual attribute: (a) small; (b) midsize; (c) large

```
CONCEPT StructureX
   SUB-TYPE-OF: Traction plus Fallout Structure;
   ATTRIBUTES:
      angularity: low_angle;
      laminaeShape: wavy lamination;
      thickness: large;
END CONCEPT StructureX;
```

Fig. 7. Description of a visual entity corresponding to a sedimentary structure

The nominal values of the visual attribute *laminae shape* are *planar*, *tangential*, *sigmoidal cross-strata*, *wavy lamination*, *truncated wavy lamination*, *trough cross-strata*, and *horizontal lamination*. These values represent the shape of the bedding in relation to the base line of the sedimentary structure. The pictorial representation for some of the values of the laminae shape are depicted in figure 5-{(a), (b), (c) and (d)}.

The nominal values of the visual attribute *thickness* are *small*, *midsize*, and *large*. They are nominal values for the height of the all structure. Structures from 1 centimetre to 5 centimetres are classified as small thickness, from 5 to 10 centimeters are midsize and over than 10 centimetres structures are classified as large thickness. The pictorial representation of the thickness attribute values is depicted inf Figure 6-{(a), (b) and (c)}.

The visual description of the sedimentary structure, presented in Fig.3-a, and its attributes using the primitives is depicted in Fig.7. The pictorial content, which represents the visual features of the sedimentary structure that is aggregated to the domain ontology are the icons presented in the Fig.4-b, Fig.5-d and Fig.6-c.

4 Discussion

The Stratigraphy domain ontology is not complete so far. There are many details in the domain which require a refinement of the constructed model.

The elicitation of the visual attributes of the sedimentary structures is a complex task, since these attributes are mostly part of the tacit knowledge of the expert that is hardly externalised. Moreover, sedimentary structures are the result of some natural process, which can show a wide range of variations in transportation media, intensity and duration, resulting in a large amount of different structures with slight distinctions from one to another. Capturing all the nuances of the visual differences can happen to be a challenge in terms of knowledge acquisition that we are still dealing with. We believe that the complete population of our models will be reach only through a cooperative work inside the Geological community.

However, the approach presented in this paper has some advantages when compared with those presented in section 2:

- The icons formalize the visual knowledge used by experts in their activities in a more straightforward way;
- We can organize the concepts by their visual aspects as the Geologists are used to do, which would not be possible without a formal representation of these visual aspects;
- The formalisation of the pictorial content allows someone to query a knowledge base for domain concepts, using the visual features of some image to drive the search.

5 Conclusion

This paper proposes a set of primitives to formalise concepts in imagistic domain that require pictorial content to complement the conceptual content in order to be correctly defined.

We have proposed a visual language whose elements can be combined to build the definition of the visual content of the concepts in a restricted domain. The elements were conceived based on the idea of free-rides and associated to a textual translation of the main representative aspects.

The visual attributes of the concepts were also applied to organise the domain in a hierarchy of objects. The chosen organisation reflects as well the organisation of the genetic processes that produced the represented objects in the Nature, thus being useful to be further applied in the interpretation of these processes from the visual characteristics of the objects.

Although the language is strongly connected with the domain, the general approach applied in its construction can be replicated to build visual representations in other image-based applications.

Future work is to develop experiments to collect and evaluate the feedback from the geologists about the proposed model.

Acknowledgement

This work is also supported by the Brazilian Council of Research CNPq through the founding Universal Program.

References

1. Yip, K., Zhao, F.: Spatial aggregation: Theory and applications. Journal of Artificial Intelligence Research 5, 1–26 (1996)
2. Mastella, L.S., Abel, M., Lamb, L.C., Ros, L.F.D.: Uma ontologia temporal para modelagem de conhecimento sobre ordenação de eventos. In: Encontro Nacional de Inteligência Artificial (2005)
3. Gruber, T.R.: A translation approach to portable ontology specifications. Knowledge Acquisition 5, 199–220 (1993)
4. Abel, M.: Estudo da Perícia em Petrografia Sedimentar e sua Importância para a Engenharia do Conhecimento. PhD thesis, Universidade Federal do Rio Grande do Sul (July 2001)
5. Fávera, J.C.D.: Fundamentos de Estratigrafia Moderna. Editora da Universidade do Estado do Rio de Janeiro, Rio de Janeiro, Brasil (2001)
6. Harnad, S.: The symbol grounding problem. Physica 42, 335–346 (1990)
7. Hudelot, C., Maillot, N., Thonnat, M.: Symbol grounding for semantic image interpretation: From image data to semantics. In: International Conference on Computer Vision Workshops, p. 1875 (2005)
8. Santin, C.E.: Construtos ontológicos para representação simbólica de conhecimento visual. Master's thesis, Universidade Federal do Rio Grande do Sul (January 2008)
9. Fiorini, S.R.: S-chart: Um arcabouço para interpretação visual de gráficos. Master's thesis, Universidade Federal do Rio Grande do Sul (2009)

10. Silva, L.A.L., Mastella, L.S., Abel, M., Gallante, R.M., Ros, L.F.D.: An ontology-based approach for visual knowledge: Image annotation and interpretation. In: Workshop on Ontologies and their Applications, in XVII Brazilian Symposium on Artificial Intelligence (SBIA), São Luis, Brazil (Setember - October 2004)
11. Liu, Y., Zhang, J., Tjondronegoro, D., Geve, S.: A shape ontology framework for bird classification. In: DICTA 2007: Proceedings of the 9th Biennial Conference of the Australian Pattern Recognition Society on Digital Image Computing Techniques and Applications, December 2007, pp. 478–484 (2007)
12. Bertini, M., Bimbo, A.D., Torniai, C., Grana, C., Cucchiara, R.: Dynamic pictorial ontologies for video digital libraries annotation. In: MS 2007: Workshop on multimedia information retrieval on The many faces of multimedia semantics, pp. 47–56. ACM, New York (2007)
13. Shimojima, A.: Operational constraints in diagrammatic reasoning. Logical Reasoning with Diagrams, 27–48 (1996)
14. Guizzardi, G., Pires, L.F., Sinderen, M.J.V.: On the role of domain ontologies in the design of domain-specific visual modeling languages. In: Second Workshop on DomainSpecific Visual Languages, 17th Annual ACM Conference on ObjectOriented Programming, Systems, Languages, and Applications (2002)
15. Lucchi, F.R.: Sedimentographica: photographic atlas of sedimentary structures, 2nd edn. Columbia University (1995)
16. Schreiber, G., Akkermans, H., Anjewierden, A., de Hoog, R., Shadbolt, N., de Velde, W.V., Wielinga, B.J.: Knowledge Engineering and Management - The CommonKADS Methodology. MIT Press, Cambridge (2000)

Using a Foundational Ontology for Reengineering a Software Enterprise Ontology

Monalessa Perini Barcellos[1,2] and Ricardo de Almeida Falbo[1]

[1] Ontology and Conceptual Modeling Research Group (NEMO), Federal University of Espírito Santo, Brazil
[2] COPPE, Federal University of Rio de Janeiro, Brazil
{monalessa, falbo}@inf.ufes.br

Abstract. The knowledge about software organizations is considerably relevant to software engineers. The use of a common vocabulary for representing the useful knowledge about software organizations involved in software projects is important for several reasons, such as to support knowledge reuse and to allow communication and interoperability between tools. Domain ontologies can be used to define a common vocabulary for sharing and reuse of knowledge about some domain. Foundational ontologies can be used for evaluating and re-designing domain ontologies, giving to these real-world semantics. This paper presents an evaluating of a Software Enterprise Ontology that was reengineered using the *Unified Foundation Ontology* (UFO) as basis.

Keywords: Foundational Ontology, Enterprise Ontology, Ontology Reengineering.

1 Introduction

Software engineering is a knowledge-intensive activity and there are many types of knowledge that are useful for software engineers. One of the main obstacles for capturing, searching and reusing this knowledge is the lack of a common conceptualizations, what can be provided by ontologies.

In context of Software Engineering Environments, ontologies have been used, among others, for building Domain Oriented Software Development Environments (DOSDEs) [1]. DOSDEs support software engineers in their tasks by providing useful knowledge during the software development process.

By building several DOSDEs for different domains, researches have noticed that other kinds of knowledge beyond domain knowledge could also be useful for software projects, mainly knowledge about the organizations involved in this context. Based on that, Villela et al. [2] defined a Software Enterprise Ontology (SEO) that was used as basis for building what they call Enterprise Oriented Software Development Environments (EOSDEs).

During the development of a research work considering high maturity aspects on software organizations supported by EOSDEs (characterized by the highest levels of maturity models like CMMI), we envision the need for building a measurement ontology. Since ideally domain ontologies should be grounded in foundational ontologies [3, 4], we decided to develop our measurement ontology taking as basis the

C.A. Heuser and G. Pernul (Eds.): ER 2009 Workshops, LNCS 5833, pp. 179–188, 2009.

Unified Foundational Ontology (UFO) [3, 5]. UFO has been used to evaluate, re-design and integrate (meta) models of conceptual modeling languages, as well as to evaluate, re-design and give real-world semantics to domain ontologies [5]. Besides, we noticed that in order to talk about measurement in high maturity software organizations, it was also necessary to use concepts regarding software organizations. Then we decided to use some concepts of SEO [2]. However, we had several problems, mainly due to implicit ontological commitments, as well as to real-world situations that were not addressed by SEO. Before integrating the ontologies, we decided to carry out an evaluation and reengineering of SEO by mapping its concepts to UFO. This allowed us to solve conceptual problems, making SEO more truthful to the domain it represents, and making explicit some ontological commitments that were implicit.

This paper presents the SEO evaluation and reengineering and it is organized as follows: Section 2 presents a brief discussion about ontologies, and presents relevant parts of UFO to this work and the fragment of SEO considered; in Section 3, we discuss the evaluation and reengineering of SEO; Section 4 presents related works; and in Section 5, we conclude presenting the final considerations of the work done.

2 Ontologies

In the context of Philosophy, ontology is a particular system of categories accounting for a certain vision of the world, independent of a particular language. Otherwise, for Computer Science communities, ontologies refer to an engineering artifact, constituted by a specific vocabulary used to describe a certain reality and by a set of explicit assumptions regarding the intended meaning of the vocabulary words [4].

In the context of the use of ontologies in Computer Science, Guarino [4] states that computational ontologies mix philosophical, cognitive and linguistic aspects, and ignoring their interdisciplinary nature makes the ontologies less useful. To capture this, Guarino says that, ideally, domain ontologies should be built based on Foundational Ontologies. Foundational ontologies are theoretically well-founded and domain- independent systems of categories, which describe very general concepts like object, event, action etc [5]. Due to its soundness, foundational ontologies can be used to improve the quality of conceptual models, including domain ontologies [8].

UFO [3, 5] is a foundational ontology that has been developed based on a number of theories from Formal Ontology, Philosophical Logics, Philosophy of Language, Linguistics and Cognitive Psychology. It is composed by three main parts. UFO-A is an ontology of *endurants*, and it is the core of UFO. A fundamental distinction in UFO-A is between *Particulars (Individuals)* and *Universals (Types)*. Particulars are entities that exist in reality possessing a unique identity, while Universals are patterns of features, which can be realized in a number of different particulars [3]. UFO-B is an ontology of *perdurants* (events). The main distinction between perdurants and endurants is that endurants are wholly present or not, while perdurants happen in time [5]. UFO-C is an ontology of social entities (both endurants and perdurants) built on the top of UFO-A and UFO-B. One of its main distinctions is between *agents* and *objects*. Agents are capable of performing actions with some intention, while objects only participate in events [5].

For the purpose of this paper, concepts from parts A and C of UFO are more important, since the analyzed fragment of SEO does not talk about events. Due to

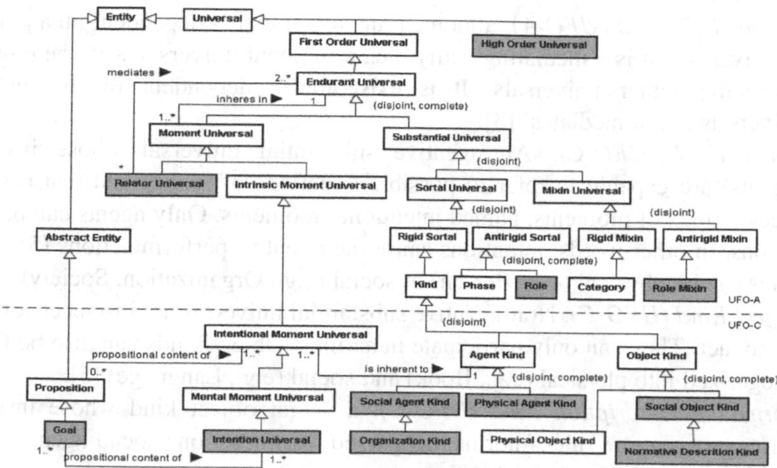

Fig. 1. An UFO Fragment

space limitations, it is impossible to discuss here all the distinctions made in those parts of UFO. So, Figure 1 presents some concepts that are important for this paper. The concepts that are directly used here are shown shaded in grey. It is worthwhile to point out that, since we used UFO concepts to ground concepts of a domain ontology, concepts related to Universals were more important. Following some of the concepts shown in Figure 1 are described.

- *First Order Universal (UFO-A):* universals whose instances are particulars.
- *High Order Universal (UFO-A):* universals whose instances are universals.
- *Kind (UFO-A):* a substance sortal[1] universal that supplies a principle of identity for its instances (rigid sortals) [3].
- *Role (UFO-A):* a possible role that a substance sortal can play along its history. An entity plays a role in a certain context, demarcated by its relations with other entities [3].
- *Role Mixin (UFO-A):* Anti-rigid mixin[2] that represents abstractions of common properties of roles [3].
- *Intrinsic Moment Universal (UFO-A):* a moment universal[3] that is dependent on a single universal.

[1] Substantials are entities that persist in time, keeping their identity. Substantial universals are patterns of features that can be realized in a number of different substantials. Some of them are sortal (sortal universals), thus providing a principle of individualization, persistence and identity. Others are merely characterizing (said mixin universals) [3].

[2] Mixins are dispersive universals, covering many concepts with different principles of identity. Anti-rigid mixins are mixins of witch patterns of features does not apply necessarily to all its instances [3].

[3] The word Moment in UFO-A is derived from the German term *Momente* and it bears no relation to the notion of time instant. It is related to the ways things are. An important feature that characterizes all moments is that they can only exist in other individuals. Thus, moment universals can only exist in other universals [3].

- *Relator Universal (UFO-A)*: a moment universal that is dependent on a plurality of universals. It is a mediating entity, i.e., a moment universal with the power of connecting other universals. It is existentially dependent of the endurant universals that it mediates[4] [3].
- *Agent Kind (UFO-C):* An agentive substantial universal whose instances (agents) are capable to refer to possible situations of reality and that can bear special kinds of moments, named intentional moments. Only agents can perform actions. In other words, intentions cause the agent to perform actions [5]. Agent kinds can be physical (e.g., Person) or social (e.g., Organization, Society).
- *Object Kind (UFO-C)*: Non-agentive substantial universal. Its instances (objects) do not act. They can only participate in actions. Object kinds can also be further categorized into physical (e.g., Book) and social (e.g., Language) [5].
- *Normative Description Kind (UFO-C)*: a social object kind whose instances define one or more rules / norms recognized by at least one social agent and that can define nominal universals [5].
- *Intention Universal (UFO-C):* a type of intentionality designating "intending something" [5].
- *Goal (UFO-C):* The propositional content of an intention [5].

In this paper, the distinctions made in UFO are shown in the concepts of the SEO as stereotypes, indicating that they are subtypes of concepts of UFO, in an approach analogous to the one defined in [3]. As pointed out by Guizzardi [3], the ontological interpretation of a UML class is of an endurant universal. Thus for simplicity, the term universal (or kind, depending on the situation) was omitted from the corresponding stereotypes. For example, the stereotype <<*social object*>> designates the UFO's concept of Social Object Kind, shown in Figure 1. When a concept is not stereotyped then it presents the same stereotype of its super-type in the model.

The Software Enterprise Ontology (SEO) considered in this paper was developed by Villela et al. [2] to establish a common vocabulary for software engineers to talk about the organizations involved in software projects. Figure 2 presents the fragment of SEO considered in this paper. According to [2], an *Organization* is an organized group of people working together for the fulfillment of a mission. Mission is the organization's purpose in a social or economic system. An organization is divided into *Functions, Organization Units* and *Committees. Organization Units* can be structured by *Positions.* An *Agent* represents a profile that allows the organization to accomplish its mission and it can be a function or a position. A *Committee* is a group of people with a specific goal, which usually works together for a period of time until a specific goal is achieved. Organizations have *Participations* on *Projects.* A *Business Agreement* is an agreement between two or more organizations which establishes a business relationship. Finally, *Objectives* are statements on the results to be reached and may be applied to the organizations, organizational units or positions.

[4] A mediation is a formal relation that takes place between a relator universal and the endurant universals it mediates [3].

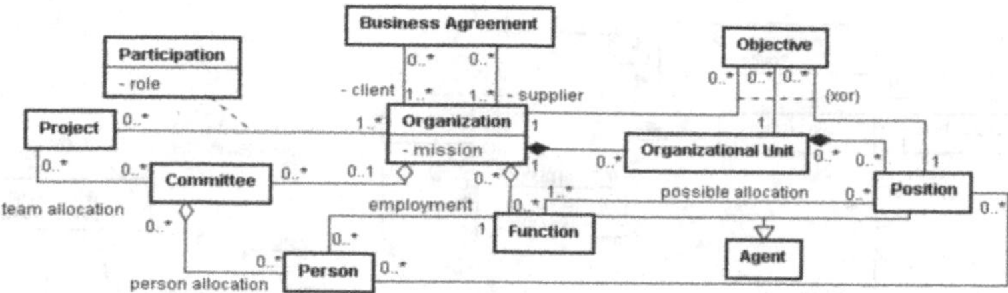

Fig. 2. An SEO Fragment [2]

3 Reengineering of the Software Enterprise Ontology

In this section we present a part of the reengineered Software Enterprise Ontology. Due to lack of space, some aspects were not shown here. We tried to include here the most relevant aspects considering the use of UFO.

As mentioned in section 2, UFO defines that agents are capable to perform actions with some intention. The analyzed fragment of SEO (Fig. 2) does not agree with this conceptualization. Since Agent is a generalization of *Position* and *Function*, it is better characterized as a Normative Description. So, we modified the original term Agent to *Profile*, as shown in Figure 3. Still considering the concept of agent in UFO, we can say that organizations, organizational units and teams are social agents, whereas people are physical agents. Moreover, agents have intentions expressed by goals. Thus, the concept *Intention* was included in the new version of SEO, where an intention is the purpose which actions are planned and performed for, and *Goal* is the propositional content of an intention [5]. The main intention of a social agent is its *Mission*. In order to maintain the alignment with the terms used in UFO, we replaced the term *Objective* by *Goal*. Furthermore, the perception that a team is also an agent drove us to the conceptualization that teams also have intentions and goals. Otherwise, the perception that a position is a normative description showed that it does not make any sense to associate goals to a position.

In the original version of SEO, an organization is defined as "an organized group of people working together for the fulfillment of a mission". However, people are kinds and they exist independently of organizations. In fact, people start playing the role of *Human Resource* within an organization when they are employed in it. So, an *Organization* is better characterized as a social agent which employs human resources for performing actions to achieve its goals.

Organizations can be divided into organizational units. An *Organizational Unit* can be defined as a grouping of human resources (the human resources allotted in it), goals and intentions, established according to the content homogeneity and alignment to the organization's goals. Similarly, a *Team* (term used as a substitute for the term *Committee* from the original version) is a grouping of human resources established with a specific purpose.

In the original version of SEO, an objective could be exclusively of one (and only one) organization or organizational unit. Analyzing this relation from the UFO perspective, we noticed that the singleness and exclusivity restrictions were not covering some situations. In the real-world it is possible, for example, that an

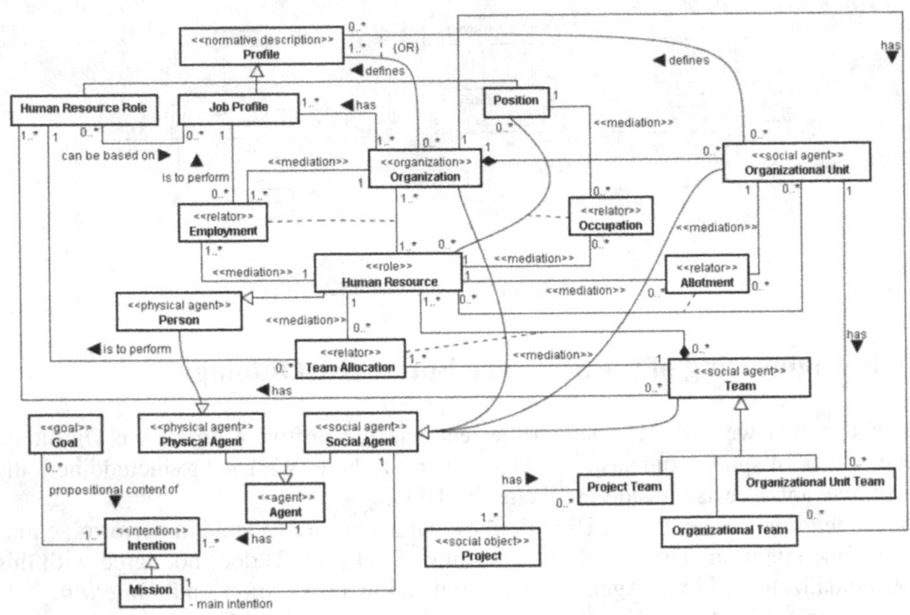

Fig. 3. Fragment of the new version of SEO that includes Agents and Human Resources

organization has the goal of "decreasing 10% product defect ratio" and this would be also a goal of some (or all) of its organizational units. Then, those restrictions were abandoned. In the new version, despite an intention is inherent to only one agent, it is possible to get different intentions when goals are the same. In other words, one goal can be the propositional content of intentions of different agents.

Turning the discussion to the concept of team, with the new conceptions that were adopted, we concluded that the original model was not appropriate when it determined that teams (committees) are parts of organizations and can be allocated to projects. Thus, in the new vision, a team can be established to fulfill a purpose in the context of a project, an organization or an organizational unit, being, respectively, a *Project Team*, an *Organizational Team* or an *Organizational Unit Team*.

As it was said before, people play the role of human resources of an organization when they are employed in it. The relation "*employment*" between Organization and Human Resource is a material relation[5] universal and, therefore, there is a relator universal (*Employment*) whose instances are individuals capable of connecting instances of both these entities. This relator is directly related to the record of the events, which establish the employment of a person in an organization, and thus it has as properties, among others, the 'start date' and 'finish date' of the employment. Moreover, this relator is associated with *Job Profile*, indicating that a human resource fulfils a job profile when it is employed in an organization (e.g., a human resource employed in the job profile of a system analyst in an organization).

---

[5] Material relations have material structure on their own, and their relata are mediated by individuals that are called relators [3].

Similar situations happen with the other relations of the Human Resource role, namely *Team Allocation*, *Occupation* and *Allotment*. The first one records the occurrence of the event of allocation of a human resource to a team, where it plays a human resource role (e.g., a human resource allocated as a project manager in a project team). The second refers to the occupation of a position by a human resource (e.g., a human resource occupying the position of portfolio manager). At last, the relator *Allotment* records the event of allotting a human resource in an organizational unit (e. g., a human resource allotted in the development systems unit). *Job Profile, Human Resource Role* and *Position* are descriptions of *profiles* that are needed for acting in specific contexts, as said before, and are, therefore, normative descriptions.

The creation of these relators is an important change in SEO, driven by the use of UFO. These concepts were included because they not only connect other entities, but also define a set of characteristics of the relationship owners (and not do so to the connected entities), allowing better coverage of real-world aspects, such as the perception that all these relators record events (broadly speaking, a relator can be seen as a static representation of an event) and, consequently, they have temporal properties. For example, an employment, despite of being represented statically, fundamentally deals with an event that has a beginning and an end, and it involves a human resource occupying a job profile in an organization. Do not representing *employment* as a concept (as was done in the original version of the SEO) would not allow the identification of crucial information, such as when the employment began and when it was finished. Furthermore, in the original version of SEO, a person occupies only a job profile and, in fact, an employment is to perform a job profile, but a human resource can occupy several job profiles, since it can have several employments, as it is modeled in the new version of SEO.

Analogous reflection can be done for Team Allocation, Allotment and Occupation. Moreover, it is possible to identify that a team allocation involves a human resource playing a human resource role in the team, what allows that the same human resource to be allocated to several roles in the same team (e.g., a human resource can play the designer and programmer roles in the same project team). In the original version of SEO, a person is allocated to a team without identifying her role. Consequently, it was not possible either to identify her specific functions and responsibilities in the team, nor to allocate her to different roles in the same team.

During the reengineering of SEO, several restrictions were identified and, since the models did not capture several of them, we defined axioms to make them explicit. For instance, the following axiom holds: if a human resource *hr* occupies, in a time window *[t1, t2]*, a position *p* in an organization *org*, then this time window must be contained in the time window *[t3, t4]* in which *hr* is employed in *org*.

$$(\forall hr \in Human\ Resource, p \in Position, org \in Organization)\ (occupation\ (hr, p, t1, t2) \wedge defines(org, p)) \rightarrow employment\ (hr, org, t3, t4) \wedge (t1 \geq t3 \wedge t2 \leq t4)$$

Figure 4 presents the fragment of the new version of the SEO that deals with projects.

In the original version of the SEO, the participation of organizations in projects was represented by the relation *Participation*, which has the property *role* that indicates the kind of their participation. Moreover, there was the concept of *Business Agreement* that is defined as an agreement between two or more organizations (clients

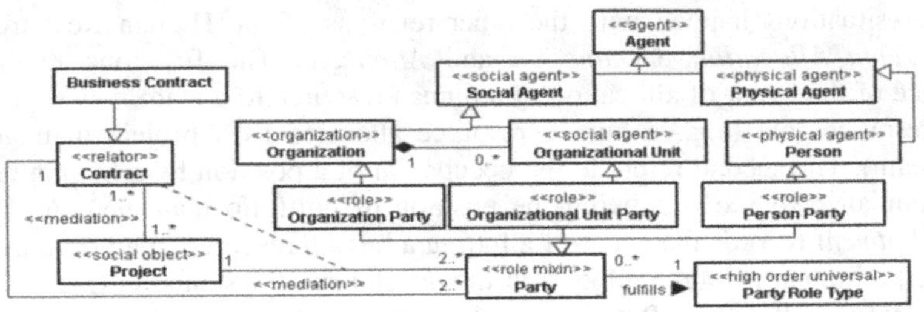

Fig. 4. Fragment of the new version of the SEO that deals with projects

and suppliers), as shown in Fig. 2. This model has several problems. First, it allows that a project has only one participation, what does not seem to make sense in reality. Projects typically involve at least two parties, playing different roles. Second, the model is negligent concerning the participation of the same organization by different ways in the same project. Can the same organization participate playing different roles? Third, only organizations can participate in projects, making it impossible to represent situations in which a person contracts (or is contracted to develop) a project, nor situations in which an organizational unit contracts or is contracted in a project context. Fourth, there is no relationship between projects and business agreement. Moreover, business agreements have two types of roles predefined: client and supplier, which do not seem to be related with the roles defined in participations.

Trying to solve these problems, in the new version of SEO, the participation on projects and the business agreement were treated by the introduction of the concepts *Party* and *Contract. Party* is a form of acting in a project that may be carried out by organizations, organizational units or people. A contract is an agreement established between parties. When a contract has business features, it is a *Business Contract.*

Analyzing the concept *Party* at the light of UFO, we can say that it is a role that can be played by different and disjunctive kinds. In other words, *Party* is a role mixin. According to [3], modeling situations like that has been a recurring problem on the literature. Fig. 5 presents two ontologically incorrect models that could be proposed, if important ontological distinctions of the UFO would not be taken into account.

In Fig. 5(a), the role *Party* is defined as a super-type of Person, Organization and Organizational Unit. Ontologically this model is not correct, because it assumes that all instances of Person, Organization and Organizational Unit are necessarily parties, what does not occur in the real-world, since a person still is a person even if she is not a party. The same happens with an organization and an organizational unit.

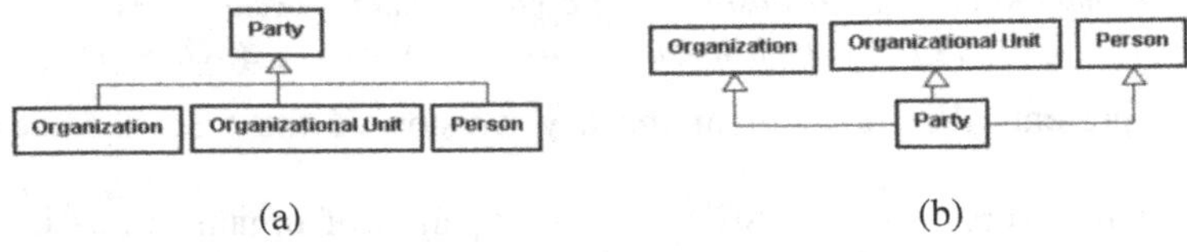

Fig. 5. Ontologically incorrect models

In Fig. 5(b), the role *Party* is defined as a subtype of Organization, Organizational Unit and Person. Ontologically this model is neither correct, because it indicates that Party has identity principles that are common to Organization, Organizational Unit and Person, what does not occur in the real-world, since it is not possible that a party is at the same time an organization, an organizational unit and a person.

In order to solve this problem, Guizzardi proposes the pattern illustrated in Fig. 6.

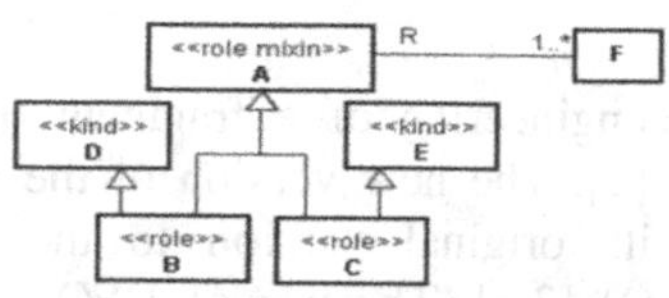

Fig. 6. Pattern for roles with multiple and disjunctive kinds [3]

Applying this pattern to model the concept *Party* in SEO, Party is a role mixin and, thus, does not have direct instances and includes different types of roles: *Organization Party, Organizational Unit Party* and *Person Party*. These roles are disjunctive. For example, the organization *O* can be an instance of Organization Party, what means that, in a given moment, *O* plays the role of Organization Party. Organization, Organizational Unit and Person are the substantial sortals that supply the identity principles to instances of Organization Party, Organizational Unit Party and Person Party, respectively. Party is relationally dependent on Project, since it is a role that an organization, organizational unit or person can play in a project context.

Finally, for solving the problem of roles in business agreement and roles in projects, we introduced the concept of *Party Role Type* that, as it is pointed out by its name, identifies the role that some parties play. *Party Role Type* is a high order universal, so its instances are first order universals, such as Client, Supplier and Partner. The singleness of *Party* to *Project* and *Party* to *Party Role Type* allows the participation of the same agent (an organization, an organizational unit or a person) as different parties in contracts of the same project. Moreover, we established that a project has at least two parties, as it is shown in Fig. 4.

4 Related Works

Some works that deal with evaluation and improvement of conceptual models based on ontological foundations have been developed in the last years. Guarino and Welty [7] developed the OntoClean methodology, which aims to provide guidance on which kinds of ontological decisions need to be made, and on how these decisions can be evaluated based on general ontological notions drawn from philosophical ontology. In [8] Welty et al. report the results of experiments that measure the advantages achieved from the use of ontologies improved based on OntoClean. Finally, Silva et al. [9] applied a technique that is based on OntoClean, called VERONTO (ONTOlogical VERification) to improve analysis patterns in the geographic domain.

As OntoClean, UFO is being used to evaluate, re-design and give real-world semantics to domain ontologies [5]. Moreover, UFO makes other distinctions and provides more guidelines to evaluate conceptual models than OntoClean does. Considering the use of UFO for this purpose, Guizzardi et al. [5] reengineered a Software Process Ontology, while Falbo and Nardi [6] evolved a Software Requirements Ontology. In both cases, ontological problems were identified and solved using UFO, such as in this work.

5 Conclusion

This paper presented the reengineering of a fragment of the Software Enterprise Ontology (SEO) defined in [2]. The new version of the ontology was obtained by mapping the concepts of its original version to the concepts of the Unified Foundational Ontology (UFO) [3,5]. The use of UFO allowed identifying several problems and driving the reengineering of the ontology, making explicit ontological commitments that were implicit and elucidating conceptual mistakes.

The need for reengineering SEO was identified during works that aimed reusing some of its concepts, what could not be done because there were problems and limitations related to real-world semantics. This situation corroborate, as argued by Guizzardi et al. [5], that the use of the UFO contributes to achieve quality attributes needed for domain ontologies. The changes done in SEO allowed the reuse and integration of it in the context of a measurement ontology which considers high maturity aspects and that is in final phase of building.

References

[1] Oliveira, K., Zlot, F., Rocha, A.R., et al.: Domain Oriented Software Development Environments. Journal of Systems and Software 72(2), 145–161 (2004)

[2] Villela, K., Rocha, A.R., Travassos, G.H., et al.: The Use of an Enterprise Ontology to Support knowledge Management in Software Development Environments. Journal of the Brazilian Computer Society 11(2), 45–59 (2005)

[3] Guizzardi, G.: Ontological Foundations for Structural Conceptual Models. Universal Press, The Netherlands (2005)

[4] Guarino, N.: Formal Ontology and Information Systems. In: Proceedings of International Conference in Formal Ontology and Information Systems, pp. 3–15 (1998)

[5] Guizzardi, G., Falbo, R.A., Guizzardi, R.S.S.: Grounding Software Domain Ontologies in the Unified Foundational Ontology (UFO): The case of the ODE Software Process Ontology. In: Proceedings of the XI Iberoamerican Workshop on Requirements Engineering and Software Environments, pp. 244–251 (2008)

[6] Falbo, R.A., Nardi, J.C.: Evolving a Software Requirements Ontology. In: XXXIV Conferencia Latinoamericana de Informática, Santa Fe, Argentina, pp. 300–309 (2008)

[7] Guarino, N., Welty, C.: Evaluating Ontological Decisions with OntoClean. Communications of the ACM 45(2), 61–65 (2002)

[8] Welty, C., Mahindru, R., Chu-Carroll, J.: Evaluating Ontology Cleaning. In: Proceedings of the National Conference on Artificial Intelligence, pp. 311–316 (2004)

[9] Silva, E.O., Lisboa, F.J., Oliveira, A.P., Gonçalves, G.S.: Improving Analysis Patterns in the Geographic Domain Using Ontological Meta-properties. In: Proceedings of International Conference on Enterprise Information Systems, pp. 256–261 (2008)

Multi-level Conceptual Modeling and OWL

Bernd Neumayr and Michael Schrefl

Department of Business Informatics - Data & Knowledge Engineering
Johannes Kepler University Linz, Altenberger Straße 69, 4040 Linz, Austria
{neumayr,schrefl}@dke.uni-linz.ac.at

Abstract. Ontological metamodeling or multilevel-modeling refers to describing complex domains at multiple levels of abstraction, especially in domains where the borderline between individuals and classes is not clear cut. Punning in OWL2 provides decideable metamodeling support by allowing to use one symbol both as identifier of a class as well as of an individual. In conceptual modeling more powerful approaches to ontological metamodeling exist: materialization, potency-based deep instantiation, and m-objects/m-relationships. These approaches not only support to treat classes as individuals but also to describe domain concepts with members at multiple levels of abstraction. Based on a mapping from m-objects/m-relationships to OWL we show how to transfer these ideas from conceptual modeling to ontology engineering. Therefore we have to combine closed world and open world reasoning. We provide semantic-preserving mappings from m-objects and m-relationships to the decideable fragment of OWL, extended by integrity constraints, and sketch basic tool support for applying this approach.

1 Introduction

Modeling domain objects at multiple levels of abstraction has received increased attention over the last years. It is nowadays also referred to as *ontological* multilevel- or meta-modeling to contrast it from *linguistic* metamodeling [1]. While linguistic metamodeling is used to define or extend modeling languages, this paper addresses ontological multilevel-modeling for modeling complex domains where the borderline between classes and instances is not clear cut.

OWL supports metamodeling in two flavours. Metamodeling in OWL Full allows to treat classes as individuals but is not decideable. OWL2 supports a very basic, but decideable approach to metamodeling called punning or contextual semantics [2]: one symbol can be used to refer both to a class as well as to an individual; the decision whether a symbol is interpreted as class, property, or individual is context-dependent. E.g. the symbol Car interpreted as class refers to the set of physical entities that belong to product category *car*, while Car interpreted as individual can be classified as ProductCategory and has assigned a value for property taxRate.

In conceptual modeling more powerful approaches to ontological metamodeling exist: materialization[3], potency-based deep instantiation[4], and m-objects/

C.A. Heuser and G. Pernul (Eds.): ER 2009 Workshops, LNCS 5833, pp. 189–199, 2009.

m-relationships[5]. These approaches not only support to treat classes as individuals but also to describe domain concepts with members at multiple levels of abstraction. For example, domain concept *Product* has members at different levels of abstraction: *Car* at level *category*, *Porsche911CarreraS* at level *model* and *MyPorsche911CarreraS* at level *physical entity*.

In [5] we introduced multi-level objects (m-objects) and multi-level relationships (m-relationships). The basic ideas of this approach are (i) to encapsulate the different levels of abstractions that relate to a single domain concept (e.g., the level descriptions, category, brand, model, physicalEntity that relate to single domain concept, *car*) into a single m-object Car, and (ii) to represent classification, aggregation and multiple generalization hierarchies by a single *concretization* hierarchy. The m-object/m-relationship approach supports modular and redundancy-free multi-level models. It allows to address sets of member objects at different levels for querying and describing their common properties. It supports heterogenous level-hierarchies and multiple relationship abstractions.

Previous work has shown how to bridge the gap between MDA and OWL [6], discussed the relation between ontologies and (meta)modeling [7], and sketched how to represent materialization with Description Logics [8]. Herein, we continue the latter line of work for the more powerful m-object/m-relationships approach by introducing a detailed mapping from m-objects/m-relationships to OWL.

Our mapping from m-objects/m-relationships to OWL is motivated as follows: (1) For conceptual modeling: to provide querying facilities and decideable consistency checking for multi-level models using OWL reasoners. (2) For ontology engineering: to extend the metamodeling features of OWL2 (punning) to objects that represent classes at multiple levels of abstraction, and still remain within the decideable and first-order fragment of OWL. In the latter case, the outcome of the mapping serves as a basis for multi-level ontology engineering and can be augmented and combined with further OWL axioms and ontologies. Some of the consistency criteria of m-objects/m-relationships have to be evaluated as integrity constraints. For this we rely on an existing approach [9] that allows to combine open world and closed world reasoning.

The paper is organized as follows: Section 2 introduces a running example and reviews the m-object/m-relationship approach. Section 3 shows how to map m-objects to OWL and how to check consistency of and to query multi-level models using OWL reasoners. Section 4 shows how to map m-relationships to OWL and shows how m-relationships can be queried using OWL reasoners. Section 5, which concludes the paper, briefly describes appropriate tool support.

2 Multi-level Conceptual Modeling

In [5] we introduced *multi-level objects* (m-objects) and *multi-level relationships* (m-relationships) and compared them to previous work in the field of conceptual modeling, especially materialization[3], potency-based deep instantiation[4], and powertypes [10]. In this section we review the structural definitions, consistency criteria, and query functionality of m-objects and m-relationships as a basis for the mapping to OWL, provided in Sections 3 and 4.

Example 1 (Sample Problem). A product catalog is described at three levels of abstraction: *category*, *model*, and *physical entity*. Each *product category* has associated a *tax rate*, each *product model* has a *list price*, each *physical entity* has a *serial number*. Book editions, i.e. objects at level *model* that belong to *product category book*, additionally have an *author*. In addition to *books*, our company sells products of category *car*. Cars differ from books in that they are described at an additional level, namely *brand*, and in that they have additional attributes: *maxSpeed* at level *product model* and *mileage* at level *physical entity*. Our company further keeps track of *companies* that produce these products. Companies are likewise described at multiple levels: *industrial sector*, *enterprise*, and *factory*. Producers of cars belong to a specific *industrial sector*, namely *car manufacturer*. To track quality problems, our company also associates with each *physical entity* of product category *car* the *factory* at which it was produced. Thereby, this *factory* must belong to the *enterprise*, which produces the respective *car model*. The example is shown in Fig. 1.

An m-object encapsulates and arranges abstraction levels in a linear order from the most abstract to the most concrete one. Thereby, it describes itself and the common properties of the objects at each level of the concretization hierarchy beneath itself. An m-object that concretizes another m-object, its parent, inherits all levels except for the top-level of its parent. It may specialize the inherited levels and introduce new levels. An m-object specifies concrete values for the properties of the top-level. This top-level has a special role in that it describes the m-object itself. All other levels describe common properties of m-objects beneath itself. Thus an m-object plays multiple roles. It represents an object described by factual properties and represents multiple classes, one for each level beneath itself, e.g., m-object Car represents a *product category*, and represents three classes, namely those of all *car brands*, all *car models* and all *car physical entites*. Such a class collects all descendants of the m-object at the respective level. It can be used in extensional constraints and as entry-point for queries. In this way the concretization hierarchy of m-objects represents at the same time multiple generalization hierarchies of classes, one for each level, e.g., the class of all *product models* generalizes the class of all *car models*, the class of all *product physical entities* generalizes the class of all *car physical entities*.

We now give a formal definition of m-objects and their arrangement in concretization hierarchies. An m-object $o = (L_o, A_o, p_o, l_o, d_o, v_o)$ consists of a set of levels L_o, taken from a universe of levels L, and a set of attributes A_o, taken from a universe of attributes A. The levels L_o are organized in a linear order, as defined by partial function parent $p_o : L_o \to L_o$, which associates with each level its parent level. Each attribute is associated with one level, defined by function $l_o : A_o \to L_o$, and has a domain, defined by function $d_o : A_o \to D$ (where D is a universe of data types). Optionally, an attribute has a value from its domain, defined by partial function $v_o : A_o \to V$, where V is a universe of data values, and $v_o(a) \in d_o(a)$ iff $v_o(a)$ is defined. The top-level of o is denoted by $\hat{l}_o$ and the set of attributes associated with the top-level by $\hat{A}_o$. Further, we say that an m-object is at level l if l is its top-level. By O we denote the set of

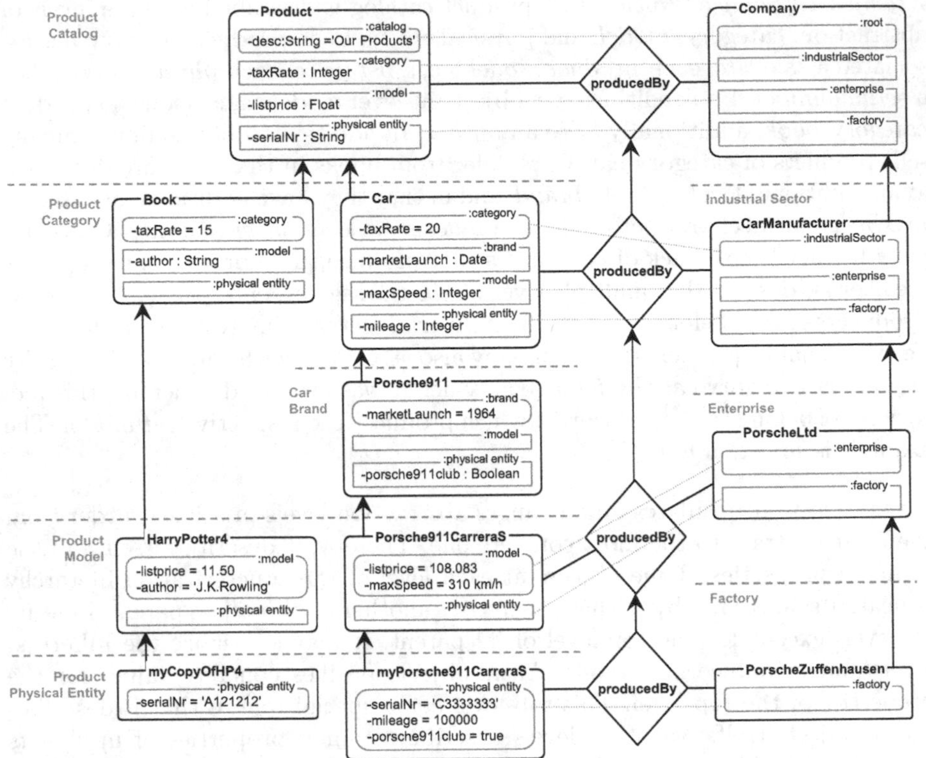

Fig. 1. Product catalog (see Example 1, only partly shown) modeled with m-objects (inherited attributes not shown) and m-relationships

m-objects. A *concretization hierarchy* of a set of m-objects O is defined by an acyclic relation $H \subseteq O \times O$. If $(o, o') \in H$ then o is a direct *concretization of* o'. By H^+ and H^* we denote the transitive, resp. transitive-reflexive, closure of H. If $(o, o') \in H^+$ then o is a direct or indirect *concretization of* o'. The class of descendant m-objects of m-object $o \in O$ at level l is denoted as $o\langle l \rangle$.

An m-object o is a *consistent concretization* of another m-object o' iff (I) each level of o', except for the top-level, is also a level of o (*level containment*), (II) the top level of o' is not a level of o, (III) all attributes of o', except for the attributes of the top-level, also exist in o (*attribute containment*), (IV) the relative order of common levels of o and o' is the same (*level order compatibility*), and (V) common attributes are associated with the same level, have the same domain, and the same value, if defined. Each level and attribute is introduced at only one m-object (*unique induction*).

M-relationships are analogous to m-objects in that they describe relationships between m-objects at multiple levels of abstraction; they have the following features: (1) M-relationships at different abstraction levels can be arranged in concretization hierarchies, similar to m-objects (e.g. *Car-producedBy-CarManufacturer*, that is m-relationship *producedBy* between *Car* and *Car-Manufacturer* (see Fig. 1), concretizes *Product-producedBy-Company*). (2) An m-relationship represents different abstraction levels of a relationship, namely one relationship occurence and multiple relationship classes. Such a relationship class collects all descending m-relationships that connect m-objects at the respective levels. E.g., m-relationship *MyPorsche911CarreraS-producedBy-PorscheZuffenhausen* is member of *Car-producedBy-CarManufacturer* at connection level *(Physical,Entity)*. (3) An m-relationship implies extensional constraints for its concretizations at multiple levels. E.g. concretizations of *producedBy* between *Car* and *CarManufacturer* only connect concretizations of *Car* and *CarManufacturer*. (4) M-relationships can cope with heterogenous hierarchies (e.g. additional level *brand* at category *Car*) and (5) M-relationships can be exploited for querying and navigating.

We now give a formal definition of m-relationships and their arrangement in concretization hierarchies. An m-relationship $r = (s_r, t_r, C_r)$ links a source m-object $s_r \in O$ with a target m-object $t_r \in O$. It connects one or more pairs of levels of source m-object s_r, $L_{s_r} \subseteq L$, to levels of target m-object t_r, $L_{t_r} \subseteq L$, as specified by $C_r \subseteq (L_{s_r} \times L_{t_r})$. The top-connection-level of an m-relationship r is given by the top-levels of its source and target, i.e. $(\hat{l}_{s_r}, \hat{l}_{t_r})$. The order of connection levels is implicitly given by the level order of its source and target.

M-relationships are organized in a concretization hierarchy, defined by an acyclic relation $H_R \subseteq R \times R$, which forms a forest (set of trees). R is the set of all m-relationships. When an m-relationship r concretizes another m-relationship r', $(r, r') \in H_R$, it does not contain the top-connection-level of r' and shares all other connection-levels of r'. It may introduce additional connection-levels for a level that is not already part of the m-objects connected by r'. The concretizes-relationship between r and r' comprises instantiation (between the top connection-level of r and the second-top-level of r' and specialization concerning all other connection levels that are shared by r and r'. M-relationship r concretizes the source object or the target object of r', or concretizes both objects related by r: $((s_r, s_{r'}) \in H^+ \wedge (t_r, t_{r'}) \in H^*) \vee ((s_r, s_{r'}) \in H^* \wedge (t_r, t_{r'}) \in H^+)$ (*source and/or target concretization*).

We allow navigating m-relationships at higher abstraction levels. By o->$r\langle l \rangle$ we denote the set of target m-objects at level l reached by traversing relationship r, or a concretization of r, from source m-object o. For example, to query which IndustrialSector produces MyPorsche911CarreraS (see Fig. 1), we write MyPorsche911CarreraS−>Product-producedBy-Company⟨IndustrialSector⟩ and get CarManufacturer as result.

3 Mapping M-Objects to OWL

The basic goals of mapping m-objects to OWL are (1) to preserve their semantics, fulfilling their basic goals and requirements, and (2) to provide an OWL-representation that allows OWL reasoners (a) to detect inconsistencies and (b) to execute queries at the different levels of abstraction. In this section we describe each step of the mapping from m-objects to OWL and exemplify the more interesting steps by showing mapping-output for m-object *Car* (see Fig. 1). The mapping of the full example is available at `http://www.dke.jku.at/research/projects/multilevel.html`. The mapping procedure is summarized in Algorithm 1. To improve readability, we use Description Logic syntax and for brevity we do not introduce entities (i.e. individuals, classes, and properties) explicitly. Also note, that Description Logic syntax does not differentiate between data properties and object properties.

Since the m-object approach makes the *closed world assumption* (CWA), which OWL does not, the mapping approach has to assure that axioms that are meant to be interpreted as integrity constraints do not lead to unwanted inferences, e.g., wrong classifications. For this we rely on an existing approach [9] that allows to designate certain TBox axioms as integrity constraints. In our mappings such axioms are marked by '`IC:`'. For TBox (schema) reasoning these axioms are treated as usual, but for ABox (data) reasoning, they are treated only as checks and do not derive additional information.

Concretization hierarchies of m-objects are mapped to OWL by representing *each m-object as individual*, e.g., Car, and *each abstraction level as primitive class*, e.g., Category. Each m-object is assigned to an abstraction level by a class assertion. Each m-object with a parent m-object is connected to this parent using functional property concretize (see Algorithm 1, lines 5 – 6). E.g., individual *Car* is member of class *Category* and conretizes *Product*:

1: Category(Car)
2: concretize(Car, Product)

Values of top-level attributes, i.e. attribute values that describe the m-object itself (a.k.a. *own-slots*), are represented as property assertions (see Algorithm 1, line 7). E.g. product category *Car* has assigned a *taxRate* of *20*:

3: taxRate(Car, 20)

As explained in Section 2, a level of an m-object is seen as a class that collects all direct and indirect concretizations of the m-object at the respective level. Such a class can be used (1) as entry point for queries, (2) to define extensional constraints, and (3) to define and refine common characteristics of its members. The class of individuals that belong to abstraction level l and that are direct or indirect concretizations of individual o, i.e. $o\langle l \rangle$, corresponds to class expression $(\exists \mathsf{concretize\_t}.\{[o]\} \sqcap [l])$. concretize_t is defined as transitive super-property of concretize. E.g., the class of all car models, $Car\langle Model \rangle$, corresponds to class expression $(\exists \mathsf{concretize\_t}.\{\mathsf{Car}\} \sqcap \mathsf{Model})$.

Thus, a concretization hierarchy implicitly introduces *multiple subsumption hierarchies*, one for each level, as explained in Section 2. (See Fig. 2 for the resulting classes and subclasses in our example). E.g., from (concretize(Car, Product)) an OWL reasoner infers that $(\exists \mathsf{concretize\_t}.\{\mathsf{Product}\} \sqcap \mathsf{Model})$ subsumes $(\exists \mathsf{concretize\_t}.\{\mathsf{Car}\} \sqcap \mathsf{Model})$ and that $(\exists \mathsf{concretize\_t}.\{\mathsf{Product}\} \sqcap \mathsf{PhysicalEntity})$

subsumes ($\exists$concretize_t.{Car} $\sqcap$ PhysicalEntity). This powerful feature of OWL facilitates inheritance between these classes as well as consistency checks.

Common characteristics of the members of a certain level of an m-object are defined by subclass axioms with the respective class expression at the left hand side. Attributes, in particular, are represented by data properties and respective value and number restrictions (see Algorithm 1, lines 8 – 9). Values of attributes might be shared with m-objects at lower levels (see Algorithm 1, line 10), no example shown. To preserve the semantics of the m-object approach, these axioms are to be interpreted as integrity constraints. E.g., *Car⟨brand⟩s* have an attribute marketLaunch, *Car⟨model⟩s* have an attribute maxSpeed (in addition to attribute listPrice from *Product⟨model⟩*), and *Car⟨physicalEntity⟩s* have an attribute mileage (in addition to serialNr from *Product⟨physicalEntity⟩*):

4: IC:$\exists$concretize_t.{Car} $\sqcap$ Brand $\sqsubseteq$ $\forall$marketLaunch.Date $\sqcap$ =1 marketLaunch.$\top$
5: IC:$\exists$concretize_t.{Car} $\sqcap$ Model $\sqsubseteq$ $\forall$maxSpeed.Integer $\sqcap$ =1 maxSpeed.$\top$
6: IC:$\exists$concretize_t.{Car} $\sqcap$ PhysicalEntity $\sqsubseteq$ $\forall$mileage.Integer $\sqcap$ =1 mileage.$\top$

A level l of an m-object o ensures that concretizations of o at lower levels also concretize a concretization of o at level l (see Algorithm 1, lines 11 – 12). This allows *stable upward navigation* and supports *heterogenous level hierarchies* by allowing that every m-object may introduce new abstraction levels for its descendants, that do not apply for descendants of other m-objects. E.g., all *Car Models* belong to a *Car Brand*:

7: IC:$\exists$concretize_t.{Car} $\sqcap$ Model $\sqsubseteq$ $\exists$concretize_t.($\exists$concretize_t.{Car} $\sqcap$ Brand)

The mapping also ensures that each attribute is inducted at only one level of one m-object (see Algorithm 1, lines 13 – 14), and that each level is inducted at only one m-object (see Algorithm 1, lines 15 – 16); no examples shown.

To ensure that an m-object belongs to one abstraction level at most, all levels are pairwise disjoint (see Algorithm 1, line 17). E.g., an individual at level *Brand* cannot, at the same time, be at level *Model*:

8: Brand $\sqcap$ Model $\sqsubseteq$ $\bot$

The m-object approach makes the *unique name assumption*. Thus we state that each pair of m-objects (cartesian product) is a member of the inequality predicate (see Algorithm 1, line 18). E.g., the symbols Car and Porsche911 refer to different individuals:

9: Car $\not\approx$ Porsche911

4 Mapping M-Relationships to OWL

There are basically two alternative representations of m-relationships in OWL, as properties or as individuals. At first sight a mapping to property assertions (*property approach*) seems intuitive. However, since property assertions do not have identifiers it is not possible to directly represent concretization links. Thus

Algorithm 1. Mapping M-Objects and M-Relationships to OWL

Input: a set O of m-objects $o = (L_o, A_o, p_o, l_o, d_o, v_o)$, a concretization hierarchy H, and a universe of levels L, as described on page 191. A set R of m-relationships $r = (s_r, t_r, C_r)$ and a concretization hierarchy H_R, as described on page 193.
Output: a set of OWL axioms.

▷ *Mapping M-Objects to OWL*

assert: $\top \sqsubseteq \leqslant 1\ \mathsf{concretize}$
assert: $\mathsf{concretize} \sqsubseteq \mathsf{concretize\_t}$
assert: $\mathsf{concretize\_t}^{+} \sqsubseteq \mathsf{concretize\_t}$
for all $o \in O$ **do**
  assert: $[\hat{l}_o]([o])$
  if $\exists o' : (o, o') \in H$ **then assert:** $\mathsf{concretize}([o],[o'])$
  for all $a \in \hat{A}_o : v_o(a)$ is defined **do assert:** $[a]([o],[v_o(a)])$
  for all $a \in (A_o \setminus \hat{A}_o)$ **do**
    assert IC: $\exists \mathsf{concretize}.\{[o]\} \sqcap [l_o(a)] \sqsubseteq \forall [a].[d_o(a)] \sqcap {=}1\ [a].\top$
    if $v_o(a)$ is defined **then assert:** $[l_o(a)] \sqcap \exists \mathsf{concretize}.\{[o]\} \sqsubseteq \exists [a].\{[v_o(a)]\}$
  for all $(l, l') \in P_o : l' \neq \hat{l}_o \wedge (\nexists o' \in O : (o, o') \in H \wedge (l, l') \in P_{o'})$ **do**
    assert IC: $\exists \mathsf{concretize\_t}.\{[o]\} \sqcap [l] \sqsubseteq \exists \mathsf{concretize\_t}.(\exists \mathsf{concretize\_t}\{[o]\} \sqcap [l'])$
  for all $a \in A_o : \nexists o' \in O : (o, o') \in H \wedge a \in A_{o'}$ **do**
    assert IC: $\exists [a].\top \sqsubseteq (\exists \mathsf{concretize\_t}.\{[o]\} \sqcup \{[o]\}) \sqcap [l_o(a)]$
  for all $l \in L_o : l \neq \hat{l}_o \wedge (\nexists o' \in O : (o, o') \in H \wedge l \in L_{o'})$ **do**
    assert IC: $[l] \sqsubseteq \exists \mathsf{concretize\_t}.\{[o]\}$
for all $l \in L, l' \in (L \setminus \{l\})$ **do assert:** $[l] \sqcap [l'] \sqsubseteq \bot$
for all $o \in o, o' \in (O \setminus \{o\})$ **do assert:** $[o] \not\approx [o']$

▷ *Mapping M-Relationships to OWL*

assert: $\top \sqsubseteq \leqslant 1 source \sqcap \leqslant 1 target$
for all $r \in R$ **do**
  if $\exists r' : (r, r') \in H_R$ **then assert:** $\mathsf{concretize}([r],[r'])$
  assert: $\mathsf{source}([r],[s_r])$
  assert: $\mathsf{target}([r],[t_r])$
  assert IC: $\exists \mathsf{concretize\_t}.\{[r]\} \sqsubseteq (\forall \mathsf{source}.(\exists \mathsf{concretize\_t}.\{[s_r]\} \sqcup \{[s_r]\}) \sqcap \forall \mathsf{target}.\exists \mathsf{concretize\_t}.\{[t_r]\}) \sqcup (\forall \mathsf{source}.\exists \mathsf{concretize\_t}.\{[s_r]\} \sqcap \forall \mathsf{target}.(\exists \mathsf{concretize\_t}.\{[t_r]\} \sqcup \{[t_r]\}))$
  for all $(l, l') \in C_r : l \neq \hat{l}_{s_r} \vee l' \neq \hat{l}_{t_r}$ **do**
    assert IC: $\exists \mathsf{concretize\_t}.\{[r]\} \sqcap (\exists \mathsf{source}.\exists \mathsf{concretize\_t}.[l] \sqcup \exists \mathsf{target}.\exists \mathsf{concretize\_t}.[l']) \sqsubseteq \exists \mathsf{concretize\_t}.(\exists \mathsf{concretize\_t}.\{[r]\} \sqcap \exists \mathsf{source}.[l] \sqcap \exists \mathsf{target}.[l'])$
  for all $r' \in R : r \neq r'$ **do assert:** $[r] \not\approx [r']$

Notation: '$[o]$' denotes a variable that is to be substituted by its actual value. '**assert:** ' adds the subsequent OWL axiom to the mapping output. '`IC:`' denotes an OWL axiom that is to be interpreted as integrity constraint.

it would be necessary to represent each connection-level of an m-relationship as property and redundantly represent a concretization-link between two m-relationships by several sub-property-axioms. To avoid this redundancy we

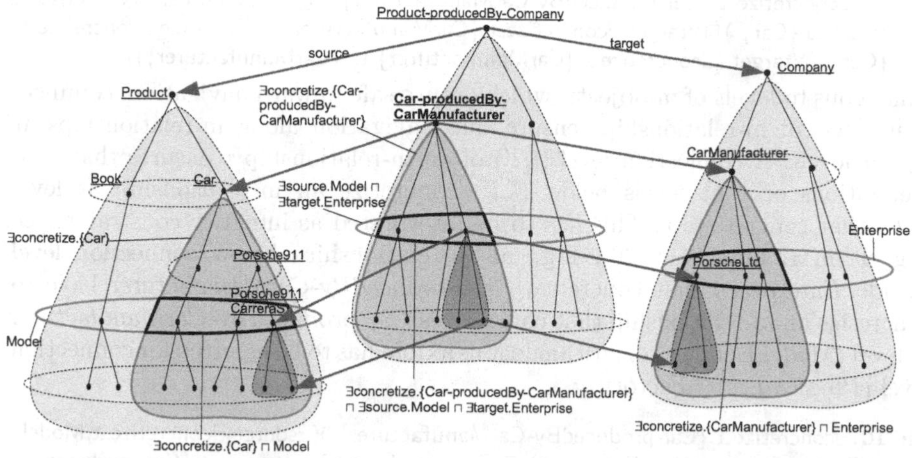

Fig. 2. Sets and subsets of m-objects and m-relationships. M-relationship *Car-producedBy-CarManufacturer* and its members, domain, and range at level *(model, enterprise)* are emphasized.

suggest to map m-relationships to individuals (*objectification approach*) which allows to directly represent concretization-links between m-relationships. We again describe each step of the mapping and exemplify them based on m-relationship *producedBy* between *Car* and *CarManufacturer* from Example 1 (see Fig. 1). The mapping procedure is summarized in Algorithm 1 from line 19 onwards.

The *objectification approach* represents each m-relationship as individual that is linked to its parent relationship and to its source- and target-m-objects by property assertions, using functional properties concretize, source, and target, respectively (see Algorithm 1, lines 21 – 23). E.g., m-relationship *producedBy* between *Car* and *CarManufacturer*, named *Car-producedBy-CarManufacturer*, concretizes m-relationship *Product-producedBy-Company*, its source is *Car* and its target is *CarManufacturer*:

```
concretize(Car-producedBy-CarManufacturer, Product-producedBy-Company)
source(Car-producedBy-CarManufacturer, Car)
target(Car-producedBy-CarManufacturer, CarManufacturer)
```

A m-relationship constrains domain and range of its concretizations as defined in Section 2 (*source- or target level concretization*) (see Algorithm 1, line 24). E.g., all direct or indirect concretizations of *Car-producedBy-CarManufacturer* have a direct or indirect concretization of *Car*, or *Car* itself, as source, and a direct or indirect concretization of *CarManufacturer*, or *CarManufacturer* itself, as target. Either *Car* or *CarManufacturer* must be concretized:

13: IC: $\exists$concretize_t.{Car-producedBy-CarManufacturer} $\sqsubseteq$ ($\forall$source.($\exists$concretize_t. {Car} $\sqcup$ {Car}) $\sqcap\forall$target.$\exists$concretize_t.{CarManufacturer})$\sqcup$ ($\forall$source.$\exists$concretize_t. {Car} $\sqcap\forall$target.($\exists$concretize_t.{CarManufacturer} $\sqcup$ {CarManufacturer}))

Analogous to levels of m-objects, which ensure safe upward navigation, connnection levels of m-relationships ensure safe navigation along m-relationships at higher levels. A connection level (l, l') of an m-relationship r ensures that concretizations of r at levels below (l, l') concretize an m-relationship at level (l, l') that concretizes r. This has to be interpreted as integrity constraint (see Algorithm 1, lines 25 – 26). E.g., all m-relationships below connection level *(Model,Enterprise)* that concretize *Car-producedBy-CarManufacturer* have to concretize an m-relationship that concretizes *Car-producedBy-CarManufacturer* at level *(Model,Enterprise)* (An analogous axiom has to be asserted for connection level *(PhysicalEntity,Factory)*):

14: IC: $\exists$concretize_t.{Car-producedBy-CarManufacturer}$\sqcap$($\exists$source.$\exists$concretize_t.Model $\sqcup$ $\exists$target.$\exists$concretize_t.Enterprise) $\sqsubseteq$ $\exists$concretize_t.($\exists$concretize_t.{Car-producedBy-CarManufacturer} $\sqcap\exists$source.Model $\sqcap\exists$target.Enterprise)

To enforce the *unique name assumption* we assert that each pair of m-relationships (cartesian product) belongs to the inequality predicate (see Algorithm 1, line 27). E.g., *Car-producedBy-CarManufacturer* and *Product-producedBy-Company* refer to different individuals:

15: Car-producedBy-CarManufacturer $\not\approx$ Product-producedBy-Company

OWL reasoners can be employed to navigate m-relationships (as defined in Sect. 2) by class expressions. For example, to query which IndustrialSector might produce MyPorsche911CarreraS (see Fig. 1) we ask for the members of class (IndustrialSector $\sqcap$ $\exists$target$^-$.(($\exists$concretize_t.{Product-producedBy-Company} $\sqcup$ {Product-producedBy-Company}) $\sqcap$ $\exists$source. ($\exists$concretize_t$^-$.{MyPorsche911CarreraS} $\sqcup$ {MyPorsche911CarreraS}))).

5 Conclusion

In this paper we showed how to apply the m-object/m-relationship approach to ontology engineering with OWL. To represent integrity constraints (closed world constraints) we applied an existing approach [9] that allows to combine open world and closed world reasoning.

In regard to tool support, we are extending the ontology editor Protégé with a plugin for modeling with m-objects and m-relationships. Multilevel-models designed with this Protégé extension can then - using an additional export plug-in - be mapped to OWL axioms (and integrity constraints) as described in the previous chapters. The result of this mapping may be freely augmented with open world OWL axioms and thereby integrated with ordinary OWL ontologies. The interested reader is referred to the project website at `http://www.dke.jku.at/research/projects/multilevel.html`.

References

1. Atkinson, C., Kühne, T.: Model-Driven Development: A Metamodeling Foundation. IEEE Software 20(5), 36–41 (2003)
2. Motik, B.: On the properties of metamodeling in OWL. In: Gil, Y., Motta, E., Benjamins, V.R., Musen, M.A. (eds.) ISWC 2005. LNCS, vol. 3729, pp. 548–562. Springer, Heidelberg (2005)
3. Pirotte, A., Zimányi, E., Massart, D., Yakusheva, T.: Materialization: A powerful and ubiquitous abstraction pattern. In: VLDB, pp. 630–641 (1994)
4. Atkinson, C., Kühne, T.: The essence of multilevel metamodeling. In: Gogolla, M., Kobryn, C. (eds.) UML 2001. LNCS, vol. 2185, pp. 19–33. Springer, Heidelberg (2001)
5. Neumayr, B., Grün, K., Schrefl, M.: Multi-level domain modeling with m-objects and m-relationships. In: APCCM (2009)
6. Gasevic, D., Djuric, D., Devedzic, V.: Bridging MDA and OWL ontologies. J. Web Eng. 4(2), 118–143 (2005)
7. Guizzardi, G.: On Ontology, ontologies, conceptualizations, modeling languages, and (meta)models. In: DB&IS 2006, pp. 18–39 (2006)
8. Borgida, A., Brachman, R.J.: Conceptual modeling with description logics. In: Description Logic Handbook, pp. 349–372. Cambridge University Press, Cambridge (2003)
9. Motik, B., Horrocks, I., Sattler, U.: Bridging the gap between OWL and relational databases. In: WWW 2007 (2007)
10. Odell, J.J.: Power Types. In: Advanced Object-Oriented Analysis & Design Using UML, pp. 23–32. Cambridge University Press, Cambridge (1998)

Preface to QoIS 2009

Isabelle Comyn-Wattiau[1] and Bernhard Thalheim[2]

[1] CEDRIC-CNAM, France
[2] Bernhard Thalheim - Kiel University, Germany

Quality assurance is a growing research domain within the Information Systems (IS) and Conceptual Modeling (CM) disciplines. Ongoing research on quality in IS and CM is highly diverse and encompasses theoretical aspects including quality definition and quality models, and practical/empirical aspects such as the development of methods, approaches and tools for quality measurement and improvement. Current research on quality also includes quality characteristics definitions, validation instruments, methodological and development approaches to quality assurance during software and information systems development, quality monitors, quality assurance during information systems development processes and practices, quality assurance both for data and (meta)schemata, quality support for information systems data import and export, quality of query answering, and cost/benefit analysis of quality assurance processes. Quality assurance is also depending on the application area and the specific requirements in applications such as health sector, logistics, public sector, financial sector, manufacturing, services, e-commerce, software, etc. Furthermore, quality assurance must also be supported for data aggregation, ETL processes, web content management and other multi-layered applications. Quality assurance is typically requiring resources and has therefore beside its benefits a computational and economical trade-off. It is therefore also based on compromising between the value of quality data and the cost for quality assurance.

The QoIS workshop covers all areas related to information systems quality: data quality, information quality, system quality as well as model, method, process, knowledge and environment quality. The aim of the workshop is twofold. Firstly, to bring together researchers and industry developers working on various aspects of information systems quality, in order to exchange research ideas and results and discuss about them. Secondly, to promote research on information systems and conceptual model quality to the broader conceptual modeling research community attending ER 2009.

The workshop continues a tradition of quality-focused workshops at the ER conferences that started with the International Workshops on Conceptual Modeling Quality (IWCMQ'02 at ER 2002 Tampere, IWCMQ'03 at ER 2003 Chicago) and was picked up again by the International Workshops on Quality of Information Systems (QoIS'05 at ER 2005 Klagenfurt, QoIS'06 at ER 2006 Tucson, QoIS'07 at ER2007 Auckland). A result of these workshops was the Data & Knowledge Engineering special issue on conceptual model quality (Vol 55, No 3, 2005).

We selected for the workshop four papers from fourteen submitted. We are sure that these papers reflect in a very good form the current state of the art. We thank the

C.A. Heuser and G. Pernul (Eds.): ER 2009 Workshops, LNCS 5833, pp. 200–201, 2009.

program committee members and additional reviewers for their support in evaluating the papers submitted to QoIS'09. We are very thankful to the ER'09 organisation team for taking care of workshop proceedings.

Last but not least we thank the participants of QoIS'09 for having made our work useful.

Completeness in Databases with Maybe-Tuples

Fabian Panse and Norbert Ritter

University of Hamburg, Vogt-Kölln Straße 33, 22527 Hamburg, Germany
{panse,ritter}@informatik.uni-hamburg.de
http://vsis-www.informatik.uni-hamburg.de/

Abstract. Some data models use so-called *maybe tuples* to express the uncertainty, whether or not a tuple belongs to a relation. In order to assess this relation's quality the corresponding vagueness needs to be taken into account. Current metrics of quality dimensions are not designed to deal with this uncertainty and therefore need to be adapted. One major quality dimension is data completeness. In general, there are two basic ways to distinguish *maybe tuples* from *definite tuples*. First, an attribute serving as a maybe indicator (values YES or NO) can be used. Second, tuple probabilities can be specified. In this paper, the notion of data completeness is redefined w.r.t. both concepts. Thus, a more precise estimating of data quality in databases with *maybe tuples* (e.g. probabilistic databases) is enabled.

Keywords: data completeness, maybe tuple, probabilistic database.

1 Introduction

Since in databases using the three-valued logic uncertain query results can appear (e.g. resulting from operations on null values), in some cases, it is not exactly known whether a tuple belongs to a query result set or not. For indicating *possible* result tuples several data models ([1], [2] et al.) use the concept of *maybe tuples*. Additionally, as a consequence of a poor information elicitation, sometimes it is not clear, whether a tuple belongs to a database relation or not. For modeling these cases *maybe tuples* can be used, too. Besides a simple indication of *maybe tuples* a more exact specification by individual tuple probabilities as it is known from probabilistic databases (e.g. a tuple belongs to a relation with a certainty of 70 percent) is possible ([9], [3] et al.). Altogether, both types of models enable the indication of tuples which may belong to a relation with less confidence.

For estimating a database's quality, for example in order to compare different databases containing information on the same issue, in the last years various data quality dimensions have been defined. Current metrics of these dimensions do not consider the uncertainty represented by *maybe tuples*. Thus, for corresponding databases some of these metrics are insufficient. Since, data completeness is one of the relevant quality dimensions, in this paper new completeness metrics with respect to the *maybe tuple* concept are defined.

Generally, we consider completeness from a theoretical point of view and try to define it as precise and exact as possible. In reality, often some required

C.A. Heuser and G. Pernul (Eds.): ER 2009 Workshops, LNCS 5833, pp. 202–211, 2009.

information are not available and more approximate and hence more imprecise methods have to be used. Since, such a practical point of view is out of the scope of this paper, it will be considered in future work.

The paper is structured as follows: In Section 2 related work is examined. Furthermore, we discuss and correct deficiencies of current data completeness metrics w.r.t. relations without *maybe tuples*. After presenting relations with *maybe tuples* in more detail (Section 3), we introduce three approaches for extending the corrected metrics to relations with *maybe tuples* (for simple maybe indications as well as for individual tuple probabilities) in Section 4. A final comparison relates these metrics to each other and points out the most suitable one. Section 5 summarizes the paper and gives an outlook to future work.

2 Related Work

Metrics of data completeness are considered in different works (Scannapieco ([7]), Naumann ([6]), Motro ([5]) et al.), but none of them regards the uncertainty resulting from *maybe tuples*. In [6], data completeness is composed by the two measures data coverage and data density[1]. Data coverage represents the completeness of the extension and is the ratio of all stored to all actually existing entities of the modeled world. Therefore w.r.t. a single relation $\mathcal{R}$, the coverage $c(\mathcal{R})$ is the ratio of all tuples of this relation to the number of entities of the corresponding entity type $\mathcal{E}$ (equation 1). Data density represents the completeness of the stored entities (intension) and can be considered at different levels of granularity (e.g. attribute value, tuple, relation). The density of an attribute value measures the information content of this value with respect to its maximal potential information content. In existing approaches (e.g. [6]) this density is either 1 if the value is specified, or 0 if it is a null value, but another value density is possible if partial information is respected ($\Rightarrow d(t) \in [0,1]$). The density $d(t)$ of a tuple t is the average of its values' densities and the density $d(\mathcal{R})$ of a relation $\mathcal{R}$ which, in turn, is the average of its tuples' densities (equation 2).

$$c(\mathcal{R}) = \frac{|\mathcal{R}|}{|\mathcal{E}|} \qquad (1) \qquad\qquad d(\mathcal{R}) = \frac{\sum_{t\in\mathcal{R}} d(t)}{|\mathcal{R}|} \qquad (2)$$

Using these two measures, the data completeness of $\mathcal{R}$ results in:

$$comp(\mathcal{R}) = c(\mathcal{R}) \cdot d(\mathcal{R}) = \frac{\sum_{t\in\mathcal{R}} d(t)}{|\mathcal{E}|} \qquad (3)$$

2.1 Metric Deficiencies

As we show by the following example, the metrics given above (equations 1-3) are deficient for relations containing tuples which do not represent an entity of the corresponding entity type: A company is assumed to have 10 employees currently.

[1] Since this decomposition increases the interpretability of completeness, we adapt the metrics defined by Naumann in the following.

Thus, a relation *employee* contains one tuple for each of them. Additionally, the relation contains a tuple for an employee who was fired last month. Resulting from a failure of the responsible secretary, the tuple has not been deleted by now. Calculating the coverage of *employee* by equation 1, $c(emplyoee) = 11/10 = 1.1$ results. Usually, quality metrics are normalized and hence a quality value has always to be within the range $[0, 1]$. Since normalization is one of the most important requirements for an adequate quality metric ([4]), this is a deficiency which must not be underrated.

Furthermore, if completeness is used to compare two or more data sources an unsound source[2] can mistakenly be regarded as the best source. For avoiding such errors only the tuples which correctly belong to the relation have to be considered (see [5]). Given $\mathcal{R}$ is the regarded relation, and $\mathcal{E}$ is the entity type which is represented by this relation, and $m : \mathcal{E} \rightarrow \mathcal{R}$ is the mapping of the entities (extension) of $\mathcal{E}$ on tuples of $\mathcal{R}$, the relation $\mathcal{R}_C(\mathcal{E})$ (short $\mathcal{R}_C$) contains all tuples which correctly belong to $\mathcal{R}$ w.r.t. the entity type $\mathcal{E}$.

$$\mathcal{R}_C(\mathcal{E}) = \{t \mid t \in \mathcal{R} \wedge (\exists e \in \mathcal{E}) : m(e) = t\}$$

Considering this 'tuple cleaning', the metrics of data coverage $c(\mathcal{R})$ and data density $d(\mathcal{R})$ have to be adapted to:

$$c(\mathcal{R}) = \frac{|\mathcal{R}_C|}{|\mathcal{E}|} \quad (4) \qquad\qquad d(\mathcal{R}) = \frac{\sum_{t \in \mathcal{R}_C} d(t)}{|\mathcal{R}_C|} \quad (5)$$

The metric of data completeness (equation 3) has to be adapted accordingly.

3 Relations with Maybe-Tuples

In contrast to *definite tuples*, as the name already says, *maybe tuples* are tuples for which it is undefined whether they belong to the associated relation or not. *Maybe tuples* can appear in database relations as well as in (intermediate) query result sets. The appearance in database relations can be traced back to a poor information elicitation. Sometimes from the available information it cannot be certainly concluded whether an entity is part of the extension of an entity type or not. As a consequence, for representing this uncertainty, the associated tuple can neither be exlcuded from nor included into the corresponding database relation. Thus, these tuples have to be indicated as 'maybe' (see attribute M of relation $\mathcal{R}_2$ in Figure 1). In addition, if a database contains null values or values which represent partial information (e.g. interval values), during query evaluation some tuples cannot be evaluated to TRUE or FALSE. In such cases, it cannot be determined, whether or not the query condition is satisfied. Thus, these tuples are *possible* query results and have to be indicated as *maybe tuples*, too.

A relation $\mathcal{R}$ with *maybe tuples* (in the following denoted as *maybe relation*) can be lossless divided into two subrelations ($\mathcal{R} = \mathcal{R}^{\mathcal{D}} \cup \mathcal{R}^{\mathcal{M}}$): Relation $\mathcal{R}^{\mathcal{D}}$

[2] A source containing many tuples which do not correctly belong to the corresponding source's relation.

contains all tuples which definitely belong to $\mathcal{R}$ and relation $\mathcal{R}^{\mathcal{M}}$ contains all tuples which may be belong to $\mathcal{R}$. If $\mathcal{R}$ does not contain duplicates (e.g. if $\mathcal{R}$ is a database relation), the two subsets have to be disjunct ($\mathcal{R}^{\mathcal{D}} \cap \mathcal{R}^{\mathcal{M}} = \emptyset$).

The individual tuple probability $p(t)_{\mathcal{R}}$ for a tuple t of the relation $\mathcal{R}$ is defined as the probability that this tuple belongs to the associated relation. Since all tuples of the subrelation $\mathcal{R}^{\mathcal{D}}$ are definitely in $\mathcal{R}$, the individual tuple probabilities of these tuples always have to be 1. Since every *maybe tuple* only possibly belongs to the relation, its individual tuple probability has to be lower than 1. However, because these tuples can certainly not be excluded from this relation, the individual tuple probability has to be within the range $]0, 1[$.

In the following, $\mathcal{S}(\mathcal{R})$ represents the set of all possible instances and $\mathcal{R}'$ represents the real instance of the relation $\mathcal{R}$ under a closed world assumption[3]. Since all tuples of $\mathcal{R}^{\mathcal{D}}$ definitely belong to $\mathcal{R}$, each possible instance of $\mathcal{R}$ contains these tuples. In general, for every possible combination of the *maybe tuples* (the power set $(\mathcal{P}(\mathcal{R}^{\mathcal{M}}))$) one possible instance of $\mathcal{R}$ results:

$$\mathcal{S}(\mathcal{R}) = \{\mathcal{R}^{\mathcal{D}} \cup M \mid M \in \mathcal{P}(\mathcal{R}^{\mathcal{M}})\} \tag{6}$$

If $\mathcal{R}$ does not contain *maybe tuples*, all tuples of $\mathcal{R}$ are known and the real set of tuples belonging to $\mathcal{R}$ is completely described by $\mathcal{R}$ itself. As a consequence, $\mathcal{S}(\mathcal{R})$ contains just one element and the relations $\mathcal{R}$ and $\mathcal{R}'$ are equal. If, in contrast, $\mathcal{R}$ contains *maybe tuples*, the set of tuples which really belong to $\mathcal{R}$ and hence the relation $\mathcal{R}'$ are not completely known. This uncertainty can be represented by a discrete probability distribution of $\mathcal{R}'$ on the set $\mathcal{S}(\mathcal{R})$. For example, we assume a relation $\mathcal{R}$ containing one *definite tuple* t_1 and one *maybe tuple* t_2 ($p(t_2)_{\mathcal{R}} = 0.6$). The set of all possible instances is $\mathcal{S}(\mathcal{R}) = \{S_0 = \{t_1\}, S_1 = \{t_1, t_2\}\}$ and the real instance $\mathcal{R}'$ is distributed over $\mathcal{S}(\mathcal{R})$ with the probability distribution $P(\mathcal{R}' = S_0) = 0.4$ and $P(\mathcal{R}' = S_1) = 0.6$.

4 Data Completeness Regarding Maybe-Tuples

Since a *maybe tuple* only possibly belongs to a relation, for measuring data completeness this imprecision has to be taken into account. In order to demonstrate this necessity, we consider the three relations $\mathcal{R}_1$, $\mathcal{R}_2$ and $\mathcal{R}_3$ as illustrated in Figure 1. $\mathcal{R}_1$ and $\mathcal{R}_3$ are relations without *maybe tuples* containing 2 or 3 tuples respectively. Relation $\mathcal{R}_2$ contains two *definite* (the same tuples as $\mathcal{R}_1$) and one *maybe tuple*. It is obvious that the completeness of $\mathcal{R}_2$ has to be greater than the completeness of $\mathcal{R}_1$. The uncertain membership of t_3 to $\mathcal{R}_2$ is also a kind of incomplete information. Since this incompleteness can influence the output of a quality driven query answering, it is also comprehensible that the completeness of $\mathcal{R}_2$ has to be smaller than the completeness of $\mathcal{R}_3$. As a consequence, the completeness of $\mathcal{R}_2$ can be limited to $comp(\mathcal{R}_1) < comp(\mathcal{R}_2) < comp(\mathcal{R}_3)$.

[3] Totally missing tuples are ignored and uncertain memberships of *maybe tuples* are the only incomplete information. Thus, w.r.t. the calculation of all possible instances only the tuples of $\mathcal{R}^{\mathcal{D}}$ and $\mathcal{R}^{\mathcal{M}}$ are considered.

	firstname	surname
t_1	Georg	Washington
t_2	Abraham	Lincoln

$\mathcal{R}_1$

	firstname	surname	M
t_1	Georg	Washington	NO
t_2	Abraham	Lincoln	NO
t_3	Theodor	Roosevelt	YES

$\mathcal{R}_2$

	firstname	surname
t_1	Georg	Washington
t_2	Abraham	Lincoln
t_3	Theodor	Roosevelt

$\mathcal{R}_3$

Fig. 1. Completeness classification of *maybe relations*

In order to calculate an exact value for the completeness of a *maybe relation*, we introduce three different but each intuitive approaches. The first one uses the average completeness of the subrelations which can result from a so-called α-selection, the second one is based on the expectation value of the completeness of the relation's real instance, and the last one considers the uncertainty of *maybe tuples* as a lower priority. Partially, we trace our new metrics to the current ones. For distinction, the newly defined metrics of completeness, coverage and density with respect to a relation $\mathcal{R}$ and an approach Ai are denoted as $comp'_{Ai}(\mathcal{R})$, $c'_{Ai}(\mathcal{R})$ and $d'_{Ai}(\mathcal{R})$.

4.1 Approach 1 (α-Selection)

The first approach is based on the α-selection introduced by Tseng ([9]). An α-selection ($\hat{\sigma}^{\alpha}(\mathcal{R})$) selects each tuple $t \in \mathcal{R}$ which belongs to $\mathcal{R}$ with a probability $p(t)_{\mathcal{R}}$ greater or equal than $\alpha \in [0, 1]$:

$$\hat{\sigma}^{\alpha}(\mathcal{R}) = \{t \mid t \in \mathcal{R} \wedge p(t)_{\mathcal{R}} \geq \alpha\} \tag{7}$$

If an α-selection is used for a probability based tuple filtering, the completeness of the resulting subrelation depends on the value α. Since the higher α the more tuples are filtered, the completeness $comp(\hat{\sigma}^{\alpha}(\mathcal{R}))$ is monotonically decreasing (see Figure 2). Additionally, the completeness of a filtered relation $\hat{\sigma}^{\alpha}(\mathcal{R})$ is always greater or equal than the completeness of $\mathcal{R}^{\mathcal{D}}$ and always smaller or equal than the completeness of $\mathcal{R}$ if maybe indications are ignored ($\alpha = 0$).

One intuitive possibility is to esteem the completeness of a *maybe relation* $\mathcal{R}$ as the average completeness of the subrelations resulting from all possible α-selections on $\mathcal{R}$.

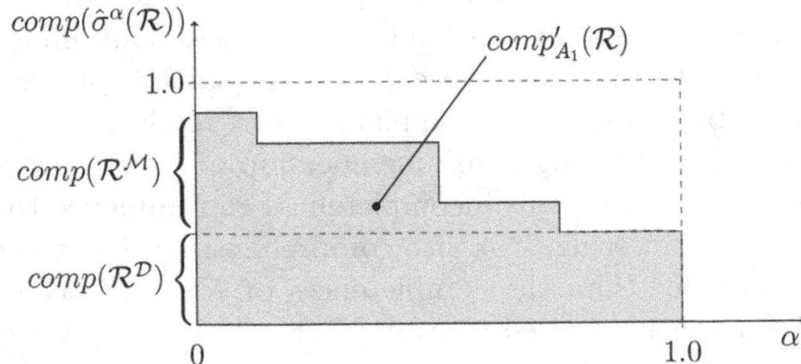

Fig. 2. Completeness of a *maybe relation* $\mathcal{R}$ w.r.t. all possible α-selections

Individual Tuple Probability: If individual tuple probabilities are given, for each α another subrelation can result from applying an α-selection. Thus, α has to be considered within the continuous range $[0,1]$ and the completeness $comp'_{A1}(\mathcal{R})$ can be defined as the integral of $comp(\hat{\sigma}^{\alpha}(\mathcal{R}))$ over α (see gray area in Figure 2):

$$comp'_{A1}(\mathcal{R}) = \int_0^1 comp(\hat{\sigma}^{\alpha}(\mathcal{R}))d\alpha \tag{8}$$

Since the coverage and the density of each subrelation are not independent of each other, by using this approach a decomposition into these two measures is not possible:

$$\int_0^1 c(\hat{\sigma}^{\alpha}(\mathcal{R})) \cdot d(\hat{\sigma}^{\alpha}(\mathcal{R}))d\alpha \neq \int_0^1 c(\hat{\sigma}^{\alpha}(\mathcal{R}))d\alpha \cdot \int_0^1 d(\hat{\sigma}^{\alpha}(\mathcal{R}))d\alpha \tag{9}$$

Simple Maybe Indication: Intuitively, in the simple case, the tuple probability of each *maybe tuple* is assumed to be 0.5. Therefore, from applying α-selections only two subrelations can result: the subrelation $\mathcal{R}^{\mathcal{D}}$, if α is within the range $]0.5, 1]$, and the whole relation $\mathcal{R} = \{\mathcal{R}^{\mathcal{D}} \cup \mathcal{R}^{\mathcal{M}}\}$ otherwise. Consequently, the completeness $comp'_{A1}(\mathcal{R})$ defined in equation 8 can be simplified to:

$$\begin{aligned} comp'_{A1}(\mathcal{R}) &= \int_0^{0.5} comp(\{\mathcal{R}^{\mathcal{D}} \cup \mathcal{R}^{\mathcal{M}}\})d\alpha + \int_{0.5}^1 comp(\mathcal{R}^{\mathcal{D}})d\alpha \\ &= comp(\mathcal{R}^{\mathcal{D}}) + \frac{1}{2}comp(\mathcal{R}^{\mathcal{M}}) \end{aligned} \tag{10}$$

4.2 Approach 2 (Expectation Value)

Another illustrative way is to calculate the completeness of $\mathcal{R}$ by using the expectation value of the completeness of $\mathcal{R}'$. As for approach 1, a decomposition of completeness into coverage and density is not possible:

$$comp'_{A3}(\mathcal{R}) = E(comp(\mathcal{R}')) = E(c(\mathcal{R}') \cdot d(\mathcal{R}')) \neq E(c(\mathcal{R}')) \cdot E(d(\mathcal{R}'))$$

Individual Tuple Probability: Defining the completeness of $\mathcal{R}$ as the expectation value of $comp(\mathcal{R}')$, the completeness[4] and probability for every possible instance of $\mathcal{R}$ have to be known.

$$\begin{aligned} E(comp(\mathcal{R}')) &= \sum_{S_i \in \mathcal{S}(\mathcal{R}_{\mathcal{C}})} P(\mathcal{R}'_{\mathcal{C}} = S_i)\, comp(S_i) \\ &= \frac{1}{|\mathcal{E}|} \sum_{S_i \in \mathcal{S}(\mathcal{R}_{\mathcal{C}})} P(\mathcal{R}'_{\mathcal{C}} = S_i) \sum_{t \in S_i} d(t) \end{aligned} \tag{11}$$

[4] Since every possible instance S_i has to be handled as a relation without *maybe tuples*, for calculating completeness the metric $comp(S_i)$ can be used.

The probability of a possible instance $S_i \in \mathcal{S}(\mathcal{R}_C)$ results from the product of the tuple probabilities of all tuples in S_i and the inverse probabilities of all tuples of $\mathcal{R}_C$ not in S_i.

$$P(\mathcal{R}'_C = S_i) = \prod_{t \in S_i} p(t)_\mathcal{R} \prod_{t \in \{\mathcal{R}_C \setminus S_i\}} (1 - p(t)_\mathcal{R})$$

Simple Maybe Indication: In the simple case, the possible instances are uniformly distributed. Thus, there exist $|\mathcal{S}(\mathcal{R}_C)| = |\mathcal{P}(\mathcal{R}^\mathcal{M})| = 2^{|\mathcal{R}_C^\mathcal{M}|}$ possible instances, and the expectation value $E(comp(\mathcal{R}'))$ and hence the completeness $comp'_{A3}(\mathcal{R})$ defined in equation 11 can be simplified to:

$$comp'_{A3}(\mathcal{R}) = E(comp(\mathcal{R}')) = \frac{1}{2^{|\mathcal{R}_C^\mathcal{M}|}} \frac{1}{|\mathcal{E}|} \sum_{S_i \in \mathcal{S}(\mathcal{R}_C)} \sum_{t \in S_i} d(t) \tag{12}$$

4.3 Approach 3 (Tuple Priorities)

In the third approach, the uncertainty resulting from *maybe tuples* is expressed by specifying lower priorities for subrelation $\mathcal{R}^\mathcal{M}$ (simple case) or each individual *maybe tuple* (exact case) respectively. In contrast to the first two approaches, completeness here can be decomposed into coverage and density. In the case of a simple maybe indication, for $\mathcal{R}$ the new metrics $comp'_{A2}(\mathcal{R})$, $c'_{A2}(\mathcal{R})$ and $d'_{A2}(\mathcal{R})$ can be traced back to $comp(\mathcal{R}^\mathcal{D})$ and $comp(\mathcal{R}^\mathcal{M})$, $c(\mathcal{R}^\mathcal{D})$ and $c(\mathcal{R}^\mathcal{M})$ or $d(\mathcal{R}^\mathcal{D})$ and $d(\mathcal{R}^\mathcal{M})$, respectively. In the exact case such a derivation is not possible. Thus, the metrics $comp'_{A2}(\mathcal{R})$, $c'_{A2}(\mathcal{R})$ and $d'_{A2}(\mathcal{R})$ have to be newly defined by regarding the individual tuple probabilities.

Simple Maybe Indication: Both, the subrelation $\mathcal{R}^D$ as well as the subrelation $\mathcal{R}^M$ cover parts of the corresponding entity type's extension. As a consequence, the coverage $c'_{A2}(\mathcal{R})$ of a relation $\mathcal{R}$ can be calculated from the coverages of these two subrelations. Assuming the probability that a *maybe tuple* belongs to a relation is equal to the probability that the *maybe tuple* does not belong to this relation, the coverage of the subrelation $\mathcal{R}^M$ is taken into account with a priority which is half as high as the priority of the coverage $c(\mathcal{R}^\mathcal{D})$:

$$c'_{A2}(\mathcal{R}) = c(\mathcal{R}^\mathcal{D}) + \frac{1}{2} c(\mathcal{R}^\mathcal{M}) = \frac{|\mathcal{R}_C^\mathcal{D}| + \frac{1}{2}|\mathcal{R}_C^\mathcal{M}|}{|\mathcal{E}|} \tag{13}$$

The densities of $\mathcal{R}^\mathcal{D}$ and $\mathcal{R}^\mathcal{M}$ are the averages of their tuples' densities (equation 5). As with the coverage, the effect of the density $d(\mathcal{R}^\mathcal{M})$ on $d'_{A2}(\mathcal{R})$ is only half as high as the effect of the *definite tuples*' densities. Since the two densities $d(\mathcal{R}^\mathcal{D})$ and $d(\mathcal{R}^\mathcal{M})$ are only relative, for the total density both have to be correlated by taking into account the associated relation's size:

$$d'_{A2}(\mathcal{R}) = \frac{|\mathcal{R}_C^\mathcal{D}| d(\mathcal{R}^\mathcal{D}) + \frac{1}{2}|\mathcal{R}_C^\mathcal{M}| d(\mathcal{R}^\mathcal{M})}{|\mathcal{R}_C^\mathcal{D}| + \frac{1}{2}|\mathcal{R}_C^\mathcal{M}|} = \frac{\sum_{t \in \mathcal{R}_C^\mathcal{D}} d(t) + \frac{1}{2} \sum_{t \in \mathcal{R}_C^\mathcal{M}} d(t)}{|\mathcal{R}_C^\mathcal{D}| + \frac{1}{2}|\mathcal{R}_C^\mathcal{M}|} \tag{14}$$

As for approach 1, the completeness $comp'_{A2}(\mathcal{R}) = c'_{A2}(\mathcal{R}) \cdot d'_{A2}(\mathcal{R})$ results in:

$$comp'_{A2}(\mathcal{R}) = comp(\mathcal{R}^{\mathcal{D}}) + \frac{1}{2}comp(\mathcal{R}^{\mathcal{M}}) \quad (15)$$

Individual Tuple Probability: If the data model supports individual tuple probabilities instead of one global maybe priority, each tuple has a different impact on the coverage and the density of $\mathcal{R}$. Considering the individual tuple probability as the degree of this impact, coverage and density are defined as:

$$c'_{A2}(\mathcal{R}) = \frac{\sum_{t\in\mathcal{R}_C} p(t)_{\mathcal{R}}}{|\mathcal{E}|} \quad (16) \qquad d'_{A2}(\mathcal{R}) = \frac{\sum_{t\in\mathcal{R}_C} p(t)_{\mathcal{R}} \cdot d(t)}{\sum_{t\in\mathcal{R}_C} p(t)_{\mathcal{R}}} \quad (17)$$

Thus, the completeness $comp'_{A2}(\mathcal{R}) = c'_{A2}(\mathcal{R}) \cdot d'_{A2}(\mathcal{R})$ results in:

$$comp'_{A2}(\mathcal{R}) = \frac{\sum_{t\in\mathcal{R}_C} p(t)_{\mathcal{R}} \cdot d(t)}{|\mathcal{E}|} \quad (18)$$

4.4 Correlated Tuples

As in most works on *maybe relations*, dependencies between tuples have not been addressed so far. Since in reality data is often correlated, a complete independence among tuples is a simplistic assumption which distorts the representation of the modeled world. Therefore, in some newer proposals ([8] et al.) probabilistic data models are extended by representing such dependencies. Since tuple dependencies restrict the set of all possible instances of a relation $\mathcal{R}$, these dependencies are completely represented by the set $\mathcal{S}(\mathcal{R})$. For example, relation $\mathcal{R}$ contains one *definite* tuple t_1 and two *maybe tuples* t_2 and t_3. A tuple dependency defines that either both *maybe tuples* belong to $\mathcal{R}$ or none of them. As a consequence, instead of four possible instances $S(\mathcal{R}) = \{\{t_1\}, \{t_1, t_2\}, \{t_1, t_3\}, \{t_1, t_2, t_3\}\}$ only two possible instances $S(\mathcal{R}) = \{\{t_1\}, \{t_1, t_2, t_3\}\}$ exist. Thus, it is obvious, that our completeness metrics which are based on the expectation value of $\mathcal{R}'$ can be used in models with tuple dependencies without any adaption.

In general, the total probability of each tuple t is (independent of correlations) always $p(t)_{\mathcal{R}}$. Thus, it does not matter in which way this probability is distributed on the possible instances. As a consequence, the completeness of a *maybe relation* is generally independent from tuple correlations and the metrics of the other two approaches do not need to be adapted to such cases, too.

4.5 Comparison of Proposed Approaches

In the approaches outlined above, we defined metrics for calculating completeness of *maybe relations*. The next step is to compare these metrics to each other and try to determine which of them is most suitable. In general, all these completeness metrics supply the same results whether tuple correlations exist or not.

This fact enhances the certainty that the resulting value is actually an adequate representation of the completeness of the considered *maybe relation*.

Regarding the requirements proposed by Heinrich ([4]) the metrics of all approaches satisfy the requirements of normalization, interval scale and adaptivity. Furthermore, the input parameters and hence the feasability of all approaches are equal. Thus, the most severe differences w.r.t. these requirements are related to the interpretability. The first approach is most suitable for illustrating the completeness of a *maybe relation*, for example on the basis of graphics as seen in Figure 2 (property $\mathcal{A}$). In contrast, the two other approaches are more abstract. The benefit of the third approach is its simplicity (see complexity below), but from a probabilistic theory point of view the concept of the second one is still more apposite (property $\mathcal{B}$). However, in contrast to the other two approaches, the third one enables a decomposition of completeness into coverage and density (property $\mathcal{C}$), which in turn improves its interpretability. Additionally, both completeness metrics of approach 3, for the simple maybe indication as well as for an indication by individual tuple probabilities are comprehensible in an easy way (property $\mathcal{D}$). In the second approach the metric for a simple maybe indication can only be derived from those of the exact case by a substitution of the value 0.5 for every tuple probability. Hence approach 2 has a poor interpretability.

Another important factor is the complexity of the individual metrics. Given a relation $\mathcal{R}$ with n *definite* and m *maybe tuples*, w.r.t. the simple case, the complexity of all metrics is equal ($\mathcal{O}(n+m)$). In the exact case, at the worst in approach 1 each *maybe tuple* has another probability and the completeness of $m+1$ subrelations have to be calculated ($\mathcal{O}(max(m^2, nm))$). In approach 2 the completeness of 2^m possible instances is required if there are no tuple correlations ($\mathcal{O}(2^m(n+m))$). In approach 3 only the completeness of a single relation is needed ($\mathcal{O}(n+m)$). The complexities w.r.t. both cases (simple and exact) as well as the mentioned benefits and drawbacks of all approaches with respect to the interpretability are summarized in the following table:

	property $\mathcal{A}$	property $\mathcal{B}$	property $\mathcal{C}$	property $\mathcal{D}$	complexity: simple case	complexity: exact case
Approach 1:	+	○	-	○	$\mathcal{O}(n+m)$	$\mathcal{O}(max(m^2, nm))$
Approach 2:	○	+	-	-	$\mathcal{O}(n+m)$	$\mathcal{O}(2^m(n+m))$
Approach 3:	○	○	+	+	$\mathcal{O}(n+m)$	$\mathcal{O}(n+m)$

Regarding its minor complexity, in databases with individual tuple probabilities, the metric of approach 3 is most suitable. At a first sight (without considerations on implementation- or application domain specific details), in databases with just a simple maybe indication all metrics can be assumed to be equivalently suitable.

5 Conclusion

Since current metrics of data completeness are not usable for estimating the completeness of *maybe relations*, we have used the metric defined by Naumann and extended it for handling the vagueness resulting from the *maybe tuple* concept. Further, we have identified two cases. In the first case, *maybe tuples* are only indicated as 'maybe'. In the second, more exact case, every tuple is indicated by a probability of its own.

We have considered completeness from three different perspectives and have therefore introduced three corresponding approaches in order to measure this quality dimension. The resulting metrics supply the same results whether or not tuple correlations exist. In general, even though all resulting completeness values are an adequate representation of this quality dimension, each of the three approaches (and hence each of the corresponding metrics) has its benefits as well as its drawbacks. In contrast to the other two approaches, the approach based on tuple priorities enables a decomposition of completeness into coverage and density, which in turn increases the interpretability of the resulting values. Furthermore, its completeness metrics have by far the lowest complexity. Thus, we favor the usage of the metrics resulting from this approach.

So far, we have considered completeness only from a theoretical point of view. In reality such an exact calculation is often impossible because important information (e.g. $|\mathcal{E}|$) is missing. Thus, in future work these approaches have to be considered from a more practical (and hence vaguer) point of view, too.

Besides completeness, other quality dimensions are influenced by the possibility of *maybe tuples*. Especially quality dimensions for which the quality of a relation is derived from the qualities of its tuples (e.g. accuracy, currency) are affected. As for completeness, the *maybe tuples* have to be considered with a minor emphasis. The lower the probability of a tuple, the lower the influence of this tuple on the quality of the associated relation has to be.

References

1. Biskup, J.: Extending the Relational Algebra for Relations with Maybe Tuples and Existential and Universal Null Values. Fundam. Inform. 7(1), 129–150 (1984)
2. DeMichiel, L.G.: Resolving Database Incompatibility: An Approach to Performing Relational Operations over Mismatched Domains. IEEE Trans. Knowl. Data Eng. 1(4), 485–493 (1989)
3. Fuhr, N., et al.: A Probabilistic Relational Algebra for the Integration of Information Retrieval and Database Systems. ACM Trans. Inf. Syst. 15(1), 32–66 (1997)
4. Heinrich, B., et al.: Metrics for Measuring Data Quality - Foundations for an Economic Data Quality Management. In: ICSOFT (ISDM/EHST/DC), pp. 87–94 (2007)
5. Motro, A., Rakov, I.: Estimating the Quality of Databases. In: Andreasen, T., Christiansen, H., Larsen, H.L. (eds.) FQAS 1998. LNCS (LNAI), vol. 1495, pp. 298–307. Springer, Heidelberg (1998)
6. Naumann, F., et al.: Completeness of integrated information sources. Inf. Syst. 29(7), 583–615 (2004)
7. Scannapieco, M., Batini, C.: Completeness in the relational model: a comprehensive framework. In: IQ, pp. 333–345 (2004)
8. Sen, P., Deshpande, A.: Representing and Querying Correlated Tuples in Probabilistic Databases. In: ICDE, pp. 596–605 (2007)
9. Tseng, F.S.-C., et al.: Answering Heterogeneous Database Queries with Degrees of Uncertainty. Distributed and Parallel Databases 1(3), 281–302 (1993)

Modeling, Measuring and Monitoring the Quality of Information

Hendrik Decker[1,*] and Davide Martinenghi[2,**]

[1] Instituto Tecnológico de Informática, Valencia, Spain
[2] Politecnico di Milano, Dipart. Elettronica e Informazione, Milano, Italy

Abstract. Semantic properties that reflect quality criteria can be modeled by integrity constraints. Violated instances of constraints may serve as a basis for measuring quality. Such measures also serve for monitoring and controlling quality impairment across changes.

1 Introduction

A pragmatic notion of information is to define it as the data that are stored in the database backend of an information system. Quality of information then is the quality of stored data. But how to model and measure such quality? Vaguely speaking, quality of data is the level of its semantic correctness, i.e., the degree of truthfulness by which the data reflect the intents of the database designer and the purposes of the database users.

From database theory [5], it is well-known that the quality, i.e., the semantic correctness of stored data can be modeled declaratively by invariant conditions called integrity constraints. Thus, if quality is modeled by integrity constraints, violations of constraints reflect a lack of data quality.

The DBMS supports the prevention of quality deterioration by enforcing the integrity constraints imposed on the database. Usually, that is done at update time, by automatically checking constraints that are potentially violated by the update. If so, then the update either is rejected, or the integrity violation in the updated state is repaired, if possible, by another update.

In spite of automated integrity checking methods, the quality of stored data often is compromised and can be severely damaged in practice. Such impairments of data quality typically are caused by integrity violations that somehow find their way into the database. There are many ways by which that may occur. For instance, constraints may be added to the schema without being checked for violations by legacy data. Or, whenever integrity checking is switched off temporarily, e.g., for uploading a backup, or for boosting availability, any amount of integrity violation may enter into the database. Data integrity and quality may also suffer from the integration of databases.

* Partially supported by FEDER and the Spanish MEC grant TIN2006-14738-C02-01.
** Supported by the ERC-funded IDEAS Advanced Grants project SeCo.

C.A. Heuser and G. Pernul (Eds.): ER 2009 Workshops, LNCS 5833, pp. 212–221, 2009.

The basic idea of quantifying the quality of stored data by measuring their integrity is the following. Constraint violations can be quantified simply by counting them. Possible refinements are to not just count violated constraints, or violated instances (later called 'cases') of constraints, but to assign and aggregate specific application-dependent weights to different violated cases. Also the sets of violated cases themselves can be taken as metrics, based on the partial ordering of set inclusion.

As soon as constraint violations are quantified, the impairment of data quality becomes measurable. In fact, it seems more feasible to implement a metric of impaired quality, i.e., semantic inconsistency, which is quantifiable, than of quality, i.e., semantic consistency, the objectification of which tends to be evasive.

In section 2, we recapitulate some background of database integrity. In section 3, we revisit the definition of 'cases', originally used for inconsistency-tolerant integrity checking [2]. Cases are instances of constraints that are apt to model the quality of information in a more differentiated manner than universally quantified constraints. In section 4, we define and discuss several inconsistency metrics. Some are based on counting violated cases of integrity or comparing sets of such cases, others on a measure of inconsistency in [4]. In section 5, we show that, apart from providing quality metrics, measures of integrity violations also may serve to control and contain impaired data quality across updates.

2 Background

We use notations and terminology that are common in the logic databases community [5]. Only databases with finite domains are considered.

2.1 Databases, Updates, Constraints

An *atom* is an expression of the form $p(a_1, ..., a_n)$, where p is a predicate of arity n ($n \geq 0$); the a_i, called *arguments*, are either constants or variables. A *literal* is either an atom A or a negated atom $\sim A$. A *fact* is an atom where all arguments are constants. A *database clause* is either a *fact*, the predicate of which corresponds to a relational table and the arguments of which correspond to column values in that table, or a formula of the form $A \leftarrow B$, where the *head* A is an atom and the *body* B is a conjunction of literals; all variables in $A \leftarrow B$ are implicitly quantified universally in front of the formula. A *database* is a finite set of database clauses. It is *definite* if there is no negated atom in the body of any of its clauses.

An *update* is a finite set of database clauses to be inserted or deleted. The writeset of each committed database transaction can be described as an update. For an update U of a database state D, we denote the 'updated database', where all inserts in U are added to D and all deletes in U are removed from D, by D^U.

An *integrity constraint* (in short, *constraint*) is a first-order predicate logic sentence which, for convenience, we assume to be always represented in *prenex* form, i.e. all *quantifiers* of variables are in front of a quantifier-free *matrix*. Constraints are often represented as *denials*, i.e., formulas of the form $\leftarrow B$, where the body B states what must not hold. Implicitly, each variable in B is universally quantified at the front of $\leftarrow B$. A denial is *definite* if there is no negated atom in its body. An *integrity theory* is a finite set of constraints.

Constraints can be read as *necessary* conditions for the quality of stored data. If all intended application semantics are expressed by constraints, then the integrity theory represents a complete set of conditions that, when satisfied, is also *sufficient* for ensuring the quality of the stored data.

The DBMS is supposed to ensure that the database satisfies its integrity theory at all times, i.e., that all constraints are logically true consequences of each state. To achieve this, database theory requires that, for each update U, the 'old' database D, i.e., the state to be updated by U, must satisfy all constraints, such that integrity checking can focus on those constraints that are possibly affected by the update. If those constraints remain satisfied, then the 'new' state D^U reached by committing U also satisfies all constraints.

From now on, let D, IC, I, U always stand for a database, an integrity theory, a constraint and, resp., an update. For convenience, we write $D(I) = \mathit{true}$ (resp., $D(I) = \mathit{false}$) if I is satisfied (resp., violated) in D. Similarly, $D(IC) = \mathit{true}$ (resp., $D(IC) = \mathit{false}$) means that all constraints in IC are satisfied in D (resp., at least one constraint in IC is violated in D).

2.2 Integrity Checking

If quality is described by constraints, the quality of stored data can be monitored by checking integrity for each update that could potentially violate it.

In definition 1, below, we revisit a previous definition of integrity checking [2]. It abstracts away from any technical detail of how checking is done. It describes each integrity checking method $\mathcal{M}$ as a mapping that takes as input a database D, and integrity theory IC and an update U, and outputs either *sat* or *vio*. If $\mathcal{M}$ is sound, $\mathcal{M}(D, IC, U) = \mathit{sat}$ indicates that $D^U(IC) = \mathit{true}$, i.e., U does not violate integrity. If $\mathcal{M}$ is also complete, then $\mathcal{M}(D, IC, U) = \mathit{vio}$ indicates that $D^U(IC) = \mathit{false}$. Also the output *vio* of an incomplete method may mean that the update would violate integrity; but it may as well mean that either further checking is needed for determining the integrity status of D^U, or, if there are not enough resources to do so, then U should be cautiously rejected.

Definition 1. (*Sound and complete integrity checking*)
Let $\mathcal{M}$ be a method for integrity checking. $\mathcal{M}$ is called *sound* or, resp., *complete* if, for each (*D,IC,U*) such that *D*(*IC*)=*true,* (1) or, resp., (2) holds.

$$\text{If } \mathcal{M}(D, IC, U) = \mathit{sat} \text{ then } D^U(IC) = \mathit{true}. \tag{1}$$

$$\text{If } D^U(IC) = \mathit{true} \text{ then } \mathcal{M}(D, IC, U) = \mathit{sat}. \tag{2}$$

Quality maintenance by integrity checking would tend to be too expensive, unless some simplification method were used [1]. Simplification essentially means that, for an update U, it suffices to check only those instances of constraints that are potentially violated by U. That idea is the basis for virtually all methods for integrity checking methods proposed in the literature or used in practice.

Example 1. Let $p(ID, Name, TelNo)$ be a relation with predicate p about the persons registered in the customers database of some telephone company (column names are self-explaining). Let I be the denial constraint

$$I = \leftarrow (p(x, y_1, z_1) \wedge p(x, y_2, z_2) \wedge (y_1 \neq y_2 \vee z_1 \neq z_2)).$$

I states that no two customers with the same ID x may have different names y_1, y_2, nor different numbers z_1, z_2, i.e., I requires that all customer *IDs* be unique.

Now, let U be an update that inserts $p(7, joe, 345)$. Usually, integrity then is checked by evaluating the following instance I' of I:

$$I' = \leftarrow (p(7, joe, 345) \wedge p(7, y_2, z_2) \wedge (joe \neq y_2 \vee 345 \neq z_2)).$$

Since U makes $p(7, joe, 345)$ true, I' can be simplified to

$$I'_s = \leftarrow (p(7, y_2, z_2) \wedge (joe \neq y_2 \vee 345 \neq z_2)).$$

The simplification I'_s asks if there is any customer with ID=7 whose name is not *joe* or whose number is not 345. Its evaluation essentially amounts to a simple search in the table of p, in order to see if there is any customer with ID=7 but with name other than *joe* or number other than 345. If so, U is rejected; if not, U can be committed. Clearly, this simplified evaluation is significantly cheaper than to evaluate I as it stands. The latter would amount to a possibly voluminous join of the entire p relation with itself.

In principle, also the following instance I'' of I would have to be evaluated:

$$I'' = \leftarrow (p(7, y_1, z_1) \wedge p(7, joe, 345) \wedge (y_1 \neq joe \vee z_1 \neq 345))$$

I'' is obtained by resolving $p(x, y_2, z_2)$ in I with the update $p(7, joe, 345)$. But no evaluation of I'' is needed since I'' is logically equivalent to I'.

3 Modeling the Quality of Information by Cases

Instead of coarsely modeling quality by universally quantified constraints, we want to be able to distinguish between different cases of violations of a constraint. Thus, a more differentiated quality metric can be obtained: the less/more cases of constraints are violated, the better/worse is the quality of data.

Essentially, a 'case' of a constraint I is an instance of I obtained by consistently substituting the $\forall$-quantified variables in I that are not 'governed' by any $\exists$-quantified variable with terms of the database language. A variable x is said *govern* a variable y if the quantifier of x occurs left of the quantifier of y.

Example 2. Each of I, I' and I'' in Example 1 is a case of I. As a counterexample, let $J = \exists x \, \forall y \, emp(y) \rightarrow sup(x, y)$. In J, x governs y. The constraint J requires that there is an individual x who is superior of all employees y. J is potentially violated by each insertion of an employee tuple and each deletion of a superior tuple. However, no instantiation of y with the value of a tuple to be inserted to *emp* can be used for simplifying the evaluation of J, nor any instantiation of x or y with the values of a tuple to be deleted from *sup*.

Assume that a new employee e is to be inserted into the database. Then, it would be wrong to only check the instance $\exists x\, emp(e) \rightarrow sup(x, e)$ of the constraint above, since there might well be some superior x of e who would satisfy that instance, but that particular x may not be a superior of all other employees.

As opposed to J in Example 2, I in Example 1 can be decomposed into a set of individual cases, where some or all variables are bound to corresponding values of customers. For insertions of new customers, the cases to be evaluated for checking I always have a form analogous to I' or I''.

In general, each constraint I is logically equivalent to the conjunction of all of its cases, i.e., I is satisfied if and only if each of its cases is satisfied. If the leftmost quantifier of I is an $\exists$, then the only case of I is I itself, modulo renamings of variables. The following definition formalizes this idea.

Definition 2. (*Cases* [2])
Let I be a constraint of the form QW, where Q is the (possibly implicit) vector of all $\forall$ quantifiers of variables in I that are not governed by any $\exists$-quantified variable in I, and W is a well-formed formula. Further, x be a variable in I.

a) x is called a *global variable* in I if x occurs in Q.

b) If ς is a substitution of the global variables in I, then $\underline{\forall}(W\varsigma)$ is called a *case* of I, where $\underline{\forall}$ denotes (a possibly implicit) universal closure. If ς substitutes each global variable by a constant, then $\underline{\forall}(W\varsigma)$ is called a *basic case* of I.

c) Let $\mathsf{Cas}(IC)$ be the set of all cases of all $I \in IC$. *Let* $\mathsf{SatCas}(D, IC)$, resp., $\mathsf{VioCas}(D, IC)$, be the set of all $C \in \mathsf{Cas}(IC)$ such that $D(C) = true$, resp., *false*. Let $\mathsf{VioBas}(D, IC)$ be the set of all basic cases in $\mathsf{VioCas}(D, IC)$ modulo variants.

Example 3. In example 1, both I' and I'' are non-basic cases of I. The constraint $\leftarrow (p(7, sue, 345) \wedge (joe \neq sue \vee 345 \neq 345))$ is a basic case of the simplification I'_s. In example 2, J itself is the only case of J, up to renamings of variables.

Typically, the size of $\mathsf{SatCas}(D, IC)$ by far exceeds the size of $\mathsf{VioCas}(D, IC)$, which reflects the amount of damaged quality in D. Thus, it is more reasonable to assess $\mathsf{VioCas}(D, IC)$, rather than $\mathsf{SatCas}(D, IC)$, for measuring the quality impairment in D with regard to IC. This thought motivates the following section.

4 Measuring Quality

We want to measure quality by measuring inconsistency. In 4.1, we axiomatize inconsistency metrics. In 4.2 and 4.3, we discuss examples of such metrics.

4.1 Axiomatizing Inconsistency Metrics

Let $\preccurlyeq$ symbolize an ordering that is antisymmetric, reflexive and transitive. For expressions E, E', let $E \prec E'$ denote that $E \preccurlyeq E'$ and $E \neq E'$.

Definition 3. We say that $(\mu, \preccurlyeq)$ is an *inconsistency metric* (in short, a *metric*) if μ is a mapping that takes tuples (D, IC) as input, and outputs a value in some lattice partially ordered by $\preccurlyeq$. Moreover, for each (D, IC) such that $D(IC) = sat$, each database D' and each integrity theory IC', the *violation-is-bad* property (3) holds:

$$\text{If } \ D'(IC') = \mathit{false} \ \text{ then } \ \mu(D, IC) \prec \mu(D', IC')\,. \tag{3}$$

Occasionally, we identify a metric $(\mu, \preccurlyeq)$ with μ, if $\preccurlyeq$ is understood.

Clearly, (3) ensures that the quality of any database with a non-zero amount of violation is always lower than for any consistent database.

For merely formal reasons, no more properties of $(\mu, \preccurlyeq)$ are needed here. Yet, we mention the following two, which can be useful and can be appreciated as desirable, even though they are not mandatory for our purposes.

First, we consider the *satisfaction-is-best* property of $(\mu, \preccurlyeq)$. It requires that, for each pair (D, IC) and each pair (D', IC'), the following holds.

$$\mathit{If} \ \ D(IC) = \mathit{true} \ \ \mathit{then} \ \ \mu(D, IC) \preccurlyeq \mu(D', IC')\,. \tag{4}$$

Clearly, (4) ensures that the quality of each (D, IC) such that $D(IC) = \mathit{true}$ is always highest. If (4) is not imposed, then the violation-is-bad property (3) still guarantees that no database state that satisfies integrity is of lower quality than any state that violates integrity.

The second property that could be required in addition to those in definition 3 presupposes that the lattice that is the range of μ is not only ordered by $\preccurlyeq$, but also is an algebra with some addition $\oplus$. Thus, it becomes desirable that μ be *additive*, i.e., that μ fulfills the following triangle inequality:

$$\mu(D, IC \cup IC') \leq \mu(D, IC) \oplus \mu(D, IC')\,. \tag{5}$$

4.2 Inconsistency Metrics Based on Cases

In this subsection, we define and discuss several metrics that comply with Definition 3. Four of them, already introduced in [3], recur on $\mathsf{VioBas}(D, IC)$ (Def. 2c).

Let $|\,.\,|$ denote the set cardinality operator, and $\overline{\mathsf{VioBas}}(D, IC)$ be the set of all cases in $\mathsf{VioBas}(D,IC)$ that are not subsumed by any other case in that set.

The lack of quality in databases can be reflected by counting and comparing sets of violated basic constraints. Example 4 illustrates why sets $\overline{\mathsf{VioBas}}(D, IC)$ are potentially interesting refinements of sets $\mathsf{VioBas}(D, IC)$. Intuitively, redundant cases in $\mathsf{VioBas}(D, IC)$ are eliminated in $\overline{\mathsf{VioBas}}(D, IC)$.

Example 4. Let $D = \{p(a),\ p(b),\ q(b,b),\ q(c,a),\ q(c,b)\}$ and $IC = \{\leftarrow q(x,x),$ $\leftarrow p(x), q(x,y)\}$. Obviously, $\mathsf{VioBas}(D,IC) = \{\leftarrow q(b,b),\ \ \leftarrow p(b), q(b,b)\}$. Since $\leftarrow q(b,b)$ logically entails $\leftarrow p(b), q(b,b)$, $\mathsf{VioBas}(D, IC)$ is redundant. So, the question arises whether $\leftarrow q(b,b)$ should count as the only violated case or whether $\leftarrow p(b), q(b,b)$ should count too. In fact, it can be argued that the presence or absence of $p(b)$ in the database makes no difference wrt. quality, since the critical fact is $q(b,b)$: if it would be deleted, then both cases in $\mathsf{VioBas}(D, IC)$

would disappear. Hence, there is a single reason for integrity violation in (D, IC). Therefore, an inconsistency metric should not take the presence or absence of $p(b)$ into account, as done by $\mathsf{VioBas}(D, IC)$. Obviously, $\overline{\mathsf{VioBas}}(D, IC)$ is independent of the presence or absence of $p(b)$.

$\mathsf{VioBas}(D, IC)$ and $\overline{\mathsf{VioBas}}(D, IC)$ and their cardinalities are partially or, resp., totally ordered by $\subseteq$, and, resp., $\leq$. Thus, the four metrics $(\nu, \subseteq)$, $(\underline{\nu}, \leq)$, $(\overline{\nu}, \subseteq)$, $(\underline{\overline{\nu}}, \leq)$ are defined by the following equations.

$$\nu(D, IC) = \mathsf{VioBas}(D, IC) \qquad \underline{\nu}(D, IC) = |\mathsf{VioBas}(D, IC)|$$

$$\overline{\nu}(D, IC) = \overline{\mathsf{VioBas}}(D, IC) \qquad \underline{\overline{\nu}}(D, IC) = |\overline{\mathsf{VioBas}}(D, IC)|$$

Clearly, each of these four structures is a metric according to Definition 3, and each of them has the satisfaction-is-best property. Also, it is easy to see that each is additive ($\oplus$ is $\cup$ for ν and $\overline{\nu}$, and $+$ for $\underline{\nu}$ and $\underline{\overline{\nu}}$).

Example 5. Let $D = \{q(a, a),\ r(b)\}$, $IC = \{\leftarrow q(x, x),\ \leftarrow q(x, y), r(y)\}$ and $U = \{\mathit{insert}\, r(a)\}$. Clearly, $\nu(D, IC) = \overline{\nu}(D, IC) = \overline{\nu}(D^U, IC) = \{\leftarrow q(a, a)\}$ and $\nu(D^U, IC) = \{\leftarrow q(a, a),\ \leftarrow q(a, a), r(a)\}$ and $\underline{\nu}(D, IC) = \underline{\overline{\nu}}(D, IC) = \underline{\overline{\nu}}(D^U, IC) = 1$ and $\underline{\nu}(D^U, IC) = 2$.

4.3 The Inconsistency Measure by Hunter and Grant

The inconsistency measure proposed in [4] also serves as a quality metric. It is based on quasi-classical *qc-models*. Roughly, a qc-model QC of (D, IC) is a set of ground literals that, together with a consequence relation $\models_s$, behaves like a Herbrand model of $D \cup IC$, except that $(QC, \models_s)$ interprets negative literals as atoms with distinguished 'negative' predicates and interprets each disjunction $L_1 \vee ... \vee L_n$ of ground literals L_i $(1 \leq i \leq n,\ n > 0)$ as follows.

$$QC \models_s L_1 \vee ... \vee L_n \text{ iff } QC \models_s L_1 \text{ or} ... \text{or } QC \models_s L_n \text{ and } QC \models_s \overline{L_i} \text{ implies } QC \models_s \otimes(L_1 \vee ... \vee L_n, L_i),$$

where, $\overline{L_i} = A$ if $L_i = {\sim}A$, and $\overline{L_i} = {\sim}A$ if $L_i = A$, for each atom A, and $\otimes(L_1 \vee ... \vee L_n, L_i)$ is the disjunction obtained by dropping L_i from $L_1 \vee ... \vee L_n$. Thus, each (D, IC) has a qc-model, even if $D(IC)$=*false*. See [4] for more details.

The *conflict set* of a qc-model QC is defined in [4] as the set of atoms A such that both A and ${\sim}A$ is in QC. For each (D, IC), the conflict set of each minimal qc-model of (D, IC) is the same. Thus, a metric η (say) is given by this set, and another one, $\underline{\eta}$, by its cardinality. The measure in [4] is defined by the ratio of $\underline{\eta}(D, IC)$ over the size of the underlying Herbrand base.

Example 6. For (D, IC, U) as in example 5, a minimal qc-model of (D, IC) consists of $q(a, a)$, $r(b)$, all negative ground literals that match ${\sim}q(x, x)$, all negative ground literals that match ${\sim}q(x, b)$ and all negative ground literals that match ${\sim}r(y)$ except ${\sim}r(b)$. Thus, $\eta(D, IC) = \{q(a, a)\}$ and $\underline{\eta}(D, IC) = 1$. Similarly, $\eta(D^U, IC) = \{q(a, a),\ r(a)\}$ and $\underline{\eta}(D^U, IC) = 2$.

5 Monitoring Impaired Quality by Integrity Checking

In section 4, we have seen that the quality of stored data can be quantified by measuring sets of violated cases of constraints. In this section, we show how such metrics also serve to monitor and control impaired quality, i.e., to prevent the degradation of integrity across updates. More precisely, we are going to see that each inconsistency metric induces a sound integrity checking method that is able to tolerate extant impairments of quality, to preserve all satisfied cases across updates, and to reject any update that would violate any satisfied case.

There is an essential difference between integrity checking methods that do and that don't tolerate extant constraint violations, i.e., impaired quality. The latter, as defined in 1, require that $D(IC) = true$ holds, i.e., that integrity be totally satisfied before each update. By contrast, inconsistency-tolerant integrity checking does not expect the total satisfaction of all integrity constraints. Rather, it prevents that the quality degrades across updates, even if the quality requirements expressed by the integrity theory are not fully complied with.

All integrity checking methods in the literature require that integrity be totally satisfied before each update. Fortunately, that requirement can be waived for many (though not all) methods, as shown in [2]. Below, we revisit an inconsistency-tolerant definition of integrity checking that does not rely on the requirement of total integrity. It characterizes integrity checking methods that can preserve all satisfied cases of constraints, while violated cases are put up with.

Definition 4. (*inconsistency-tolerant integrity checking*)
Let $\mathcal{M}$ be a method for integrity checking. $\mathcal{M}$ is called *sound,* resp., *complete wrt. case-based inconsistency tolerance* if, for each (D, IC, U) and each $C \in \mathsf{SatCas}(D, IC)$, (6) or, resp., (7) holds.

$$\text{If } \mathcal{M}(D, IC, U) = sat \text{ then } D^U(C) = true. \tag{6}$$

$$\text{If } D^U(C) = true \text{ then } \mathcal{M}(D, IC, U) = sat. \tag{7}$$

The quality of data is easily compromised. Thus, the tolerance of inconsistency, i.e., of logically modeled quality impairment, is indispensable for monitoring quality by integrity checking. Hence, some form of inconsistency tolerance should be provided also from any integrity checking method based on inconsistency metrics. Such methods are going to be defined below.

Clearly, each metric for quantifying the inconsistency, i.e., lack of quality, of databases serves for comparing the values measured in consecutive states. Thus, each such metric induces a method defined by accepting the update if quality impairment does not increase, and repelling it or raise a warning if it does.

Definition 5 below captures this kind of methods. For any given μ, each such method accepts an update only if the integrity violation measured by μ does not increase, while tolerating extant violations of integrity. The preceding 'only if' becomes an 'if and only if' for complete measure-based methods.

Definition 5. (*measure-based integrity checking*)
Let $\mathcal{M}$ be a method for integrity checking and $(\mu, \preccurlyeq)$ be a metric. $\mathcal{M}$ is called *sound*, resp., *complete wrt. measure-based inconsistency tolerance* if, for each triple (D, IC, U), (8) or, resp., (9) holds.

$$\text{If } \mathcal{M}(D, IC, U) = sat \text{ then } \mu(D^U, IC) \preccurlyeq \mu(D, IC). \tag{8}$$

$$\text{If } \mu(D^U, IC) \preccurlyeq \mu(D, IC) \text{ then } \mathcal{M}(D, IC, U) = sat. \tag{9}$$

If (8) holds, then $\mathcal{M}$ is also called *measure-based*, and, in particular, *μ-based.*

Measure-based methods constitute an approach to inconsistency tolerance that does not refer to cases. The following result, which follows from the definitions, provides examples of measure-based methods.

Theorem 1. Each method $\mathcal{M}$ that is sound wrt. case-based inconsistency tolerance is ν-based and $\underline{\nu}$-based. If $\mathcal{M}$ is also complete wrt. case-based inconsistency tolerance, then $\mathcal{M} = \mathcal{M}^{\nu}$. □

Note that definition 5 does not give any hints on how $\mathcal{M}$ computes its output. However, for each metric μ, a μ-based method can be defined, as done below. For that purpose, let us first define, for each metric $(\mu, \preccurlyeq)$, a mapping M^{μ} for consistency checking, as follows.

$$\mathcal{M}^{\mu}(D, IC, U) = sat \text{ iff } \mu(D^U, IC) \preccurlyeq \mu(D, IC). \tag{10}$$

Clearly, the output of $\mathcal{M}^{\mu}$ is determined entirely by the output of μ: the inconsistency before and after the update is measured and compared. If it does not increase, $\mathcal{M}$ outputs *sat*, otherwise *vio*.

The following result asserts that each $\mathcal{M}^{\mu}$ for which (10) holds is indeed a method in the sense of definition 1.

Theorem 2. Let μ be an inconsistency metric. Then, $\mathcal{M}^{\mu}$ is a sound integrity checking method and also a method that is sound and complete wrt. measure-based inconsistency tolerance.

Proof. Let (D, IC, U) be a triple such that $D(IC) = true$ and $\mathcal{M}(D, IC, U) = sat$. By definition 4, we have to show that $D^U(IC) = true$. Suppose that $D^U(IC) = false$. By definition 5, $\mathcal{M}(D,IC,U) = sat$ entails $\mu(D^U, IC) \preccurlyeq \mu(D, IC)$. From $D^U(IC) = false$, however, the contradiction $\mu(D, IC) \prec \mu(D^U, IC)$ would be entailed by the violation-is-bad property (3). Hence, $\mathcal{M}^{\mu}$ is a sound method. By its definition (10), $\mathcal{M}^{\mu}$ obviously fulfills (8) and (9). □

Example 7 shows that the converse of Theorem 1 does not hold, i.e., not each measure-based method is sound wrt. case-based inconsistency tolerance.

Example 7. Let $D = \{p(a)\}$, $IC = \{\leftarrow p(x), \leftarrow q(x)\}$ and $U = \{delete\ p(a), insert\ q(b)\}$. Clearly, $\mathsf{VioBas}(D, IC) = \{\leftarrow p(a)\}$, $\mathsf{VioBas}(D^U, IC) = \{\leftarrow q(b)\}$, i.e, $\mathsf{VioBas}(D, IC)$ and $\mathsf{VioBas}(D^U, IC)$ have the same size but different contents. Thus, $\underline{\nu}(D^U, IC) = \underline{\nu}(D, IC)$, and $\mathcal{M}^{\underline{\nu}}(D, IC, U) = sat$. However, it follows from

$\leftarrow q(b) \in \mathsf{SatCas}(D, IC) \cap \mathsf{VioCas}(D^U, IC)$ that $\mathcal{M}^{\underline{\nu}}$ is not a case-based inconsistency-tolerant method. However, $\mathcal{M}^{\nu}(D, IC, U) = vio$ holds. More generally, it is easy to see that $\mathcal{M}^{\nu}$ is sound wrt. case-based inconsistency tolerance.

Example 8 shows that there is a method that is sound and complete wrt. measure-based inconsistency tolerance but not complete wrt. case-based inconsistency tolerance.

Example 8. Let $D = \{p(x) \leftarrow q(x),\ \ p(x) \leftarrow r(x),\ \ q(a)\}$, $IC = \{\leftarrow p(x)\}$ $U = \{insert\ r(a)\}$, and δ be the metric defined by a count of distinct derivation paths of violated cases. Then, the δ-based method $\mathcal{M}^{\delta}$ diagnoses an increase of inconsistency caused by U, hence $\mathcal{M}^{\delta}(D, IC, U) = vio$. By Theorem 2, $\mathcal{M}^{\delta}$ is sound and complete wrt. measure-based inconsistency tolerance. However, $\mathcal{M}^{\delta}$ is not complete wrt. case-based inconsistency tolerance. In fact, for each method $\mathcal{M}$ that is complete wrt. case-based inconsistency tolerance, $\mathcal{M}$(D,IC,U) $= sat$, since $\mathsf{VioCas}(D, IC) = \mathsf{VioCas}(D^U, IC) = \{\leftarrow p(a)\}$.

6 Conclusion

We have proposed to model the quality of stored data by integrity constraints, and to measure the lack of data quality by assessing sets of violated instances of constraints. Such measurements also serve to monitor and control the increase of unavoidable quality impairment. Further research is needed wrt. assigning application-specific weights to violated cases, for obtaining more considerate quality metrics, and wrt. efficient implementations of inconsistency metrics.

References

1. Christiansen, H., Martinenghi, D.: On simplification of database integrity constraints. Fundam. Inform. 71(4), 371–417 (2006)
2. Decker, H., Martinenghi, D.: A relaxed approach to integrity and inconsistency in databases. In: Hermann, M., Voronkov, A. (eds.) LPAR 2006. LNCS (LNAI), vol. 4246, pp. 287–301. Springer, Heidelberg (2006)
3. Decker, H., Martinenghi, D.: Classifying integrity checking methods with regard to inconsistency tolerance. In: Proc. 10th PPDP, pp. 195–204. ACM Press, New York (2008)
4. Grant, J., Hunter, A.: Measuring inconsistency in knowledgebases. J. Intelligent Information Systems 27(2), 159–184 (2006)
5. Ramakrishnan, R., Gehrke, J.: Database Management Systems, 3rd edn. McGraw-Hill, New York (2003)

Evaluating the Functionality of Conceptual Models

Kashif Mehmood[1,2] and Samira Si-Said Cherfi[1]

[1] CEDRIC-CNAM, 292 Rue Saint Martin, F-75141 Paris Cedex 03, France
[2] ESSEC, Avenue Bernard Hirsch B.P. 50105, 95021 Cergy-Pontoise Cedex, France
Kashif.Mehmood@essec.fr, samira.cherfi@cnam.fr

Abstract. Conceptual models serve as the blueprints of information systems and their quality plays decisive role in the success of the end system. It has been witnessed that majority of the IS change-requests results due to deficient functionalities in the information systems. Therefore, a good analysis and design method should ensure that conceptual models are functionally correct and complete, as they are the communicating mediator between the users and the development team. Conceptual model is said to be functionally complete if it represents all the relevant features of the application domain and covers all the specified requirements. Our approach evaluates the functional aspects on multiple levels of granularity in addition to providing the corrective actions or transformation for improvement. This approach has been empirically validated by practitioners through a survey.

Keywords: Conceptual Model Quality, Functional Quality, Quality Metrics, Quality Evaluation, Quality Improvement.

1 Introduction

An information system is designed to answer the user's requirements. Therefore, users expect it to deliver all the required functionalities, for which it has been designed, correctly. It has now been widely agreed that the quality of the end-system depends on the quality of the Conceptual Models (CM). These CMs are designed as part of the analysis phase and are the basis for further design and implementation. Thus, if there are errors and deficiencies in the CMs then they are propagated along the development process. These errors are more expensive to fix once the system is developed and deployed. For these reasons, different methodologies propose different methods and guidelines to ensure a certain degree of quality to the produced deliverables. These guidelines aim to make the developed models correct, consistent and complete with respect to the specified requirements. Some of these criteria are relatively simple and easy to check such as syntactic correctness since it relates to the used notation. Whereas, some characteristics are more difficult to verify and to ensure. For example, verifying that the designed CMs cover all the user requirements or verifying that the different designed models are consistent are more difficult. Indeed, a "good" conceptual model should respond to some characteristics that we

C.A. Heuser and G. Pernul (Eds.): ER 2009 Workshops, LNCS 5833, pp. 222–231, 2009.

refer to in this article as functionality. A conceptual model, with respect to functional quality, should: (i) cover all the requirements by proposing suitable functions, (ii) not propose functions out of the system scope, (iii) be consistent, (iv) reuse common functions if possible and (v) be reliable. However, the translation of the requirements into conceptual model depends heavily on the degree of expertise of the analyst. For this reason, we propose a quality approach to help the evaluation and improvement of the functional quality of conceptual models.

In this paper, we propose a quality model for functional quality. This model proposes a set of quality attributes for functional quality and a set of metrics to measure these attributes. Moreover, our quality evaluation is enriched with corrective actions provided to the designer, leading to a quality guided modeling process. The rest of the paper is organized as follows. Section 2 is a brief state-of-the-art. Section 3 describes our quality model for functional quality. A first validation based on a survey is described in Section 4. Section 5 concludes and mentions future research directions.

2 Literature Review

Research in software quality is rather mature and has produced several standards such ISO 9126 [1]. This standard defines a set of six characteristics to describe and to evaluate software quality. These characteristics are Functionality, Reliability, Usability, Efficiency, Maintainability and Portability. More precisely, Functionality is defined as a set of attributes that expect the existence of a set of functions satisfying the stated requirements.

In the domain of conceptual modeling, research on quality evaluation is rather young. The first structured approach dates back to the contribution of [2]. They were the pioneers in proposing quality criteria relevant to conceptual schema evaluation (completeness, correctness, minimality, expressiveness, readability, self-explanation, extensibility, normality). In [3], the quality of schemas is evaluated along three dimensions: syntax, semantics and pragmatics. Syntactic quality refers to the degree of correspondence between the conceptual schema and its representation. The semantic quality refers to the degree of correspondence between the conceptual schema and the real world. Finally, the pragmatic quality defines the degree of correspondence between a conceptual schema and its interpretation, which can be defined as the degree to which the schema can be understood. [4] identified the quality framework proposed in [3] to be the only one having a theoretical basis and an empirical foundation. In the context of Business Process Reengineering (BPR), the authors in [5] proposed a four dimensional framework for evaluating models and tools. Functionality is listed as the first dimension in their framework and is defined by the following criteria: expressiveness, structuring, formal/methodological support and relevance of concepts. The second dimension is ease of use. The third dimension is BPR trajectory and finally a general dimension related to tool price and customer support. [6] have reviewed existing frameworks on conceptual modeling quality and found lack of generalizability among the frameworks and lack of collaboration between researchers and practitioners.

The authors in [7] stepped ahead and reviewed cognitive mapping techniques to improve the quality of conceptual models.

To summarize, we could say that conceptual modeling is still considered as an art, which is poorly supported, by methods and tools. Our vision of functionality considers both the requirements and their coverage in the future system, and the modeling principles emerging from analysis and design practices. We argue that functionality of conceptual models cannot be evaluated on the same attributes as those used for software functionality evaluation as the software is not at hand yet. Moreover, as the conceptual modeling occurs early in the development process, it is important to make the emphasis on both the coverage of requirements and the fulfillment of good analysis and design principle.

This paper is a step forward in the description of conceptual models functionality. Moreover, we propose metrics to measure functionality and a set of transformation rules to improve it. These proposals are inserted in a generic quality model that can be applied not only to functionality but also to other criteria. The next section describes this quality model.

3 Quality Model for Functionality

Most of the end-users evaluate their Information systems (IS) based on its functionality. It has been noticed that majority of the IS change-requests results due to deficient functionalities in the information systems such as the lack of desired functionality within a system etc. Similarly, Studies show that defect detection in the early stages of the application development can be 33 times more cost effective than testing done at the end of development [8]. Therefore, it is very effective to catch the defects much earlier in the design phase. Conceptual models serves as the communicating mediator between the end user and the development team. Hence if the conceptual models are scanned for defects and the defects be corrected then it is likely to reduce the number of change requests for the end system. In this paper, we propose a feedback driven quality approach for the functional aspects of the conceptual models. We tend to detect and correct the functionality driven errors in the earlier stage of designing. For this, we have proposed a set of attributes that can evaluate the different aspects of conceptual models with respect to user's desired functionality. Our approach is unique in a way that it not only detects the errors but also provides corrective suggestions or transformations for their rectification.

3.1 Meta-model

In this article, we propose a meta-model driven solution that is generic and simple (Fig. 1). The meta-model starts by defining a quality goal that in this article is about improving the functional aspects of the conceptual models. Then respective quality dimensions are identified and similarly goal specific quality attributes and the corresponding metrics are identified. Once the metrics are calculated, corresponding transformations can be used for improving the quality of the conceptual model based on the quality goal.

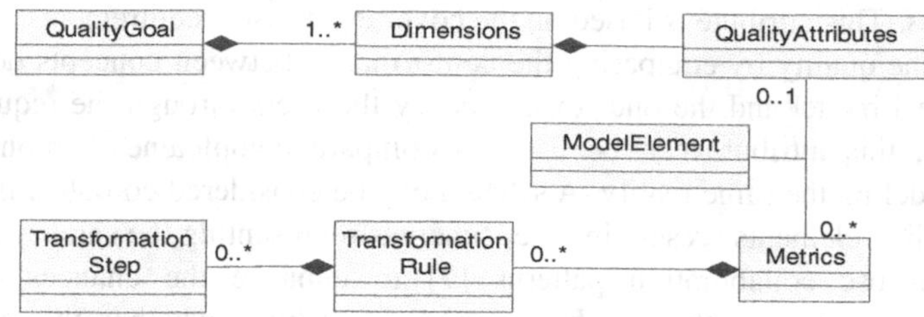

Fig. 1. The underlying meta-model

3.2 Defining a Quality Model for Functionality

In this article, we are interested in evaluating the quality of conceptual model with respect to its functional aspects. Thus, the quality goal is to "evaluate the functionality of the conceptual model". We can use the meta-model in figure-1 to instantiate our goal specific model for evaluation. One of the possible models for the problem in question can be that of Figure-2. The first level of the tree is the functionality dimension. It is composed of six attributes (second level). The third level proposes metrics to evaluate these attributes. The last level suggests transformation rules to improve these quality attributes. This hierarchy relates to the quality goal "Improve the functionality".

Quality Attributes for Functionality. Our solution suggests employing the functionality and its constituting attributes and metrics for evaluating the conceptual model based on the above mentioned quality goal of improving the functional aspects.

Functionality dimension consists of the set of attributes responsible for evaluating the model quality based on functional aspects. These attributes are, directly or indirectly related to the functional quality of the future product and addresses issue that could lead to functional changes in the final product. Furthermore, these attributes tries to identify the key problems that can hamper the functionality of the final product. Some of the attributes that can be used for our quality goal are:

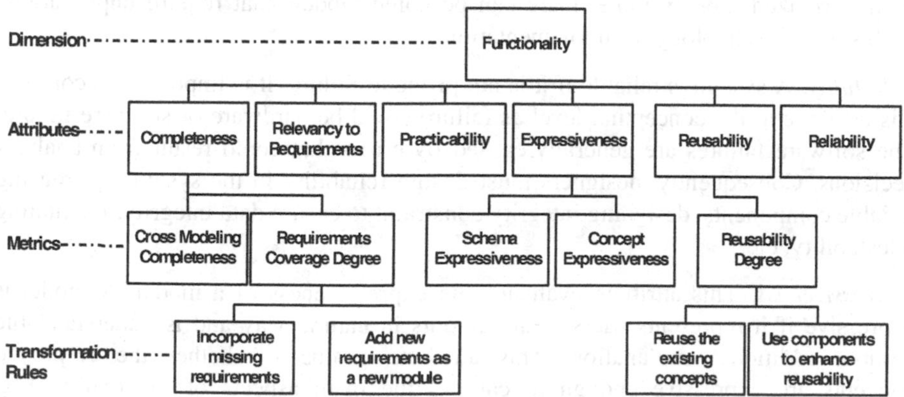

Fig. 2. A quality model for Functionality

Completeness. This attribute is based on the coverage of user requirements. It will try to evaluate the quality by comparing the conformance between concepts depicted in the conceptual model and the ones expressed by the users through the requirements. Furthermore, this attribute can be used to compare completeness among several schemas modeling the same reality. A schema can be considered complete if it covers all the modeling elements present in other schemas representing the same reality. This attribute can use collaboration patterns [9] to enhance the chances of model completeness. Moreover, this attribute can also evaluate whether the number of concepts present in the model corresponds to the number of concepts demanded by the user in their requirements.

Reusability. This attribute has been widely recognized and appreciated in the Object Oriented Paradigm. Reusability is considered a major opportunity for improving quality and productivity of systems development [3]. We choose this attribute to evaluate the quality of the model in twofold: First, to checks whether the model employs the previously developed models (e.g. use of existing modules) and secondly to check whether this model can be reused in future (for example to check if this model is specific or generic). Such an attribute will help in speeding up the process of modeling. Some studies suggest that reusability is feasible only if planned at the design stage because of loss of generalizability at subsequent stages [3]. Reusability is important in our model since it enhances the system's functional reliability because the reused component/module has been tested multiple times therefore errors and deficiencies would have been rectified during its maturity cycle.

Relevancy to requirements. This attribute is different from "Completeness" in a way that it is employed for finding the relevancy between the concepts present in the model and the ones required by the users. It will help in removing the irrelevant concepts present in the model thus will implicitly affect the complexity and functionality dimensions.

Practicability. This attribute is based on the notion of feasibility of the model. It verifies whether the model employs the concepts or elements that are realistic and can be materialized. For example, there can be some models that require unprocurable sophisticated technology for implementation.

Reliability. A system is reliable if it is not prone to failure. It is important to consider this attribute at the conceptual level as failure could be hardware or software failure. The software failures are generally caused by errors that could result from analysis decisions. Consequently, designers must design reliability in the system by; reusing reliable components, designing integrity constraint to ensure data integrity, facilitating it testability, etc.

Expressiveness. This attribute evaluates the expressiveness of a model. A model is expressive if it represents users' requirements in natural way and is understandable without additional explanation. This attribute evaluates whether the employed concepts are expressive enough to capture the main aspects of the reality. E.g. Inheritance link is more expressive than an association. So the more the expressive concepts are used, the more the schema will be expressive. Furthermore, this attribute evaluates the expressiveness by validating whether the existing notations are used to increase the expressiveness or not. For example, it can verify whether the multiplicities are defined in an ER diagram or not.

Quality Metrics Quantifying Functionality. The above mentioned quality model for functionality lists some of the metrics that can be used to quantify our quality attributes. Due to space constraints, we are listing only some of the metrics that are available for functionality. A more complete and a more formal description of metrics could be found in [12, 13].

Requirements Coverage Degree. This metric is based on notion of completeness of user requirements. It has been widely accepted that if the requirements errors are detected earlier in the designing phase then the cost of their rectification gets much lower. This metric calculate the ratio between the concepts covered by the modeling elements in the conceptual schema and the ones expressed by the users through the requirements.

Cross Modeling Completeness. This criterion is used to compare completeness among several schemas modeling the same reality. A schema is considered to be complete if it covers all the modeling elements present in the other schemas. Thus, this metric calculates the ratio between the number of concepts present in the model and the union of all the distinct concepts present in all the schemas representing the same reality.

Reusability Degree. This metric calculates the ratio between the reused concepts and the total concepts present in the model.

Reusability Degree = Reused Concepts / Total Concepts

Overall Model Reuse. This metric is adopted from Basili's[10] metric for overall system reuse. It calculates the aggregated reuse of the whole model by summing the reuse of every individual concept in the model. This metric uses the following formula for calculation:

$$\text{Reuse (Model)} = \sum \text{Reuse (Concept)}$$

Where *Reuse (Concept)* includes the count of all the ancestors of that concept and the concepts aggregated by that concept.

Coupling Between Concepts. This metric is adopted from [11]. It calculates the number of other concepts to which a concept is coupled. Low value for this metric signifies that the model is modular and promotes encapsulation.

Concept Expressiveness. It measures whether the used concepts are expressive enough to capture the main aspects of the reality. For example, Inheritance link is more expressive then association. So the more the expressive concepts are used, the more the schema will be expressive.

Schema expressiveness. It measures the expressiveness of the schema as a whole. A schema is said to be expressive when it represents users' requirements in natural way and can be easily understood without additional explanation. This metric assigns the weights of every concept and then takes the ratio between the calculated total value of the schema and the union of all the schemas describing the same reality.

Transformation Rules for Improvement. Corrective action or transformation rules are the main strength of our proposed solution. Once the quality metrics are calculated, corresponding corrective actions or transformations can be proposed to optimize the model. Due to space constraints, we are just defining two correction actions for the above mentioned two quality metrics.

Requirements coverage degree. If the metric shows that the model doesn't cover all the user requirements then the corrective actions can include the following:

1. Incorporate all the uncovered requirements.
2. If the incorporation of the missing requirements demands major modification to the model and if the model is modular then a new module can be used to address these upcoming requirements and can be interfaced with the existing model.
3. If the incorporation of the missing requirements completely changes the model then the whole model must be retested for conformance.

Reusability degree. If 'reusability degree' metric shows a very low value for reusability then some of the corrective actions could be:

1. Search the model to find the concepts for which equivalent concepts exists in the repository for reusability.
2. Decompose the model into multiple independent modules to facilitate the reusability.

4 Empirical Support

A web-based survey was used to empirically validate the quality model for functionality. The purpose of this survey was twofold:

i. To serve as a validation exercise in providing feedback from professionals including practitioners over the efficacy of our quality model.
ii. To study the general practices and views of the professionals over the quality of conceptual models. This includes the identification of attributes or factors important to professionals for evaluating the quality of conceptual models.

As mentioned above, a web-based survey was formulated to conduct this study. This was a closed survey and was accessible through a special link, provided to the invited participants only to avoid unintended participants. This was a comprehensive survey containing 42 general questions and our model specific questions. However, all the questioned were directly related to the quality of conceptual models. These questions include the two feedback questions where the participants were required to mention the quality attributes/factors that in their view are crucial to the quality of conceptual models. Moreover, they were also required to identify their practice for comparing two conceptual models representing the same reality or modeling the same problem. They were required to identify and mention up to seven attributes/properties that they think they will employ in choosing the best model with respect to their perception of quality.

The survey provides the dictionary and instant help about the definitions and details of all the terms and concepts that were present in the survey including the definitions of all the attributes in our model. Respondents were asked to classify each of our quality attributes into 'directly related to quality', 'indirectly related to quality', 'not related to quality' and 'I am not sure'.

4.1 Sample

In total 179 professionals (including IS managers, IS developers, Researchers etc.) were contacted to complete the survey. However, 57 professionals completed the survey that resulted in the response rate of 31.8%. Among the received 57 responses, three were discarded due to errors in the provided data or incomplete information. Average age of the respondents was approximately 30 years and average modeling experience was 4 years and 3 months.

Respondents belong to different organizations ranging from small organizations having less than 50 employees to as big as having more than 1000 employees. Moreover, respondents were required to select their occupation from a list of fifteen pre-defined occupations.

4.2 Data Analysis

The collected data shows that 85% of the respondents consider the imposition of quality approach on the conceptual models to directly influence the quality of the final product. However, it is interesting to note that 87% of the respondents have never used any method or approach to evaluate the quality of conceptual models. This shows that despite the appreciation of importance of implementing quality approach, professionals do not employ any methods to improve the quality. This behavior can be due to the gap between research and practice. To date there does not exist any quality framework for conceptual models that is standardized and comprehensive enough to accommodate the requirements of the practitioners. However, our proposed model is unique in a way that it is generic, simple and easy to implement. Moreover, the proposed approach follows a hierarchy of different quality levels starting from a quality goal and ending at the corrective suggestions.

As mentioned above, respondents were asked to provide feedback over the efficacy of the above mentioned attributes of our model. They were required to mark these attributes into either 'not related to quality', 'I am not sure', 'directly related to quality' or 'indirectly related to quality'. However since the last two options affirm that the attribute is related to quality therefore we have merged these two options as one to have a clear distinction between the attributes that are related to and not related to quality. The responses are summarized in the Table 1. All the values are in percentages of the responses and are rounded off to the nearest tenth digit. Table-1 should be read as, for example, 75.9 % of the respondents think that 'completeness' is related to quality against 7.4% that think 'completeness' is not related to quality. Similarly, 13% of the respondents declare their inability to categorize 'completeness' in any of four classes.

Table 1. Respondents' feedback on the quality model for Functionality

Attributes	NOT Related to Quality	Related to Quality	Not answered	I am not sure
Completeness	7.4	75.9	3.7	13
Reusability	16.7	64.8	3.7	14.8
Relevancy	3.7	83.3	3.7	9.3
Reliability	11.1	79.6	3.7	5.6
Practicability	9.3	77.8	3.7	9.3
Expressiveness	0	74.1	3.7	22.2

After viewing the above feedback, we can say that the attributes in the functionality dimension are well identified and represent the attributes and factors required by professionals. However, respondents have also identified some attributes that they think are important to quality such as Validity and Degree of abstraction. These attributes will be incorporated in our approach after validation.

5 Conclusion and Implications for Further Research

The functional aspects of conceptual models cannot be evaluated on the same attributes as the ones used for software functionality as the software is not at hand yet. We propose to optimize the overall quality of the IS by ensuring the functional aspects of the conceptual models during the analysis and design phase. In this paper, we have addressed the concept of functionality at the conceptual level. Our approach emphasizes the coverage of requirements and the fulfillment of good analysis and design principles. Our main contribution is a model for evaluating and improving the functionality of conceptual schema. This model has been instantiated from our meta-model and defines a set of quality attributes for functionality refinement. The strength of our approach lies in the post evaluation feedback in the form of corrective actions or transformations. The functionality model has been empirically validated by the professionals through a web based survey. The empirical results show that the respondents consider the identified attributes to be related to functional quality. Our meta-model can also be used to evaluate the conceptual models on other user specified quality goals.

Future directions of this work include:

- The extension and enrichment of the current quality model;
- The development of an environment implementing the proposed quality approach;

References

1. Software Engineering - Product quality - Part 1: Quality model 2001(ISO/IEC 9126)
2. Batini, C., Ceri, S., Navathe, C.: Conceptual Database Design: An Entity-Relationship approach, p. 496. Benjamin/Cummings Publishing Company Inc. (1992)
3. Lindland, O.I., Sindre, G., Sølvberg, A.: Understanding Quality in Conceptual Modeling. IEEE Software 11(2), 42–49 (1994)

4. Maes, A., Poels, G.: Evaluating Quality of Conceptual Modeling Scripts Based on User Perceptions. Data & Knowledge Engineering 63(3), 701–724 (2007)
5. Teeuw, B., Van Den Berg, H.: On the Quality of Conceptual Models. In: Embley, D.W. (ed.) ER 1997. LNCS, vol. 1331. Springer, Heidelberg (1997)
6. Moody, D.L.: Theoretical and Practical Issues in Evaluating the Quality of Conceptual Models: Current State and Future Directions. Data & Knowledge Engineering 55, 243–276 (2005)
7. Siau, K., Tan, X.: Improving the quality of conceptual modeling using cognitive mapping techniques. Data & Knowledge Engineering 55(3), 343–365 (2005)
8. Walrad, C., Moss, E.: Measurement: The Key to Application Development Quality. IBM Systems Journal 32(3), 445–460 (1993)
9. Ribbert, M., Niehaves, B., Dreiling, A., Holten, R.: An Epistemological Foundation of Conceptual Modeling. In: The Proceedings of the 12th European Conference on Information Systems, Turku, Finland (2004)
10. Basili, V.R., Rombach, H.D., Bailey, J., Delis, A.: Ada Reusability and Measurement. Technical Report, Institute for Advanced Computer Studies and Department of Computer Science at the University of Maryland College Park (1990)
11. Chidamber, S.R., Kemerer, C.F.: A Metrics Suite for Object Oriented Design. IEEE Transactions on Software Engineering 20(6), 476–493 (1994)
12. Cherfi, S.S., Akoka, J., Comyn-Wattiau, I.: Measuring UML Conceptual Modeling Quality-Method and Implementation. In: Pucheral, P. (ed.) The Proceedings of the BDA Conference, Collection INT, France (2002)
13. Cherfi, S.S., Akoka, J., Comyn-Wattiau, I.: A Framework for Conceptual Modeling Quality Evaluation. In: The Proceedings of ICSQ 2003 (2003)

Qbox-Services: Towards a Service-Oriented Quality Platform*

Laura González[1,2], Verónika Peralta[3,2], Mokrane Bouzeghoub[2], and Raúl Ruggia[1]

[1] Instituto de Computación, Facultad de Ingeniería, Universidad de la República
Julio Herrera y Reissing 565, 5to piso, 11300 Montevideo, Uruguay
{lauragon, ruggia}@fing.edu.uy
[2] Laboratoire PRiSM, Université de Versailles
45 avenue des Etats-unis, 78035 Versailles cedex, France
mokrane.bouzeghoub@prism.uvsq.fr
[3] Laboratoire d'Informatique, Université François Rabelais Tours
3 place Jean Jaurès, 41000 Blois, France
veronika.peralta@univ-tours.fr

Abstract. The data quality market is characterized by a sparse offer of tools, providing individual functionalities which have their own interest with respect to quality assessment. But interoperating among these tools remains a technical challenge because of the heterogeneity of their models and access patterns. On the other side, quality analysts require more and more integration facilities that allow them to consolidate and aggregate multiple quality measures acquired from different observations. The QBox platform, developed within the ANR Quadris project, aims at filling this gap by supplying a service-based integration infrastructure that allows interoperability among several quality tools and provides an OLAP-based quality model to support multidimensional analysis. This paper focuses on the architectural principles of this infrastructure and illustrates its use through specific examples of quality services.

Keywords: data quality, quality assessment tools, service-oriented architecture.

1 Introduction

Data quality management is a key problem for all kinds of public and private organizations. A large spectrum of commercial and open source tools have been proposed for dealing with data quality problems in information systems. They support many types of data sources (relational databases, XML files, text files, etc.) accessed in different ways (JDBC, ODBC, FTP, etc.) and provide different kinds of functionalities (measurement, analysis, improvement, etc.).

In general, quality tools provide general purpose functionalities that need major adaptation in order to be used with specific data; for example, reusing techniques for measuring data consistency requires the formulation of specific rules modeling data

* This work was partially financed by the "Quadris" project (ANR, France).
http://deptinfo.cnam.fr/xwiki/bin/view/QUADRIS

C.A. Heuser and G. Pernul (Eds.): ER 2009 Workshops, LNCS 5833, pp. 232–242, 2009.

relations. In addition, quality tools are generally stand-alone applications and do not provide interoperation mechanisms, which are a major need in many application domains in order to consolidate and aggregate multiple quality measures acquired from different observations. Furthermore, quality tools manage different quality concepts, at different abstraction levels, expressed with ad-hoc terminology. These limitations generate an important gap between users' quality needs (generally complex requirements combining several quality indicators) and the quality indicators that can be effectively computed from isolated tools. Many organizations experiment the problem of having several sophisticated quality tools but leading to the manual "glue" of tool results. This highlights the importance of developing a quality management platform that handles a unified catalog of quality concepts and allows the interoperation of a variety of tools.

We follow a top-down approach for assessing data quality [20][5], which is a refinement of the Goal-Question-Metric (GQM) paradigm [2]. In this approach, data quality is analyzed at three abstraction levels: (i) at conceptual level, identifying high-level quality goals; (ii) at operational level, enouncing a set of questions that characterize the way to assess each goal, and (iii) at quantitative level, defining a set of quality measures that quantify the way to answer each question and a set of measurement methods for computing them. An advanced prototype, called Qbox-Foundation [5], implements the approach, assisting the analyst in the definition of quality goals, questions and metrics. It includes a set of predefined measurement methods and provides a programming interface for implementing new ones.

In this paper we extend the approach by providing the mechanisms to locate and invoke quality assessment tools, either if they are user-defined or market supplied. Starting from the quality meta-model of the QBox-Foundation, we propose a new mechanism based on a service-oriented architecture. Functionalities of quality tools are described as abstract services. A delegation mechanism allows binding an abstract service to a specific implementation in a given external quality tool. This new architecture will then behave as a mediator between the analyst (whose view is the quality meta-model) and the external quality tools (encapsulated by abstract descriptions of the functionalities they provide).

The contributions of this paper are the following: (i) the definition of a service-based architecture, Qbox-Services, as well as its components (abstract service, adapter, delegation), (ii) elements of the implementation of Qbox-Services and (iii) a case study showing the use of the platform through some examples.

The remaining of the paper is organized as follows: Section 2 presents previous and related work. Section 3 introduces the new service-oriented approach. Section 4 describes a prototype of the platform, whose use is illustrated, in Section 5, in a case study. Finally, Section 6 presents our conclusions and future works.

2 Previous and Related Work

Qbox-Foundation has been experimented in several application scenarios (data warehousing, CRM, medical data) [5,9,6,14,1]. The main developments consist in instantiating the GQM approach in specific application contexts and programming the quality processes which evaluate the selected quality factors depending on the metrics chosen and the target objects to evaluate (data files or business processes).

The experiments performed on these applications have shown the relevance and the usefulness of the Qbox-Foundation, in particular its ability to characterize quality goals with multidimensional factors, to reuse basic measurement processes and to aggregate measurement values along defined time intervals. The refinement process of quality goals and the browsing facilities provided through quality factors and quality methods have been appreciated as powerful tools which drive quality analysts and improve productivity of quality assessment procedures.

However, the same experiments revealed some limitations of this first version of the QBox: incompleteness of quality methods and lack of connectivity to integrate functionalities of existing tools. Actually, both limitations can be seen as a problem of scalability. On the one side, developing within the QBox all possible quality methods is time and money consuming although the results might be more efficient regarding user's needs. On the other side, many tools in the market provide generic quality methods which can be applied to many data sources whatever their models and access patterns are. However these tools do not interoperate with each other and do not share a common quality model. Consequently, extending the current QBox version to achieve connectivity requirements will solve the scalability problem.

A variety of quality tools has been proposed in the last years. Many of them provide either low-level profiling functionalities (e.g. number of tuples, number of null-values) or quality-oriented functionalities (e.g. rule validation, duplicate search) [4,18,11,16]. *Datacleaner* [4] allows profiling, validating and analyzing data through the identification of string patterns, dictionary lookup, JavaScript validation rules and regular expression validation. *Talend Open Profiler* [18] generates statistics of many types. *Oracle Data Profiling* [11] allows monitoring quality metrics and discovering rules. *Joopelganger* [16] allows similarity checking in addresses lists. Other tools also allow data cleaning, standardization and duplicate elimination [17,15,8]: *Power-MatchMaker* [17] allows cleaning and eliminating duplicates in addresses. *Aggregate Profiler* [15] allows enriching data after profiling, filtering, checking for similarities and processing real-time alerts. *Open Data Quality* [8] manages data from multiple sources allowing the matching, standardization and cleaning of those data.

These tools are stand-alone applications, and have no facilities for interacting with other tools. Some development environments have been proposed for dealing with the integration of quality measurement and improvement functionalities [19,13,11,10]. They provide a graphical environment for designing data transformation operations and manage the execution and coordination of such operations. *Talend Open Studio* [19] generates code for executing the graphically-defined operations. *Pentaho DI (Kettle)* [13], *Microsoft Integration Services* [10] and *Oracle Data Quality* [11] also allows invoking executable files.

To the best of our knowledge there exists no platform allowing the invocation of remote quality tools and providing directory services for locating and invoking tools.

3 Qbox-Services Architecture

In this section we describe a service-oriented architecture, called Qbox-Services, which addresses the interoperability and scalability limitations enounced in previous section. First, we present the main components and concepts involved in the architecture and the way they interact with each other. Then, we bring some details on the discovery and invocation of quality services. In next section we describe the development of a prototype of Qbox-Services; more details can be found in [7].

3.1 General Overview and Architecture

Applying a service-oriented architecture enables to model the loosely-coupled and distributed characteristics of Qbox-Services. As this system would be Internet-distributed and highly dynamic in terms of service provision, then a SOA-based modelling appears to be an adequate approach for modelling and implementation. SOA (Service-Oriented Architecture) is a logical way of designing a software system to provide services either to end-user applications or other services distributed in a network, via published and discoverable interfaces. SOA provides a flexible infrastructure and processing environment by provisioning independent, reusable application functions as services and a robust foundation for leveraging services [12].

Qbox-Services platform consists of three main components called Qbox-Foundation, QMediator and QManagement. These components, and other important concepts involved in the architecture, are sketched in Figure 1.

The Qbox-Foundation component provides the functionalities for specifying a quality model based on the GQM paradigm [2]. It allows defining high-level quality goals and choosing a set of quality factors and metrics that characterize the way to assess each goal. Metrics are chosen, refined and specialized from a general quality meta-model, which constitutes a library of quality concepts [5]. Besides, personalization and binding functionalities help the quality expert to derive, from the generic quality meta-model, a Personalized Quality Model (PQM), i.e. a refined set of quality factors and metrics that correspond to specific quality goals and refer to specific IS objects. Details on Qbox-Foundation can be found in [5].

As a result, a PQM embeds specific quality requirements that can be answered by executing appropriate quality services. Quality services may be of three types: *measurement services* that compute quality metrics; *analysis services* that analyze a set of measures and calculate complex indicators and *improvement services* that perform data updates in order to improve quality. Examples of these types of services are the calculation of the percentage of null values, the analysis of growing/decreasing tendencies in data freshness and the elimination of duplicate tuples, respectively.

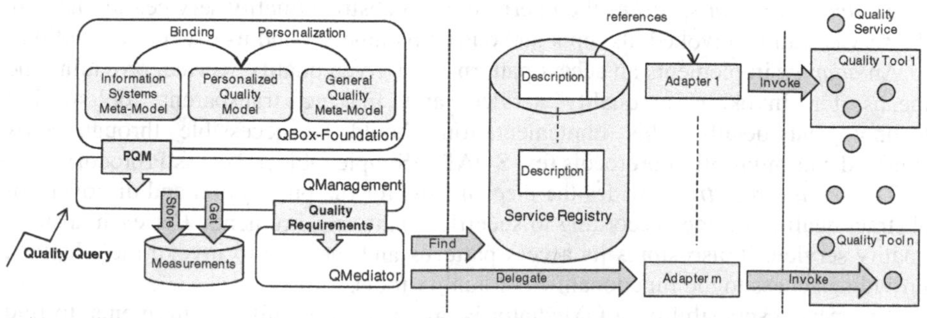

Fig. 1. Qbox-Services Architecture

The *QMediator* component provides functionalities for finding appropriate services for these requirements, enabling their execution and returning their results. To this end, it accesses a service registry that contains abstract service descriptions and access patterns of available quality services. A set of adapters implement these access patterns and invoke the quality services encapsulating technological details.

QMediator acts as a mediator between quality requirements expressed in the PQM (e.g. the need to assess the metrics of a given goal) and the available quality services (especially those calculating such metrics).

The *QManagement* component executes the required quality services for specific goals and provides an interface for analyzing results. Quality services may be periodically executed, may be punctually invoked by a user or may be triggered by another service (e.g. an analysis service that needs comparing some non-available measures). The measurement results are stored in a star-like database schema which favors the aggregation of measures, the computation of complex indicators or even the analysis of correlations among measures. Result analysis is carried out by evaluating quality queries, called *Qolap queries,* defined over the PQM. QManagement includes a decision-support interface that allows browsing in the star-like quality model and posing Qolap queries. If some of the necessary measures are not available, the corresponding services may be executed in order to obtain them. The implementation of QManagement is out of the scope of this paper.

Next sub-section presents more details on the QMediator component.

3.2 Mediation of Quality Services

QMediator acts as a mediator between quality requirements, taken from the PQM or stated by the user, and quality services listed in a service registry. We describe functionalities of quality tools as abstract quality services and propose a delegation mechanism which binds an abstract quality service to a specific implementation in a given quality tool. The fundamental concepts handled by QMediator are: quality services, abstract quality services, access patterns, adapters and service registry.

We define a *quality service* as an implementation of a *quality functionality* that can be either custom implemented or provided by external quality tools.

An *abstract quality service* is the description of the quality functionality provided by a quality service. This description could include, for example, the quality concepts it addresses (dimensions, factors and metrics) or the type of functionality it provides (measurement, analysis or improvement).

An *access pattern* specifies the interface of an abstract quality service, i.e. the way the service can be invoked, its input and output parameters and its exception handling.

An *adapter* implements an access pattern for a given quality service, providing the means for invoking a quality service and making transparent tool-specific technological details. This implementation should be accessible through SOA standard communication protocols (ex. SOAP – Simple Object Access Protocol).

The *service registry* provides the mechanisms for the publication and discovery of abstract quality services according to specific quality requirements. For each abstract quality service, it also stores its access patterns and the way to invoke the adapters, providing access to the functionalities of quality tools.

The main responsibility of QMediator is, given some quality requirements, to find and execute the quality services that best match these requirements. To this end,

QMediator has to find in the registry the abstract quality services which declare fulfilling the given quality requirements, select the abstract quality services to use (automatically or according to user input), get the corresponding access patterns from the service registry, delegate the invocation of the quality services to the adapters components, and finally, consolidate the result of the invocations in a unified result.

4 Qbox-Services Prototyping

In this section we describe the development of a prototype, based on the Web Services technology, which implements the Qbox-Services architecture presented in the previous section. The development of the prototype consisted in two main tasks: specification and implementation of the QBox-Services components and the incorporation of quality services of third-party tools, through the implementation of adapters and the publication of abstract service descriptions in a service registry.

4.1 QBox-Services Components

In order to implement the services registry component, we leveraged an open source implementation of an UDDI (Universal Description, Discovery and Integration) registry, called jUDDI. An UDDI registry provides the mechanisms for publishing and finding business entities and business services, and the means to describe them in various ways.

Taking advantage of the mechanisms provided by UDDI, we represent an abstract quality service as a business service published in the UDDI registry. Additionally, using the categorization mechanisms provided by UDDI, we specify the type of quality functionalities the services provide (measurement, improvement or analysis), the quality concept they address (i.e. the quality metrics) and the IS object types they support (relational databases, XML files, etc.). We describe access patterns using WSDL (Web Services Description Language). These descriptions, and the access points to invoke the adapters, which implement the access patterns, are stored in the UDDI registry and associated with the corresponding abstract quality services [3].

Finally, the QualityMediator and QualityManagement components were coded in Java. We also developed a RegistryService component that encapsulates the access to the UDDI Registry. All the components were deployed in a JBoss Application Server.

4.2 Incorporation of Quality Services

The prototype incorporated quality services provided by external tools, as well as custom-developed quality services. Specifically, we integrated quality services provided by the DataCleaner Project [4], an open source project which provides functionalities for profiling, validating and comparing data. The profiling functionalities are grouped in profiles: standard measures profile (highest and lowest values, number of null values, row count), number analysis profile (mean, sum, variance, etc.), time analysis profile (highest and lowest values), etc. Regarding validation, DataCleaner provides functionalities to perform validations based on dictionaries, regular expressions, ranges of values and JavaScript scripts. The DataCleaner tool is coded in Java and supports many types of data sources, e.g. files in different formats (csv, xml, txt) and relational databases (accessed via JDBC).

In order to incorporate DataCleaner quality services in our prototype, we describe their access patterns using WSDL, implement their adapters in Java and expose them as Web Services. Then, we publish this information, along with the abstract quality service descriptions, in the UDDI registry.

DataCleaner tool consists of two main components: Core and Gui (graphical user interface). The Core component provides an API for executing a specific profile (for instance standard measure profile) returning the results in a matrix. The adapters for DataCleaner functionalities call the API methods in the same way the Gui module does, as shown in Figure 2. Table 1 presents various quality services provided by the DataCleaner tool which were published in the UDDI registry.

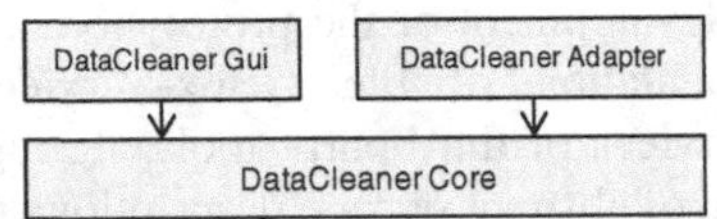

Fig. 2. Adapters accessing DataCleaner Logic

Table 1. Examples of quality services provided by DataCleaner

Quality Service		Parameters		Categorization in UDDI Registry		
Name	Description	IN	OUT	Quality Concept	Quality Functionality	IS Type
getRowCount	Returns the number of rows in a given table	t: table_name	int	any	measurement	table
getLowest-Value	Returns the lowest date value in a given column	t: table_name, c: column_name	date	any	measurement	column
getDictionary-Validation	Returns the number of not valid values in a column according to the specified dictionary	t: table_name, c: column_name, d: dictionary	int	any	measurement	column
getRegExp-Validation	Returns the number of not valid values in a column according to the specified regular expression	t: table_name, c: column_name, r: reg_expression	int	any	measurement	column

Table 2. Example of custom developed quality service

Quality Service		Parameters		Categorization in UDDI Registry		
Name	Description	IN	OUT	Quality Concept	Quality Functionality	IS Type
getSyntactic-CorrectnessRatio	Ratio of syntactically correct values in the column according to a regular expression	t: table_name, c: column_name, r: reg. expression	double	syntactic correctness ratio	measurement	table

In the same way, we develop adapters, describe access patterns and publish several functionalities for custom-build quality services. Some of these services call other services to calculate intermediate results. For instance, Table 2 shows the getSyntacticCorrectnessRatio quality service which, using the getRegExpValidation and the getRowCount services provided by DataCleaner, calculates the syntactic correctness ratio of a given column.

5 Qbox-Services Applied to CRM Data: A Case Study

In this section, we illustrate the use of Qbox-Services for evaluating data quality in a real application: a Customer Relationship Management (CRM) application in the area of financial business, in a database replication scheme[1]. The customers database is some decades old and has suffered several schema transformations, data migrations and fusions that cause most of the current quality problems. Other changes were originated by new regulations and new commercial goals, which led to the definition of new business rules (not necessarily satisfied by old data).

Several quality goals have been defined in this application context [9]. In this case study we address some of them:

- Goal 1: Obtaining and managing up-to-date information about customers
- Goal 2: Assuring that data satisfies all business validation rules
- Goal 3: Migrating data to current formats and assuring all data has allowed values
- Goal 4: Detecting and consolidating customers that are registered multiple times

These goals have been decomposed and refined in quality questions and several quality metrics have been defined for them; Table 3 lists the ones selected for this case study. The complete list of goals, questions and metrics can be found in [9].

In an initial stage, a set of measurement methods was implemented as SQL stored procedures in a MS SqlServer 2005 DBMS. The obtained measures served to diagnose the quality of the customers' database and to define and prioritize cleaning tasks. In a second stage, we evaluated the use of the QBox-Services platform as an alternative to the custom-built stored procedures.

During this second stage, we use QMediator interface to look for services which could implement each metric. In this process, we faced three situations: (i) we found one or more services that implement the metric; (ii) we did not find any service that fully implements the metric, but we did find services that could be used to implement the metric (adding some code or composing services); and (iii) we did not find any service that fully or partially implementes the metric. In the remaining of this section we present examples of these three situations and explain why the use of Qbox-Services has advantages over the approach taken during the initial stage.

When we examine in detail the *Ratio of valid addresses* metric, we realized that most of the rules that were used in the initial stage to validate addresses could be specified in terms of regular expressions. As a result, we found that the *getSyntacticCorrectnessRatio* service, described in Table 2, could be directly used to implement the metric. In this situation the advantage of using QBox-Services is clear: no coding was required and the invocation of the quality service is transparent. Depending on the complexity of the calculus this advantage can lead to faster and cheaper solutions.

Besides, in the initial stage, the *Maximum age of cells* metric was calculated by comparing the last modification date of each tuple (stored in an attribute of the *Customers* table). Although, we did not find any quality service that could be directly used to implement the metric, we noticed that the *getLowestValue* service (described in Table 1), can solve part of the problem. In this way the *Maximum age of cells* of a

[1] The name of the company is omitted for confidentiality purposes.

Table 3. Goals, questions and metrics for the case study

Goal	Question		Factor	Metric
1	1.1	How old is customer data?	Age	Maximum age of cells
2	2.1	Does customer data satisfy validation rules?	Domain integrity	Ratio of tuples satisfying validation rules
3	3.1	Are addresses valid?	Syntactic correctness	Ratio of valid addresses
	3.2	Does data belong to appropriate ranges?	Syntactic correctness	Ratio of cells belonging to the corresponding domain ranges
4	4.1	How many duplicate customers are there?	Uniqueness	Ratio of non-duplicated customers

table *T* can be easily implemented by the following formula: *now() – get LowestValue (T,A)*, being *A* the attribute storing the last modification date and *now()* a function returning the current date. So, in this situation we had to develop a simple quality service, which calls the *getLowestValue* service and computes the previous formula. Although in this case some coding was required and there was some overhead for describing and publishing the implemented quality service, we noticed some advantages of using the QBox-Services platform. First, the implementation of the quality service was not as complex as the implementation in the initial stage. Additionally, after incorporating the implemented quality service in the QBox-Services platform, this service could then be used to implement other metrics or to build more complex quality services. This encourages reusability and addresses the scalability problem stated before.

Finally, we faced situations where we did not find any method in the QBox-Services platform to implement a specific metric. In this case, we first search for available quality tools which could provide the required functionality, with the purpose of incorporating their quality services in the QBox-Services platform. Specifically, for calculating the *ratio of non-duplicated customers*, we found the *Joopelganger* tool for similarity checking and the *Aggregate Profiler* tool for duplicate detection. Although, in this last situation the solutions are more costly and time consuming, it is important to note that being able to incorporate quality services provided by different quality tools encourage not only reusability but also interoperability among tools.

6 Conclusions

In this paper, we proposed a quality assessment platform, Qbox-Services, based on service-oriented mechanisms. Its architecture allows interoperation between user-defined quality methods and external quality tools, giving the user the possibility to access a larger set of quality functionalities and reducing implementation efforts. In addition, performing quality evaluation through external services would enable to face complexities in computation and data management, as well as privacy issues.

Qbox-Services uses the Qbox-Foundation component for defining and personalizing a quality model according to user quality goals and the underlying information system. A new QMediator component was developed which, using Web Service technologies, manages a dynamic library of quality tools (catalogued in a service registry) and provides functionalities for finding quality services according to quality requirements. Adapters for several tools were also developed and registered in the service registry. The QManagement component is currently being implemented; the formalism for expressing quality queries is one of our current research directions.

Our approach was validated in a CRM application. Our experimentations showed that although we could integrate quality services provided by various open source projects (with minor technical difficulties), integrating quality services from other commercial and not commercial tools could involve a major effort. However, this effort can be worthy when we deal with complex quality services whose implementation from scratch could represent a major cost. Additionally, once the service is integrated in the platform, it could be reused and combined, so in the long run, the cost required to perform the integration can also be worth the effort.

The SOA-based approach has pros and cons. On one side it involves a rather complex design and communication management between the system components. But, on the other side, the application of SOA enables to scale up to an Internet distributed architecture and provides flexibility and robustness for an evolving environment. We believe that a key aspect of quality evaluation environments relies on the possibility of integrating external methods for measuring the quality of specific data, which may be provided "as a service" (without disclosing the code) for expert organisations. The here proposed architecture is a step forward in this direction, and intends to highlight and experiment on the main related issues. The ultimate goal is to make available the service registry allowing the incorporation of third-party adapters.

References

1. Akoka, J., Berti-Equille, L., Boucelma, O., Bouzeghoub, M., Comyn-Wattiau, I., Cosquer, M., Goasdoué-Thion, V., Kedad, Z., Nugier, S., Peralta, V., Quafafou, M., Sisaid-Cherfi, S.: Evaluation de la qualité des systèmes multisources: Une approche par les patterns. In: 4th Data and Knowledge Quality Workshop (DKQ 2008), France (2008)
2. Basili, V., Caldiera, G., Rombach, H.D.: The Goal Question Metric Approach. In: Encyclopedia of Software Engineering, vol. 1, pp. 528–532. John Wiley & Sons, Inc., Chichester (1994)
3. Colgrave, J., Januszewski, K.: Using WSDL in a UDDI Registry. Oasis Technical Note, http://www.oasis-open.org/committees/uddi-spec/doc/tn/uddi-spec-tc-tn-wsdl-v2.htm (access: 10/2008)
4. DataCleaner: The DataCleaner Project, http://eobjects.org/trac/wiki/DataCleaner (accessed 12/2008)
5. Etcheverry, L., Peralta, V., Bouzeghoub, M.: Qbox-Foundation: a Metadata Platform for Quality Measurement. In: 4ème Atelier Qualité des données et des Connaisances (QDC 2008), Nice, France (2008)
6. Etcheverry, L., Graña, M., Marotta, A., Naya, H., Raggio, V., Ruggia, R.: Enabling GWAS Meta-Analysis through data quality management. In: Microsoft eScience Workshop, Indianápolis, USA (2008)
7. Gonzalez, L.: Qbox-Services: Towards a Service-Oriented Quality Platform. Internal Report, Université de Versailles (2009), http://www.fing.edu.uy/~lauragon/qboxservices/techreport.pdf
8. Java.net: Open-DM-DQ: Open Data Quality Project, https://open-dm-dq.dev.java.net/ (accessed: 10/2008)

9. Martirena, E.: Medición de la calidad de datos: un enfoque parametrizable. Master Thesis, Universidad de la República, Uruguay (2008)
10. Microsoft SQL-Server: Integration Services, http://www.microsoft.com/sqlserver/2008/en/us/integration.aspx (accessed: 10/2008)
11. Oracle: Oracle Data Quality and Oracle Data Profiling, http://www.oracle.com/products/middleware/odi/oracle-data-quality.html/ (accessed: 10/2008)
12. Papazoglou, M., Traverso, P., Dustdar, S., Leymann, F.: Service-Oriented Computing: State of the Art and Research Challenges (November 2007)
13. Pentaho: Pentaho Commercial Open Source Business Intelligence: Kettle Project, http://kettle.pentaho.org/, http://sourceforge.net/projects/joppelganger/ (last accessed 10/2008)
14. Sastre, D., Peralta, V., Ruggia, R.: Evaluación de Calidad en una Aplicación de Data Warehousing: de la Definición de Metas a la Especificación de Métricas. In: 6th Chilean Workshop on Databases (WBD 2008), Chili (2008)
15. SourceForge: Open Source Data Quality and Profiling, http://sourceforge.net/projects/dataquality/ (accessed: 10/2008)
16. SourceForge: Joppelganger, http://sourceforge.net/projects/joppelganger/ (accessed: 10/2008)
17. SQLPower: PowerMatchMaker, http://www.sqlpower.ca/matchmaker (accessed: 10/2008)
18. Talend: Talend Open Profiler, http://www.talend.com/products-data-quality/talend-open-profiler.php (accessed: 10/2008)
19. Talend: Talend Open Studio, http://www.talend.com/products-data-quality/talend-open-studio.php (accessed: 10/2008)
20. Vassiliadis, P., Bouzeghoub, M., Quix, C.: Towards Quality-oriented Data Warehouse Usage and Evolution. Information Systems 25(2), 89–115 (2000)

Preface to RIGiM 2009

Colette Rolland[1], Eric Yu[2], Camille Salinesi[1], and Jaelson Castro[3]

[1] Université Paris 1 Panthéon - Sorbonne, France
[2] University of Toronto, Canada
[3] Universidade Federal de Pernambuco, Brazil

The use of intentional concepts, the notion of "goal" in particular, has been prominent in recent approaches to requirement engineering (RE). Goal-oriented frameworks and methods for requirements engineering (GORE) have been keynote topics in requirements engineering, conceptual modelling, and more generally in software engineering. What are the conceptual modelling foundations in these approaches? RIGiM (Requirements Intentions and Goals in Conceptual Modelling) aims to provide a forum for discussing the interplay between requirements engineering and conceptual modelling, and in particular, to investigate how goal- and intention-driven approaches help in conceptualising purposeful systems. What are the fundamental objectives and premises of requirements engineering and conceptual modelling respectively, and how can they complement each other? What are the demands on conceptual modelling from the standpoint of requirements engineering? What conceptual modelling techniques can be further taken advantage of in requirements engineering? What are the upcoming modelling challenges and issues in GORE? What are the unresolved open questions? What lessons are there to be learnt from industrial experiences? What empirical data are there to support the cost-benefit analysis when adopting GORE methods? Are there application domains or types of project settings for which goals and intentional approaches are particularly suitable or not suitable? What degree of formalization and automation, or interactivity is feasible and appropriate for what types of participants during requirements engineering?

This year, five high quality papers were accepted out of the 16 initially submitted. Two sessions entitled "modelling", and "elicitation issues" were organized to discuss these papers. The papers were allocated as follows:

Session 1: Modelling

- Clarissa Borba and Carla Silva. A Comparison of Goal-Oriented Approaches to Model Software Product Lines Variability.
- Daniel Amyot, Jennifer Horkoff, Daniel Gross, Gunter Mussbacher. A Lightweight GRL Profile for i* Modeling.

Session 2: Elicitation Issues

- Luiz Olavo Bonin. From User Goals to Service Discovery and Composition.
- Bruno Claudepierre, Selmin Nurcan. ITGIM: An Intention-Driven Approach for Analyzing the IT Governance Requirements.
- Sandra António, João Araújo, Carla Silva. Adapting the Framework i* for Software Product Lines.

C.A. Heuser and G. Pernul (Eds.): ER 2009 Workshops, LNCS 5833, p. 243, 2009.

A Comparison of Goal-Oriented Approaches to Model Software Product Lines Variability

Clarissa Borba[1] and Carla Silva[2,*]

[1] Centro de Informática, Universidade Federal de Pernambuco, 50740-540, Recife, Brazil
ccb@cin.ufpe.br

[2] Centro de Ciências Aplicadas e Educação, Universidade Federal da Paraíba, 58297-000, Rio Tinto, Brazil
ctaciana@ccae.ufpb.br

Abstract. In the requirements engineering for software product lines (SPL), feature modeling is used to capture commonalities and variabilities in system families. However, it is a great challenge to establish the relationship among features in an application and stakeholders' goals. This makes it difficult to justify why a specific feature configuration is required, for example. On the other hand, goal-oriented requirements engineering provides a natural way to identify and specify how the stakeholders' interests and concerns might be addressed by the intended system. The strength of goal modeling to represent commonalities and variabilities in early stages of software product lines development has been recognized. As a result some goal-oriented approaches for modeling requirements variability in SPL have been recently proposed. In this paper we perform a comparison among existing goal-oriented techniques and then, we propose a new extension to the i* framework to capture common and variable requirements in software product lines.

Keywords: Goal Oriented Requirements Engineering, Software Product Lines.

1 Introduction

A feature model [1] can represent commonalities and variabilities in software product lines (SPLs). Feature models can be used to define core assets from which products for individual users can be derived in a cost-effective way. Features may model parts of a system which correspond to entities, entity attributes, processes, or non-functional properties [2].

The variability of a product line has to be documented explicitly to enable a strategic reuse of requirements artifacts. This means it should relate different types of requirements, such as organizational, non-functional and functional, and keep trace among them. But a feature model is a very concise taxonomic form, in which features are modeled as symbols [3]. Thus, feature models do not capture explicitly non-functional

* This work was partially performed at Centro de Informática – Universidade Federal de Pernambuco and partially supported by CAPES Proc. PNPD-0092088 research grant.

C.A. Heuser and G. Pernul (Eds.): ER 2009 Workshops, LNCS 5833, pp. 244–253, 2009.

requirements and the positive/negative influence from these requirements to the system configuration alternatives elicited during the development of the core assets. This influence could help to choose a specific configuration for an application to meet organizational goals. Goals are states of affairs in the world that the stakeholders want to attain while features are characteristics the system must present. In this light, it seems clear that goal-oriented approaches could be used to capture features using more meaningful models, to keep trace of system features to their motivations and to reason about the implications of the system configuration alternatives to the achievement of the stakeholders' goals.

Goal models provide a natural way to identify variability at the early requirements of SPL [4, 5]. Goal-oriented approaches can be used as an effective way for discovering SPL common and variable requirements, and reduce time and costs associated with the configuration of an individual product in a product family. Since some goal-oriented approaches for SPL requirements have recently been proposed to capture the semantics of the feature model, we decided to compare these approaches taking into account expressiveness related to all the possible variability modeling provided by the feature model [1, 6]. In this paper we also propose a new extension to the i* framework [7] to capture common and variable requirements in software product lines. We present guidelines to map feature models to our approach.

This paper is organized as follows. Section 2 overviews the software product line engineering and feature modeling. Section 3 presents the goal-oriented approaches. Section 4 explains and illustrates our proposed approach. Section 5 summarizes this paper and points out future directions for our work.

2 Software Product Line and Feature Modeling

According to [8], a Software Product Line (SPL) is a set of software-intensive systems sharing a common, managed set of features satisfying the specific needs of a particular market segment or mission they are developed from a common set of core assets in a prescribed way. Feature-oriented domain analysis gathers abstract concepts of the domain and organizes them as features [1]. A feature is a system property that is relevant to some stakeholder and is used to capture commonalities or discriminate it among systems in a family. Feature modeling can be used at any stage of the software product line engineering (e.g., requirements, architecture, design) and for any kind of artifacts (e.g., code, models, documentation). At an early stage, feature modeling enables to decide which features should be supported, or not, by a product line [9].

At its essence, a product line involves core asset development (also known as Domain Engineering) and product development using the core assets (also known as Application Engineering). Commonalities, as well as the flexibility to adapt to different product requirements are captured in core assets. Those reusable assets are created during domain engineering process. During application engineering, products are either automatically or manually assembled, using the assets created during the previous process and completed with product-specific artifacts. Thus, products differ by the set of features they include to fulfill customer requirements [10].

The feature model [1, 6] describes the configuration space of a system family. An application engineer may specify a member of a system family by selecting the

Table 1. Feature Model Concepts

	Concept	Meaning	Explanation
(A)	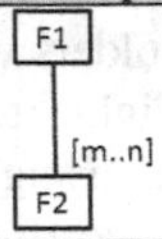 F1 [m..n] F2	Solitary Feature with cardinality [m..n]	Feature cardinalities are placed on solitary features and denoted using square brackets; for example, a feature cardinality of [1..k] indicates that at least one and at most *k* clones must be present in a concrete product.
(B)	 F1 F2	Solitary Feature with cardinality [0..1] (optional)	An optional feature may or may not be present in a concrete system. The feature cardinality is [0..1].
(C)	 F1 F2	Solitary Feature with cardinality [1..1] (mandatory)	A mandatory feature must be present exactly once in a concrete system. The feature cardinality is [1..1].
(D)	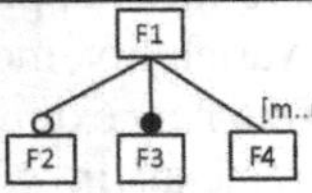 F1 [m..n] F2 F3 F4	Binary relations which includes optional, mandatory and cardinality–based like relations [16]	In binary relations its sub-features are of different types.
(E)	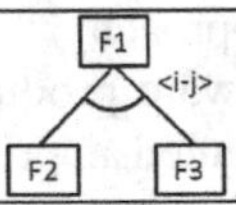 F1 <i-j> F2 F3	Feature Group with group Cardinality <i-j>	Group cardinality is an interval placed on a feature group that denotes how many grouped features can be selected from the feature group in a concrete system.
(F)	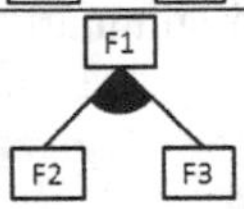 F1 F2 F3	Feature Group with group Cardinality <1-k>, k= size of the group (inclusive-or)	A group cardinality of <1..k> indicates that at least one and at most *k* features can be selected from the feature group, and that *k* is the total number of grouped features in the feature group.
(G)	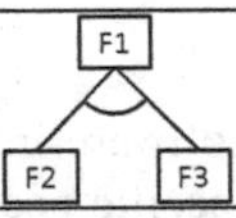 F1 F2 F3	Feature Group with group Cardinality <1-1> (exclusive-or)	A group cardinality of <1..1> indicates that exactly one feature can be selected from the feature group.

desired features from the feature model within the variability constraints defined by the model. Table 1 presents the types of features and relationships among feature that there exists in the feature modeling approach [6, 9].

To illustrate a feature model, let us consider a family of online Business to Customer (B2C) solutions presented in [2] and extended with some features of the Security profile presented in [9]. It is represented using the cardinality-based feature model [6] and a fragment of this model is shown in Figure 1. The model in Figure 1 contains one feature diagram, with ECommerce as its root feature.

The root feature has four solitary sub-features: Payment, Shipping, PasswordPolicy and Security. The filled circle symbol indicates that these mandatory sub-features have a feature cardinality of [1..1], meaning that the feature must exist once and only once. On the other hand, the empty circle symbol indicates that PaymentTypes is an optional feature with cardinality [0..1]. The cardinality of Method(String) feature is [1..*], meaning that there can be at least one clone of this feature in a concrete product. Available PaymentTypes, in this case DebitCard, PurshaseOrder and CreditCard, are members of a feature group. The filled ramification symbol denotes a

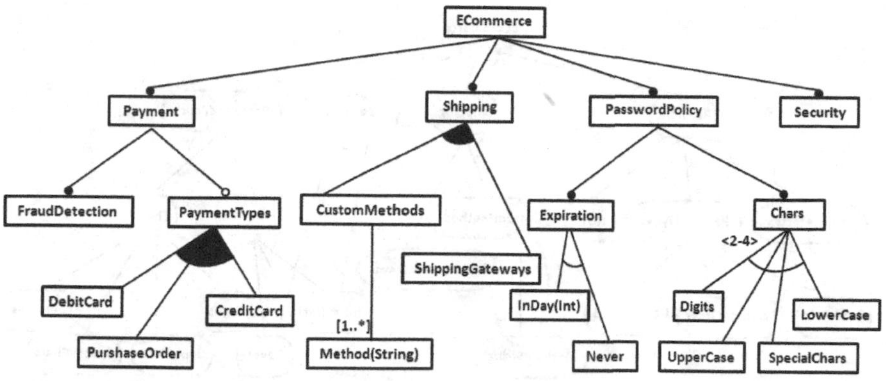

Fig. 1. ECommerce Feature Model (adapted from [2])

group with cardinality <1– k>, where k is the group size. Thus available payment types can be any non-empty subset of the three payment types. The empty ramification symbol under Expiration feature denotes a group with cardinality <1–1>, meaning that there can be only one kind of expiration of the password. The empty ramification symbol with the cardinality <2-4> under Chars feature denotes a group in with at least 2 and at most 4 sub-features can be chosen in a product configuration.

In the sequel we try to model the same ECommerce problem using different goal oriented approaches. We then discuss their limitations.

3 Goal-Oriented Approaches to Model SPL Variability

Yu and others [11] uses, as starting point, goal models that represent how both the system and its operating environment can achieve stakeholders goals. From these models, they identify which goals are assigned of the system and which are assigned to its environment. The latter are replaced by NOP (no operation) goals, meaning that they must not be mapped to features. The remaining goals must be mapped to features according to the mapping presented in Table 2. They recognize that AND/OR decompositions cannot be directly mapped to Alternative and Optional feature sets. The feature models produced by this approach are restricted to (i) features decomposed into sub-features of a single type (mandatory, optional), (ii) features cannot be grouped using general cardinality such as <i,j>. The produced feature model can be re-structured if appropriate. Figure 2 depicts the models resulting from the mapping from the feature model presented in Figure 1 to the goal modeling presented in [11]. In this representation we could add the influence of some alternatives in relation to the satisfaction of the Security softgoal. Observe that the Expiration goal can be achieved through two different ways: the password can expire in days or never expire. Each alternative contributes in a different way to satisfice the Security softgoal. The highlighted circles in Figure 2 indicate situations where concepts (A) and (E) shown in Table 1 should have been used. Unfortunately, the approach presented in [11] does not support cardinality and then, the expressivity is limited. In this example, as a possible solution, we could have used AND

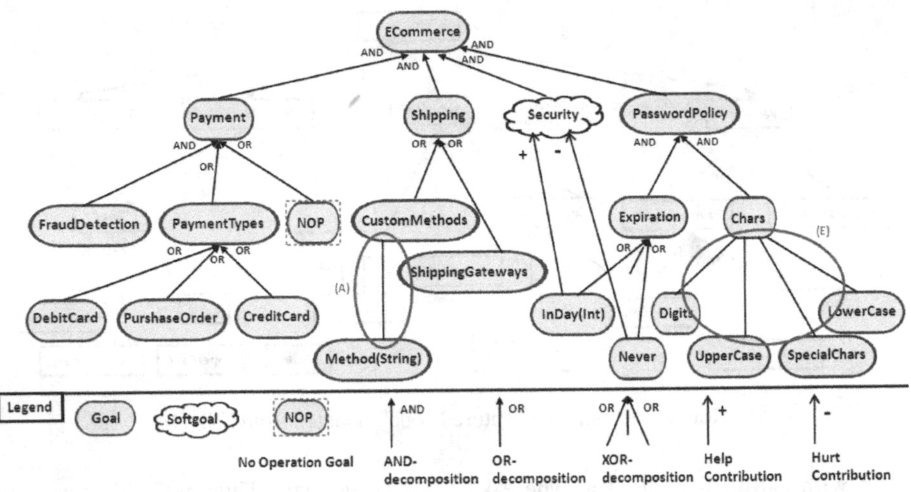

Fig. 2. ECommerce Goal Model

decomposition to reflect (A) and OR decomposition to reflect (E). However, in both cases, they not able to capture the same meaning of Figure 1.

AOV-graph is an aspect-oriented intentional model, represented by AND/OR decomposition graphs. Aspect-orientation aims at handling crosscutting concerns (concerns that cut across the boundaries of other concerns) explicitly, providing means for their systematic identification, modularization, representation and composition during all phases of the software development lifecycle [12]. AOV-graph's relationships map not just positive and negative conflicts between requirements (goals, softgoals and tasks), they map how these requirements crosscut each other and they also represent choices of different options of how a given requirement may be achieved, being a reasonable choice to represent variability. Batista and others [13] extends the AOV-graph to represent variability in SPL, resulting in the PL-AOVGraph. Table 2 summarizes the constructs mapping from feature models to PL-AOVGraph. Figure 3 illustrates the PL-AOVGraph representation of the ECommerce variability represented in Figure 1 using feature model. PL-AOVGraph models also allow the representation of the influence of alternatives for the satisficing of a softgoal. The highlighted circles in Figure 3 indicate limitations of the expressivity of PL-AOVGraph since it does not support cardinality corresponding to (A) and (E) shown in Table 1. Similar to previous case, as a possible solution, we have also used AND relations to reflect (A) and Inclusive OR relations to reflect (E). Again, in both cases, they do not represent the same meaning of Figure 1.

Silva and others [14] present how the aspectual i* modeling language can be used to both describe features in the domain engineering and in application engineering. They argue that the aspectual i* framework can be used both to describe features in the domain engineering and in application engineering. They use the aspectual i* approach to modularize variabilities and to compose variable features with common features because they believe that representing these features as aspects can reduce

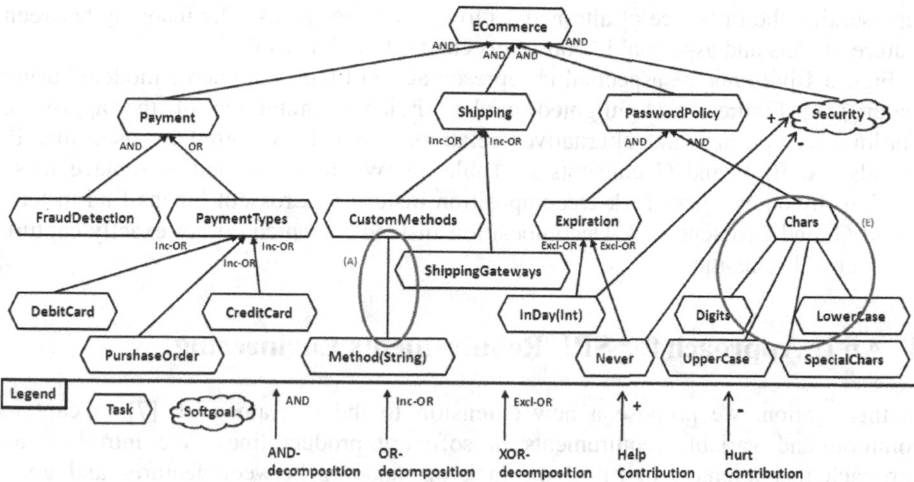

Fig. 3. ECommerce PL-AOVGraph Model

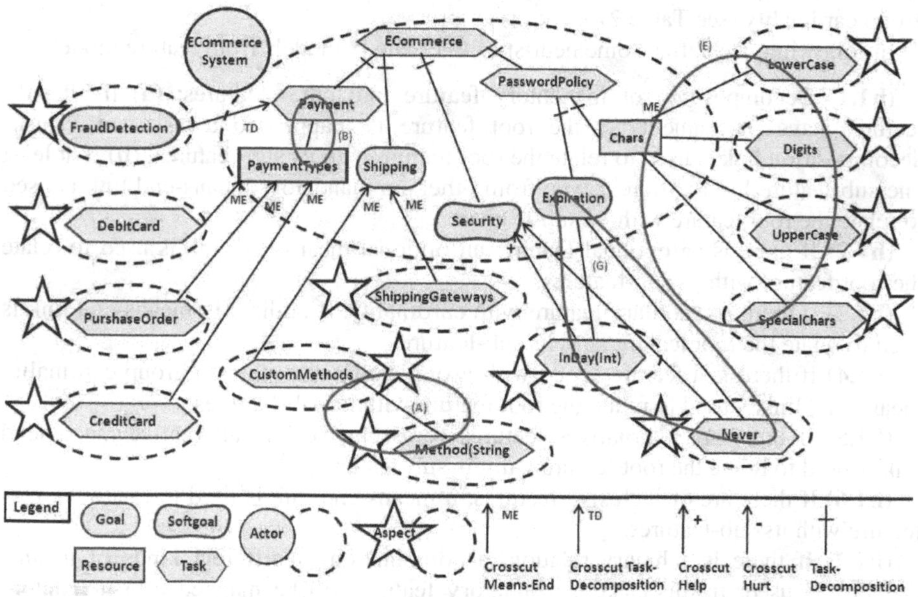

Fig. 4. ECommerce i* Aspectual Model

time and costs associated with the configuration activity. Besides, it is required an improved localization of features in software artifacts to facilitate the incremental evolution of feature functionality. Thus, these variable features could be modularized into aspects and later composed with common features in application engineering. To facilitate even more the configuration activity, aspectual i* models also allow representing the influence of alternatives to satisfice softgoals. The mapping between feature models and aspectual i* constructs can be found in Table 2.

Figure 4 presents the aspectual i* representation of the Ecommerce modeled using features in Figure 1. Highlighted circles indicate limitations of the approach. Cardinality, optional and alternatives relations are not supported by aspectual i* models (A, B, E and G concepts in Table 1). We have decided to replace these relations by other types (Task-Decomposition instead B, Crosscut Means-End instead E and G, and Crosscut Task-Decomposition instead A) which do not exactly capture the intended semantic.

4 An i* Approach for SPL Requirements Engineering

In this section, we propose a new extension to the i* framework [7] to capture common and variable requirements in software product lines. We introduce an approach that attempts to fulfill the lack of mapping between features and goal-oriented techniques discussed above. Thus, in order to generate the types of features described in Table 1, we create new types of means-end link: mandatory means-end, optional means-end, means-end cardinality, alternative means-end and means-end group cardinality (see Table 2).

In following, we define some heuristics to create i* models from feature models:

(h1.1) Decomposition of mandatory feature into other features: **(a)** if all sub-features have the same type, the root feature is mapped to a task and a task-decomposition link is used to relate the root feature with its sub-features; **(b)** if at least one sub-feature has a different type from others, a mandatory means-end link is used to relate the root feature with its sub-feature.

(h1.2) If there is an optional feature, an optional means-end link is used to relate the root feature with its sub-features.

(h1.3) If there is a solitary feature with cardinality, a cardinality means-end link is used to relate the root feature with its sub-feature.

(h1.4) If there is a feature group with group cardinality <i-j>, a group cardinality means-end link is used to relate the root feature with its sub-features.

(h1.5) If there are alternatives features (or-exclusive), an alternative means-end link is used to relate the root feature with its sub-features.

(h1.6) If there are or-inclusive features, a means-end link is used to relate the root feature with its sub-features.

(h1.7) If there is a binary relation, a combination of different kinds of means-end link is used. In this case, a mandatory feature will be mapped to a mandatory means-end link.

To illustrate our approach, we apply the mapping heuristics described previously in the Ecommerce variability represented in Figure 1. The resulting model is shown in Figure 5. This model captures the same information of the feature model presented in Figure 1. Moreover, similarly to the previous analyzed approaches, our approach also allows representing richer information such as the positive/negative contribution of some lower level tasks to achieve higher level (soft)goals of the system. For example, in Figure 5, InDay(Int) task contribute positively to satisfice Security softgoal while

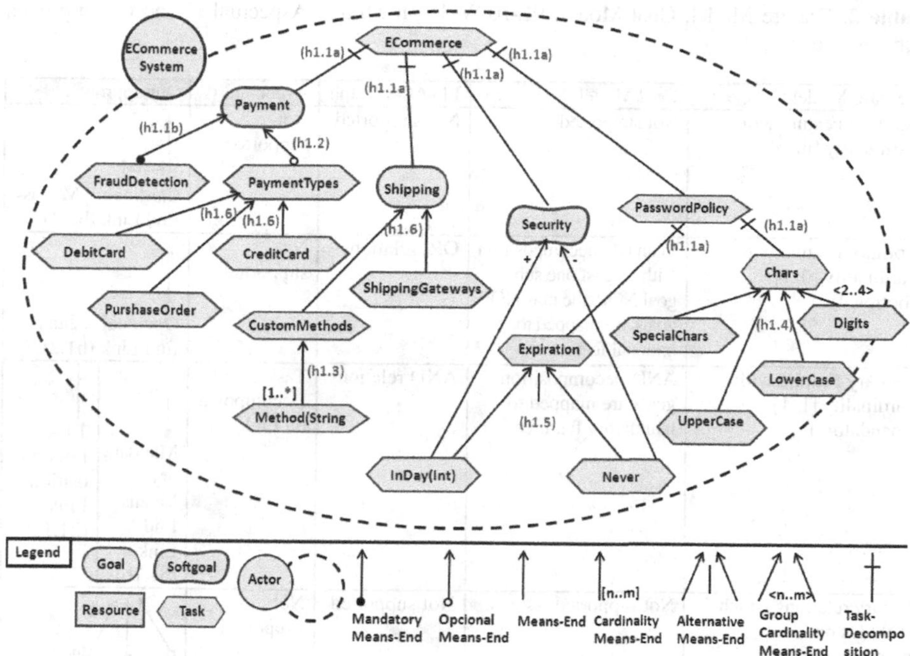

Fig. 5. ECommerce i* with our approach

Never task contribute negatively to the same softgoal. This extra information could help to choose and justify a configuration for a specific application. If Security is a requirement with high priority in an application, the InDay(Int) task should be chosen to be present in that application, instead of Never task.

5 Conclusions and Future Work

In this paper, we evaluate the expressiveness of goal-oriented techniques for modeling variability in SPL and introduced an approach to capture common and variable requirements in software product lines.

Expressiveness assesses the degree to which the SPL application domain is represented precisely in terms of the concepts offered by the modeling technique [15]. To evaluate this property, we analyzed every concept present in the feature model proposed by [6, 9].

We observe that goal models [11] and PL-AOVGraph models [13] have limited expressiveness, because they were not able to model the representation of cardinality of feature model. Aspectual i* [14] cannot represent cardinality neither alternatives of feature model. In order to solve some of the limitations discussed, we have extended the i* framework [7] by proposing some heuristics and creating new types of means-end link. Our proposal proved to be more efficient than the others because it was able to model every concept of the feature model presented in Table 1. The Table 2 shows the result of the comparison of the goal-oriented approach, PL-AOVGraph approach, Aspectual i* approach and our approach.

Table 2. Feature Model, Goal Model, PL-AOVGraph Model, Aspectual i* and our approach equivalence

Feature Model	Goal Model	PL-AOVGraph	Aspectual i*	Our approach
Solitary Feature with cardinality [m..n]	Not supported	Not supported	Not supported	 [n..m] Cardinality Means-End Link (h1.3)
Solitary Feature with cardinality [0..1] (optional)	In an OR-decomposition with at least one sub-goal NOP, the non-NOP goals are mapped to optional features.	OR relations	Not supported	 Opcional Means-End Link (h1.2)
Solitary Feature with cardinality [1..1] (mandatory)	AND-decomposition goals are mapped to mandatory features	AND relations	Task-Decomposition Link	Mandatory Means-End Link (h1.1b) Task-Decomposition Link (h1.1a)
Binary relations which includes optional, mandatory and cardinality–based like relations	Not supported	Not supported	Not supported	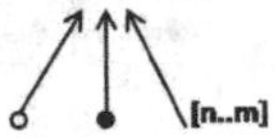 [n..m] Binary Relations Means-End Link (h1.7)
Feature Group with group Cardinality <i-j>	Not supported	Not supported	Not supported	 <n..m> Group Cardinality Means-End Link (h1.4)
Feature Group with group Cardinality <1-k>, k= size of the group (inclusive-or)	OR(inclusive)-decomposition	Inclusive-OR relations	Means-End link	 Means-End Link (h1.6)
Feature Group with group Cardinality <1-1> (exclusive-or)	Annotate the OR-decomposition with the symbol "|" to mean that this decomposition is now an OR (exclusive)-decomposition.	Exclusive-OR relations [1].	Not supported	 Alternative Means-End Link (h1.5)

[1] It is not clear how this relation is used, since there is no example in [13] illustrating this. To keep consistency, in this paper we chose to use Exclusive-OR relations (Exc-Or) because in [13] the authors mentioned its existence by only illustrated the Inclusive-OR relations (Inc-Or).

As future work we intend to (i) validate our approach by modeling larger and more complex case studies; (ii) develop a metamodel for our language and (iii) develop a tool to support our extensions to i*.

References

1. Kang, K., Cohen, S., Hess, J., Nowak, W., Peterson, S.: Feature-oriented domain analysis (FODA) feasibility study. Technical Report CMU/SEI-90-TR-21, Software Engineering Institute, Carnegie Mellon University, Pittsburgh, PA (November 1990)
2. Lau, S.: Domain Analysis of E-Commerce Systems Using Feature-Based Model Templates. Master Thesis, ECE Department, University of Waterloo, Canada (2006)
3. Czarnecki, K., Antkiewicz, M.: Mapping features to models: A template approach based on superimposed variants. In: Glück, R., Lowry, M. (eds.) GPCE 2005. LNCS, vol. 3676, pp. 422–437. Springer, Heidelberg (2005)
4. Liaskos, S., Lapouchnian, A., Yu, Y., Yu, E., Mylopoulos, J.: On Goal-based Variability Acquisition and Analysis. In: 14th IEEE Requirements Engineering (RE 2006), pp. 76–85. IEEE Press, USA (2006)
5. Mylopoulos, J., Chung, L., Liao, S., Wang, H., Yu, E.: Exploring Alternatives during Requirements Analysis. IEEE Software 18(1), 92–96 (2001)
6. Czarnecki, K., Helsen, S., Eisenecker, U.: Formalizing cardinality-based feature models and their specialization. Software Process Improvement and Practice 10(1), 7–29 (2005)
7. Yu, E.: Towards modelling and reasoning support for early-phase requirements engineering. In: 3rd IEEE Requirements Engineering (RE 1997), USA, pp. 226–235 (1997)
8. Clements, P., Northrop, L.: Software Product Lines: Practices and Patterns. Addison-Wesley, Boston (2002)
9. Czarnecki, K., Helsen, S., Eisenecker, U.: Staged configuration through specialization and multilevel configuration of feature models. Software Process: Improvement and Practice 10(2), 143–169 (2005)
10. Pohl, K., Böckle, G., van der Linden, F.J.: Software Product Line Engineering: Foundations, Principles, and Techniques. Springer, New York (2005)
11. Yu, Y., Leite, J.C.S.P., Lapouchnian, A., Mylopoulos, J.: Configuring features with stakeholder goals. In: ACM Symposium on Applied Computing (SAC), pp. 645–649. ACM Press, Fortaleza (2008)
12. Rashid, A., Moreira, A., Araújo, J.: Modularisation and Composition of Aspectual Requirements. In. 2nd Intl. Conf. on Aspect-Oriented Soft. Develop., USA, pp. 11–20 (2003)
13. Batista, T., Bastarrica, M., Soares, S., Fernandes, L.: A Marriage of MDD and Early Aspects in Software Product Line Development. In: Early Aspects Workshop at 12th International Software Product Line Conference (SPLC 2008), Limerick, Ireland, pp. 97–104 (2008)
14. Silva, C., Alencar, F., Araújo, J., Moreira, A., Castro, J.: Tailoring an Aspectual Goal-oriented Approach to Model Features. In: 20th International Conference on Software Engineering and Knowledge Engineering (SEKE 2008), pp. 472–477. Knowledge Systems Institute Graduate School, San Francisco, CA, USA (2008)
15. Estrada, H.: A service-oriented architecture for the i* Framework. Ph.D. Thesis. DSIC. Universidad Politècnica de Valencia (2008)
16. Benavides, D., Trujillo, S., Trinidad, P.: On the modularization of feature models. In: 1st European Workshop on Model Transformation. Rennes, France (2005)

A Lightweight GRL Profile for i* Modeling

Daniel Amyot[1], Jennifer Horkoff[2], Daniel Gross[3], and Gunter Mussbacher[1]

[1] SITE, University of Ottawa
{damyot,gunterm}@site.uottawa.ca
[2] Department of Computer Science, University of Toronto
jenhork@cs.utoronto.ca
[3] Faculty of Information, University of Toronto
daniel.gross@utoronto.ca

Abstract. The i* framework is a popular conceptual modeling language for capturing and analyzing socio-technical motivation and properties of complex systems in terms of actors, their intentions, and their relationships. In November 2008, the International Telecommunications Union finalized the standardization of the User Requirements Notation (URN). URN is composed of two loosely coupled yet integrated sub-languages: the Goal-oriented Requirement Language (GRL), which is an intentional modeling language based on a subset of i*, and the Use Case Map notation for representing and capturing high-level system scenarios and structures. GRL was specifically defined in a non-restrictive way to foster the development and use of different agent and/or goal modeling approaches and techniques. However, because of its permissiveness, GRL can be used in ways that deviate from conventional i* modeling guidelines. In addition, some i* concepts do not have equivalent first-class concepts in GRL. In this paper, we present a lightweight GRL profile for i* that takes advantage of GRL's extensibility features to capture missing i* concepts. The profile presents formal constraints on the use of GRL and its extensions to restrict it to an i* style. Using GRL constrained by this profile enables GRL modeling and analysis tools to be used for i* models, and ensures that resulting i* models conform to an international standard and that they can be integrated with Use Case Maps. Variants and extensions of the original i* can also be supported in a similar way. This profile is implemented in the jUCMNav modeling tool.

Keywords: Goal-oriented Requirement Language, i*, jUCMNav, OCL, profile, User Requirements Notation.

1 Introduction

The i* modeling framework [12, 13] introduced aspects of intentional and social modeling and reasoning into information system engineering methods, especially at the requirements level. Unlike traditional systems analysis methods which strive to abstract away from the people aspects of systems, i* recognizes the primacy of social actors. Actors are viewed as being intentional, i.e., they have goals, beliefs, abilities, and commitments, which must be discovered, captured and analyzed. The analysis

C.A. Heuser and G. Pernul (Eds.): ER 2009 Workshops, LNCS 5833, pp. 254–264, 2009.

focuses on how well the goals of various actors are achieved given some configuration of relationships among human and system actors, and what reconfigurations of those relationships can help actors advance their strategic interests. Such analysis supports many software and system requirements engineering activities.

The i* framework has stimulated considerable interest in a socially-motivated approach to systems modeling and design, and has led to a number of extensions and adaptations, many of which are discussed in the i* Wiki [6]. One of these adaptations was recently standardized by the International Telecommunications Union (ITU-T) as part of the User Requirements Notation (URN – Recommendation Z.151) [7]. URN combines the Goal-oriented Requirement Language (GRL) with the Use Case Map (UCM) scenario notation in a single language, with a mature and well-defined metamodel supplemented by a concrete graphical syntax.

GRL supports many of the core concepts of i*, including actors, intentional elements, dependencies, contributions, and decompositions. However, GRL also differs from i* in a number of ways, such as the following:

1) *Missing first-class concepts in GRL*: i* contains concepts that are missing from GRL. For instance, GRL has only one type of actor, whereas i* also defines the notions of roles, agents and positions.
2) *GRL permissiveness*: GRL is voluntarily permissive in how intentional elements can be linked to each other. This is meant to support the wide variety of ways people actually create goal models [5]. However, i* proposes more specific and restrictive usages of relationships. For instance, an i* contribution link cannot have a task as a destination.
3) *Additional concepts in GRL*: GRL contains additional first-class concepts such as strategies (for the analysis of GRL models), metadata, and URN links (which enable the creation of typed links between any GRL/UCM elements).

In this paper, we present a lightweight profile for GRL that enables one to create goal models in a particular i* style according to the i* Guide in [6] and Yu's work [12, 13]. We take advantage of URN links and metadata to create relationships and stereotypes (respectively) for the missing GRL concepts found in i*. We specify constraints in UML's Object Constraint Language (OCL) [9] in order to restrict the usage of GRL to commonly used i* guidelines. We say that this profile is lightweight because it uses simple extensibility mechanisms and it does not require the extension of the URN metamodel or the use of heavyweight profiling mechanisms à la UML.

We also provide tool support for this profile with the jUCMNav tool, an Eclipse plug-in for the creation, analysis, and transformation of URN models [8, 10]. jUCMNav supports the notion of metadata together with an OCL engine that can check violations of user-defined constraints [2], enabling low-cost language customization.

A profile enabling the creation of GRL models in an i* style allows i* models to follow the standard defined in Z.151, including its interchange format. In addition, the use of the jUCMNav tool for i* models provides support for the division of models into consistent views (addressing scalability), the application of various pre-defined and automated quantitative and qualitative evaluation algorithms (with easy addition of new ones), the integration with UCMs, and simple modification or addition of constraints (for handling other variants and extensions of i*).

Because GRL is a recent language, background information on its notation and metamodel is given in section 2, followed by the profile definition in section 3. Section 4 presents the support of the profile in the jUCMNav tool. Related work is briefly discussed in section 5, followed by our conclusions.

2 Goal-Oriented Requirement Language (GRL)

GRL is a graphical language that focuses primarily on goal modeling. One of GRL's major assets is to provide ways to model and reason about functional and non-functional requirements in terms of goal achievement in a social context. With GRL, the modeler is primarily concerned with exposing "why" and "for whom" certain choices for behavior and/or structure were introduced, leaving the "what" and the "how" to other languages such as UCM and UML. GRL integrates core elements of i* [12, 13] and the NFR framework [3] relevant for intentional modeling. Major benefits of GRL over other popular notations include the integration of GRL with a scenario notation, the support for qualitative and quantitative attributes, and a clear separation of GRL model elements from their graphical representation, enabling a scalable and consistent representation of multiple views/diagrams of the same goal model.

The graphical syntax of GRL (see Fig. 1) is based on the syntax of the i* language. There are three main categories of concepts in GRL: actors, intentional elements, and links. A GRL goal graph is a connected graph of intentional elements that optionally reside within an actor boundary. An actor represents a stakeholder of the system or

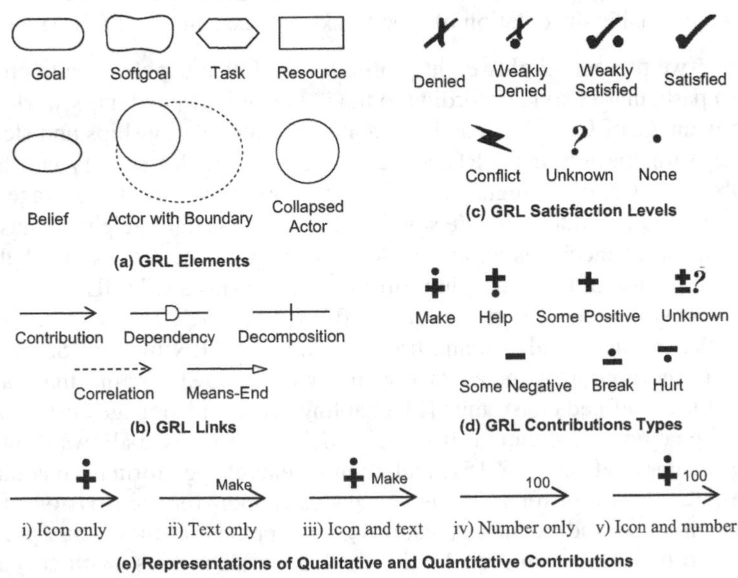

Fig. 1. Basic Elements of GRL Notation

another system. Actors are holders of intentions; they are the active entities in the system or its environment who want goals to be achieved, tasks to be performed, resources to be available and softgoals to be satisfied. A goal graph shows the high-level business goals and system goals (functional and non-functional) of interest to a stakeholder and the alternatives for achieving these goals. A goal graph also documents beliefs (rationales) important to the stakeholder.

In addition to beliefs, *intentional elements* can be softgoals, goals, tasks, and resources. *Softgoals* differentiate themselves from *goals* in that there may not exist a clear, objective measure of satisfaction for a softgoal whereas a (hard) goal is usually quantifiable. In general, softgoals are more related to non-functional requirements, whereas goals are more related to functional requirements. *Tasks* represent solutions to (or operationalizations of) goals or softgoals. In order to be achieved or completed, softgoals, goals, and tasks may require *resources* to be available.

Links (see Fig. 1b) are used to connect elements in the goal model. *Decomposition links* allow an element to be decomposed into sub-elements. AND, IOR, as well as XOR decompositions are supported. XOR and IOR decomposition links may alternatively be displayed as *means-end* links. *Contribution links* indicate desired impacts of one element on another element. A contribution link can have a qualitative contribution type (see Fig. 1d), or a quantitative contribution (integer value between -100 and 100, see Fig. 1e). *Correlation links* are similar to contribution links, but describe side effects rather than desired impacts. Finally, *dependency links* model relationships between actors (one actor depending on another actor for something).

Fig. 2 presents the metamodel of the core GRL concepts, which constitute a part of the URN metamodel from Recommendation Z.151 [7]. These concepts represent the abstract grammar of the language, independently of the notation. This metamodel also formalizes the GRL concepts and constructs introduced earlier.

In addition, GRL inherits the concepts of *URN link* (Fig. 3-left), which enable one to create a link of a user-defined type between any pair of URN model elements (GRL

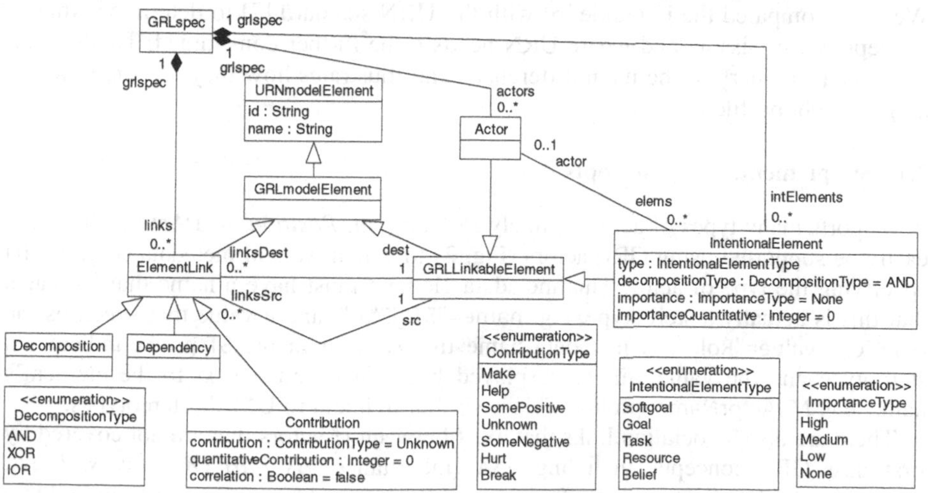

Fig. 2. GRL Metamodel – Core Concepts (from Z.151)

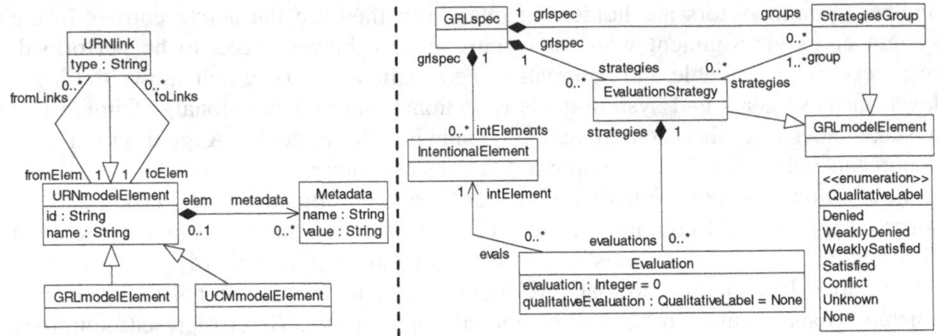

Fig. 3. GRL Metamodel – URN Links and GRL Strategies (from Z.151)

and UCM model elements alike). GRL model elements may also contain *metadata* that capture user-defined name-value pairs. These two concepts help extend the language or add precision to the model without having to change the metamodel.

Finally, a GRL model may also contain *evaluation strategies* (Fig. 3-right), which allow modelers to analyze the model for various what-if contexts. A strategy contains initial qualitative or quantitative evaluation levels attached to a subset of the intentional elements. These represent satisfaction levels that are propagated to the other intentional elements of the model through the decomposition, contribution, and dependency links. Several bottom-up evaluation algorithms (quantitative, qualitative and mixed) are proposed in the Appendix II of the URN standard [7] and are supported by jUCMNav [8, 10].

3 GRL Profile for i*

We have compared the i* Guide [6] with the URN standard [7] to determine what i* concepts were missing and where URN needs to be further constrained. This section presents a summary of the main differences and illustrates how they are supported in a lightweight profile.

3.1 Supplementary i* Concepts

i* supports many types of actors, namely *Role*, *Agent*, *Position*, and *Actor*. These can easily be supported using GRL actors (Fig. 2) to which we add metadata (Fig. 3-left) specifying the type of actor. This metadata element must have a name that indicates that this is actually a stereotype, e.g. name="ST_iStar", and a value that specifies the type, e.g. value="Role". A metadata name that starts with the ST prefix indicates a stereotype and the value will be displayed between « and » next to the element's name, e.g. MyActorName «Role». This is similar in intent to UML's stereotypes.

There are also association links that exist between i* actors that are not covered by first-class GRL concepts, including: *ISA* (inheritance), *Is Part Of*, *Covers*, *Plays*, *Occupies*, and *INS* (instance of). These could be captured by a stereotype applied to a

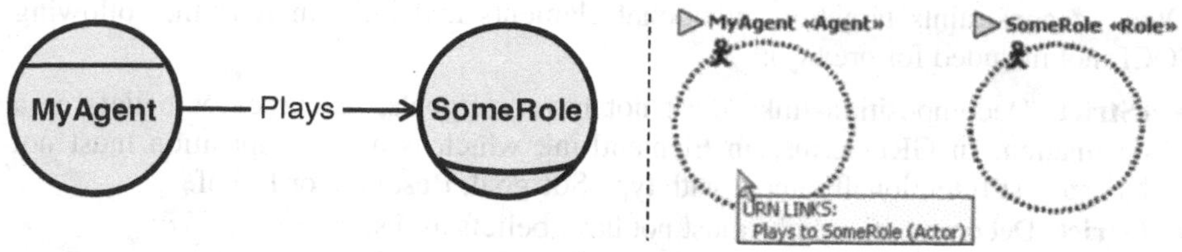

Fig. 4. Example: i* Actor Types and Association Links in GRL

dependency link, but this might be overstretching this concept. A better alternative is to use a URN link between the two actors (Fig. 3-left), where the link type corresponds to the desired i* association type. Note however that a drawback of URN links is that they do not have a visual representation in GRL diagrams (but these relationships can be exploited during analysis). The ▶symbol indicates the presence of URN links on a GRL element, and tools like jUCMNav can show the nature of these from/to links in a tool tip, as shown in Fig. 4, where an agent plays some role.

As for i* *AND/OR* contribution links, they correspond semantically to GRL AND/OR decomposition links. The later should simply be used.

Finally, GRL does not make the distinction between *Strategic Dependency* (SD) models and *Strategic Rationale* (SR) models, as i* does. GRL has one integrated model, with multiple views (diagrams) based on the concrete syntax. URN provides additional metamodel elements to support the concrete syntax (location, size, colors, etc.), including one for GRL graphs (diagram). Hence, one can associate a stereotype to a GRL graph (ST_iStar metadata with value SD or SR) to enforce this distinction.

3.2 Constraints for i* Guidelines

Restricting the use of GRL to an i* style can be achieved by defining OCL constraints on the URN metamodel presented in section 2. Such constraints can target not only the core GRL concepts but also the various extensions described in section 3.1. Rules presented here use the conventions proposed in [5] and are tagged as strict (must never be violated) or loose (should not happen, but is tolerable).

One commonly accepted rule for i* is the following: ***(Strict)*** *Contribution links must only have softgoals as destinations*. In terms of the URN metamodel, this means that the dest GRLLinkableElement of an ElementLink which is a Contribution must be an IntentionalElement with type Softgoal. The corresponding OCL invariant is:

```
context Contribution
inv SoftgoalAsContributionDestination:
    self.dest.oclIsTypeOf(IntentionalElement)
    implies
    (self.dest.oclAsType(IntentionalElement)).type =
        IntentionalElementType::Softgoal)
```

Other i* constraints targeting intentional elements and links include the following (OCL not included for brevity):

- **(Strict)** Decomposition links must not have softgoals, resources or beliefs as a destination. In GRL terms, an ElementLink which is a Decomposition must not have a dest IntentionalElement with type Softgoal, Resource or Belief.
- **(Strict)** Decomposition links must not have beliefs as a source.
- **(Loose)** Beliefs should not be the destination of element links.
- **(Loose)** AND decomposition links should only have tasks as destinations.
- **(Loose)** Means-end links (i.e., OR/IOR decomposition links in GRL) should only have goals as destinations.

Interestingly, one can also define constraints that involve the metadata/stereotypes and URN links used to add i* concepts to GRL, as explained in the previous section. For instance, the rule ***(Strict)*** *ISA (generalization) must be between two actors of the same type* can be encoded in OCL using the following constraint:

```
context Actor
inv ISAbetweenActorsOfSameType:
    self.getLinksTo('ISA')->
        forall(to | to.oclIsTypeOf(Actor) and
            ( to.oclAsType(Actor).getMetadata('ST_iStar') =
              self.getMetadata('ST_iStar') )
        )
```

The above rule states that for all the URN elements that are the targets of URN links of type ISA, each such element (`to`) must be an Actor and must have the same ST_iStar metadata value as the source Actor. The rule takes advantage of two reusable OCL helper functions defined in our framework to query URN links (`getLinksTo`) and metadata (`getMetadata`). Other rules involving metadata and URN links include:

- **(Loose)** An *Is Part Of* association should be between two actors of the same type.
- **(Strict)** A *Covers* association must be from a Position to a Role.
- **(Strict)** A *Plays* association must be from an Agent to a Role.
- **(Strict)** An *Occupies* association must be from an Agent to a Position.
- **(Strict)** An *INS* association must only be used between Agents.

Finally, similar restrictions can also target actor boundaries and dependencies (note that the other relevant situations are already covered by standard GRL constraints).

- **(Strict)** Dependency links must never completely be inside of an actor boundary. In GRL terms: For an ElementLink which is a Dependency (with source and destination GRLLinkableElements):
  - If src and dest are both Actors, then dest ≠ src
  - If src is an Actor and dest an IntentionalElement, then src ≠ dest.actor
  - If dest is an Actor and src an IntentionalElement, then src.actor ≠ dest
  - If src and dest are both IntentionalElement, then src.actor ≠ dest.actor

- **(Strict)** Dependency links in an SD model must always have a dependum, i.e., there should never be a dependency link from an actor to an actor.
- **(Strict)** SD models must not have links other than dependency and actor association links.
- **(Loose)** Dependency links in an SR model should always have a dependum.
- **(Loose)** The only links that cross actor boundaries should be dependency links.

4 Tool Support

The lightweight GRL profile for i* was implemented in the jUCMNav tool. A GUI for managing metadata and URN links is already available [10], so supporting the supplementary i* concepts from section 3.1 and Fig. 4 is already covered.

The integrated OCL-based engine for the verification of user-defined semantic rules presented [2] can also be used as is to define and check the i* constraints highlighted in section 3.2. The `SoftgoalAsContributionDestination` rule previously see is repeated in Fig. 5. The name, context, and constraint expression are essentially the same, except for the precision of metamodel packages (`grl::`) required by jUCMNav. The tool also allow for the definition of an informal description and of supplementary utility functions (such as `getLinksTo` and `getMetadata`). An OCL query expression is required to collect all the instances of a particular URN metaclass (`Contribution` here) used in the model being edited. Such a rule is created once and can then be checked against any URN model. The tool also allows for rules to be exported and imported, so modelers can share their rules.

Constraints to be checked can be selected individually. For convenience, constraints can also be grouped to ease their selection. In Fig. 6, several groups of constraints are present, including three groups related to our profile for i*: one for strict rules, one for loose rules, and one for both (as a constraint can be part of multiple groups).

The verification of constraints is done on demand via menu selection. Violations are reported in Eclipse's standard Problems view. For example, suppose an i* model

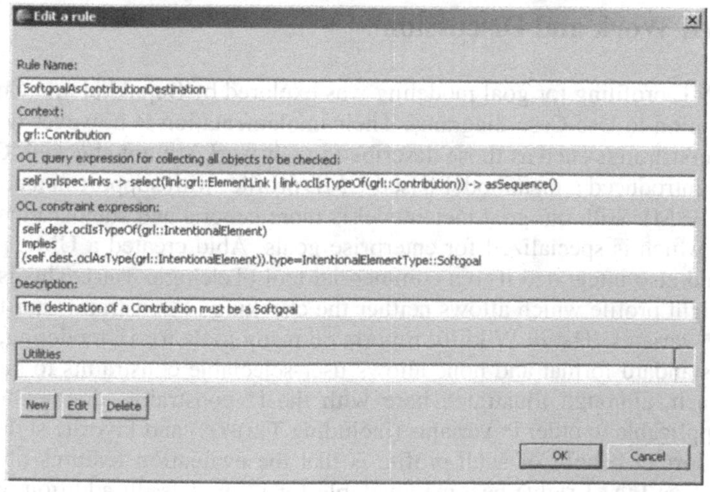

Fig. 5. Example of Constraint Definition: *SoftgoalAsContributionDestination*

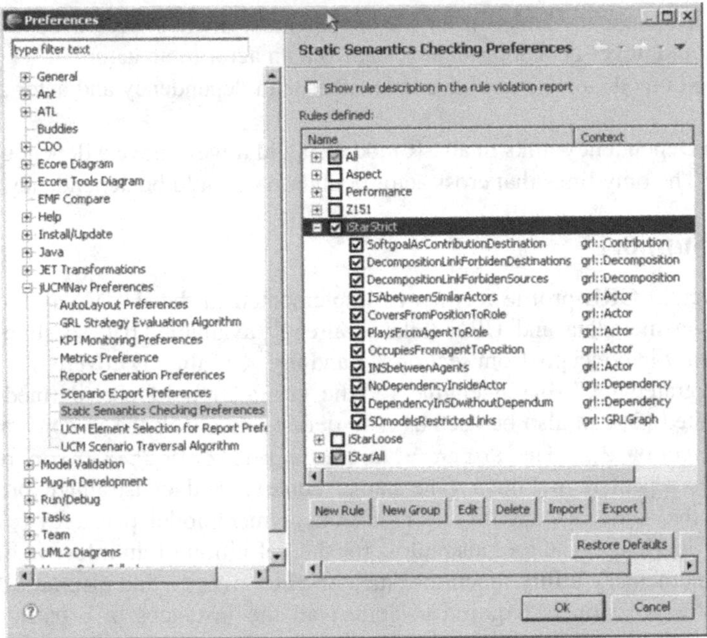

Fig. 6. Selectable Strict and Loose i* Constraints in the GRL Profile

where one of the tasks contributes to a goal, a situation that is forbidden according to the `SoftgoalAsContributionDestination` constraint. Out of the 11 strict i* constraints selected in Fig. 6 and checked against the model, one error would be detected and reported in the Problems view. Double-clicking on the error actually brings the focus of the model editor to the violating element (the contribution in this case).

5 Related Work and Discussion

General UML profiling for goal modeling was explored by Supakkul and Chung [11], with integration to Use Case diagrams. Their implementation is however not focused on i*, and constraints such as those described in our work are not checked. Grangel *et al.* [4] also introduced a metamodel-based UML profile, with support in a commercial tool (IBM RSM). Still, our goal metamodel is more general and standard than the one they used, which is specialized for enterprise goals. Abid created a UML profile for GRL [1] and also integrated it to a commercial tool (Telelogic Tau). This is however a heavyweight profile which allows neither the checking of constraints nor the exploration of i* variants. The i* Wiki [6] reports on many tools for i* modeling, but none is using a standard format and none allows user-selectable constraints to be checked. Our approach, although illustrated here with the i* constraints documented in [5], could be applicable to other i* variants (including TROPOS) and favorite styles.

One important benefit of such profile is that the evaluation features of GRL and jUCMNav (see Fig. 3-right) become available for i* models. In addition, jUCMNav allows one to add new evaluation algorithms [10], some of which could take advantage of the new stereotype and URN link information now captured. This would also permit one to compare evaluation procedures from i* and GRL. In addition to the support of UCM scenarios, jUCMNav also contains extensions of GRL not included in standard URN (e.g. for key performance indicators and aspect-oriented modeling), which are now available as a byproduct to i* models for further exploration.

6 Conclusions

We have introduced a lightweight profile for GRL that allows the representation of concepts unique to i* and restricts the usage of GRL to comply with i* guidelines. This is supported by the jUCMNav tool, where constraints can be captured in OCL and violations reported to modeler. This work casts i* onto a standardized metamodel (URN's) that enables the use of GRL-like analysis for i* models, the comparison of i* stylistic guidelines and evaluation algorithms, and a common representation of i* models. Using the underlying mechanisms presented here (metadata and URN links), language designers can prototype and explore new language features easily and at low cost before making them first-class entities in a revised metamodel.

Several directions for future work were already identified in the previous section. In addition to those, the jUCMNav tool could be improved in a number of ways, including by reporting warnings (instead of errors) for violating constraints that are loose (instead of strict). Also, more in-depth, hopefully industrial case studies to test the use of i* models with jUCMNav and GRL evaluation is required.

Acknowledgments. This work was supported by NSERC (Discovery Grants and Postgraduate Scholarships). We also thank Eric Yu for his helpful comments.

References

1. Abid, M.R.: UML Profile for Goal-oriented Modelling. Master of Computer Science Thesis, University of Ottawa, Canada (2008)
2. Amyot, D., Yan, J.B.: Flexible Verification of User-Defined Semantic Constraints in Modelling Tools. In: CASCON 2008. ACM, New York (2008)
3. Chung, L., Nixon, B.A., Yu, E., Mylopoulos, J.: Non-Functional Requirements in Software Engineering. Kluwer Academic Publishers, Dordrecht (2000)
4. Grangel, R., Chalmeta, R., Campos, C., Sommar, R., Bourey, J.-P.: A Proposal for Goal Modelling Using a UML Profile. In: Enterprise Interoperability III, pp. 679–690. Springer, Heidelberg (2008)
5. Horkoff, J., Elahi, G., Abdulhadi, S., Yu, E.: Reflective Analysis of the Syntax and Semantics of the i* Framework. In: Song, I.-Y., Piattini, M., Chen, Y.-P.P., Hartmann, S., Grandi, F., Trujillo, J., Opdahl, A.L., Ferri, F., Grifoni, P., Caschera, M.C., Rolland, C., Woo, C., Salinesi, C., Zimányi, E., Claramunt, C., Frasincar, F., Houben, G.-J., Thiran, P. (eds.) ER Workshops 2008. LNCS, vol. 5232, pp. 249–260. Springer, Heidelberg (2008)
6. i* wiki, http://istar.rwth-aachen.de (accessed, May 2009)
7. ITU-T – International Telecommunications Union: Recommendation Z.151 (11/08) User Requirements Notation (URN) – Language definition. Geneva, Switzerland (2008)

8. jUCMNav website. University of Ottawa, http://jucmnav.softwareengineering.ca/jucmnav/ (accessed, May 2009)
9. OMG – Object Management Group: Object Constraint Language Specification, 2.0 (2006)
10. Roy, J.-F., Kealey, J., Amyot, D.: Towards Integrated Tool Support for the User Requirements Notation. In: Gotzhein, R., Reed, R. (eds.) SAM 2006. LNCS, vol. 4320, pp. 198–215. Springer, Heidelberg (2006)
11. Supakkul, S., Chung, L.: A UML profile for goal-oriented and use case-driven representation of NFRs and FRs. In: Dosch, W., Lee, R.Y., Wu, C. (eds.) SERA 2005. LNCS, vol. 3647, pp. 29–41. Springer, Heidelberg (2006)
12. Yu, E.S.K.: Modelling strategic relationships for process reengineering. Ph.D. dissertation. Dept. of Computer Science, University of Toronto, Canada (1995)
13. Yu, E.: Towards Modelling and Reasoning Support for Early-Phase Requirements Engineering. In: 3rd IEEE Int. Symp. on RE, Washington, USA, pp. 226–235. IEEE CS, Los Alamitos (1997)

From User Goals to Service Discovery and Composition

Luiz Olavo Bonino da Silva Santos[1], Giancarlo Guizzardi[2], Luís Ferreira Pires[1], and Marten van Sinderen[1]

[1] Department of Computer Science, University of Twente, Enschede 7500AE, The Netherlands
[2] Ontology and Conceptual Modeling Research Group (NEMO), Universidade Federal do Espírito Santo, Vitória, Brazil

Abstract. Goals are often used to represent stakeholder's objectives. The intentionality inherited by a goal drives stakeholders to pursuit the fulfillment of their goals either by themselves or by delegating this fulfillment to third parties. In Service-Oriented Computing, service client's requirements are commonly expressed in terms of inputs, outputs, preconditions and effects, also known as IOPE. End-users, i.e., human service clients, may have difficulties to express such requirements as they would have to deal with technical issues such as the request's language, and the type, format and coding of the IOPE. This paper presents the core concepts of the Goal-Based Service Ontology (GSO) that relates goals and services. By grounding GSO in a well-founded ontology we aim at clarifying the semantics for a set of relevant domain concepts that can support specialists in defining application ontologies based on goals and services.

1 Introduction

Service-Oriented Computing (SOC) has been gaining momentum in recent years with an increase in industry adoption and research efforts. SOC has been seeing as an approach to integrate legacy and new systems with a standardized set of protocols and interfaces in a distributed manner. Among the research efforts in this area we can include the pursuit of supplying semantics to service descriptions, message exchanges and service requests. The addition of semantics aims at supporting semantic interoperability for heterogeneous systems. Ontologies are being used in the realm of SOC for providing this semantic richness [1], [2].

Even when semantically enriched, service client's requirements are commonly expressed in terms of inputs, outputs, pre-conditions and effects, also known as IOPE. In this manner, the intentionality of the service client (why he wants the service) is not clear or explicit in the mix of technological details such as input and output parameter types and restrictions to the service selection and execution.

In this paper we present an ontology-based approach to support dynamic service discovery and composition. The main element of this approach is the

C.A. Heuser and G. Pernul (Eds.): ER 2009 Workshops, LNCS 5833, pp. 265–274, 2009.

Goal-Based Service Ontology (GSO). GSO includes concepts and relationships that (represented by the Goal-Based Service Metamodel) allows domain specialists to define their goal-based service-oriented models. Clarity and an appropriate formalization of semantics are important requirements for ontologies. These requirements are especially relevant in Service-Oriented Computing (SOC) to enable complex tasks involving multiple agents. GSO aims at providing ontologically sound concepts relating concepts of SOC (e.g., Service Provider, Service Client and Service) with concepts pertinent to our goal-based approach, such as Goal and Task. Nevertheless these concepts are not sufficient for a complete domain specification. Other domain-independent concepts and relations are necessary such as Description, Agent, Intention, Material Relation, among others. In order to provide these concepts and relations and at the same time supply semantic clarity we are working towards a domain ontology for the domain of goal-based service specification making use of the foundational ontology Unified Foundational Ontology (UFO) [3]. UFO is based on formal principles derived from formal ontology in philosophy, cognitive sciences, philosophical logics and linguistics.

GSO is part of a framework to support dynamic service discovery and composition called Goal-Based Service Framework (GSF). In GSO the concept of goal is used to express the service client's intention towards a service, i.e., why the service client used the service and why the selected service is beneficial to the service client. This paper is further structured as follows. Section 2 gives an overview of the architecture of the Goal-Based Service Framework. Section 3 details and discusses the proposed Goal-Based Service Ontology. Section 4 presents an example usage scenario of GSF in the Home Health Care domain. Section 5 presents some final considerations.

2 Goal-Based Service Framework (GSF)

In our work we consider the scenario of Pervasive Computing associated with SOC technologies and concepts. In this scenario we have human agents surrounded by and interacting with a plethora of computational devices and services. This motivates the need of a platform support to tackle with the issues of service discovery and composition in an unobtrusive way.

Our framework to support dynamic service discovery and composition is based on goal modeling and assumes that the involved stakeholders (service clients, service providers, supporting platform) share the same conceptual models, i.e., the same set of domain ontologies. This requirement is necessary because the approach relies on the availability of domain-specific ontologies. The elements of this Goal-Based Service Framework (GSF) are described as follows:

- *Goal-Based Service Ontology (GSO).* This ontology defines domain- independent concepts such as service, service client, service provider, goal, task and their relations, among others. This domain independency is however limited to domains and applications within the scope of the aforementioned scenario of Pervasive and Service-Oriented Computing.

- *Goal-Based Service Metamodel (GSM).* Generated from Goal-Based Service Ontology, this metamodel represents the concepts defined in GSO and defines the language used by domain specialists to create domain specifications.
- *Domain Specification.* GSF can be used in different application domains such as Health Care, Ambient Intelligence, etc. For each of these application domains a domain specialist defines a domain specification, namely the concepts and relations relevant to the domain, goals that users can have, valid tasks in the application, etc. GSM, representing GSO concepts, provides a modeling language that enables domain specialists to define domain specifications allowing a shared knowledge about particular domains. A domain specification is composed of: *(i)* a domain ontology including domain-specific concepts, the relations among these concepts and valid goals that users of that domain can have; and *(ii)* a task ontology which uses the concepts defined in the domain ontology and provides domain-specific definitions of valid tasks and how they can be related to user's goals fulfillment.
- *Context-Aware Service platform.* The context-aware service platform supports the interaction between service providers and service clients. From the service provider's perspective, the platform supports the publication of service descriptions. From the service client's perspective, the platform provides mechanisms for service discovery, composition, invocation and monitoring, among others. Moreover, the context-aware components of our supporting platform provide user's contextual information that is used *(i)* to select which of the tasks that support a given goal will be used in the service discovery and composition procedures and, *(ii)* as input data for the discovered services. The context information gathering reduces the need of direct user input and, thus, reduces also the need of user's interaction supporting a more autonomic behavior of the platform.

A normal deployment of GSF consists in the GSO, GSM and the CA Service Platform. A second step is the addition of domain specifications by domain specialists. Service providers can start to semantically annotate their services and service descriptions based on the concepts present on these domain specifications. The service descriptions are added to the CA Service Platform by the service providers.

3 Goal-Based Service Ontology

3.1 Goal Definition

The concept of goal has several different definitions depending on the domain the term is used, e.g., Philosophy, Sports, Economy, among others. Narrowing down to the Computer Science domain, a variety of definitions of the goal concept can also be found such as in [4]. Regarding the community of Semantic Web, in the goal definition of the Web Service Modeling Ontology (WSMO) [1] a goal is closely tied to Web services, i.e., a commitment is done already in the ontological level w.r.t. the specific technology to realize services. An example of this close tie between a WSMO goal and Web services is in WSMO's goal description which includes the interface of the Web service the user would like to interact with. In our work we consider Web services as one possible technological solution for implementing services and do not limit our approach to this specific technology.

For the purposes of this framework, we define goal as the *propositional content of a service client's intention.* In other words, a service client (an intentional agent) has an intentional moment of the type *Intention.* Here, moment is used in its ontological sense of being an individual that can only exists in other individual, i.e., moments are existentially dependent of other individuals. Every intentional moment has a type and a propositional content. The propositional content is an abstract representation of a class of situations referred by that intention. In an intention, the intentional agent commits at pursuing the satisfaction of this intention. Therefore, by having a goal, a service client commits to pursue the fulfillment of that goal. Using this definition we can have that many alternative state of affairs can satisfy (in the logical sense) the goal. Belief is defined in UFO as an intentional state about a certain state of affairs in reality. Examples include my belief that the Moon orbits the Earth and, my belief that Paris is the capital of France.

3.2 Goals, Tasks and Services

Figure 1 depicts the *Goal* concept of GSO and how it is related to UFO concepts (grayed boxes). In GSO we added that a Goal is owned by a Service Client Type. This ownership relation defines a meta-commitment making that the individual instances of the Service Client Type have a goal of certain kind, i.e., let S be a service type and g a goal, we have that S owns g iff for every instance x of S there is an intention I which is an intrinsic property of x (inheres in x) and g is the propositional content of I. For example, when an individual is a patient (in this example a patient is a service client type) he/she is characterized by the goal GetMedicalTreatment.

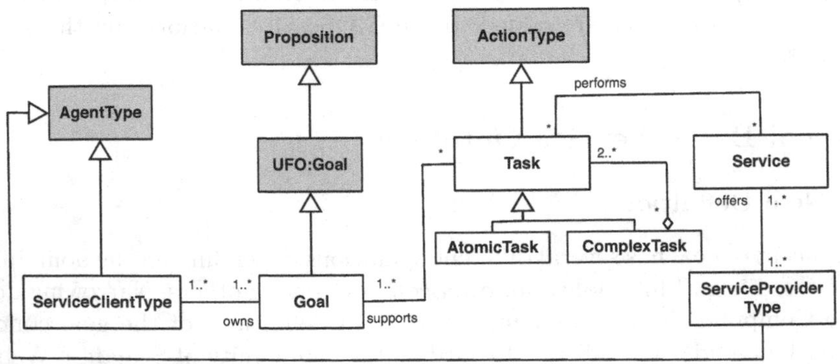

Fig. 1. Goal definition

Task in GSO is a specialization of the UFO concept of Action Type. An Action in UFO is an intentional event, i.e., an event performed by one or more agents in order to accomplish a goal. In figure 1, the relation *performs* between *Service* and *Task* (again, an Action Type) represents that instances of *Task* are executed when the associated service is executed. Finally, the relation *supports* between task and goal represents that a successful execution of that task satisfies that goal.

A domain specialist can define goals for different types of service client types in a domain. For instance, a *Doctor* (a type of service client) can be specified as having the goals *ProvideMedicalTreatment*, *KeepUpdatedWithMedicalAdvancements*, etc.

As depicted in Figure 2 (a model of instances), a Goal can be structured in two different ways, namely, in a decomposition structure (GoalANDDecomposition) and in a specialization structure (GoalORDecomposition). These two structures have different implications on goal fulfillment. In the decomposition structure, the fulfillment of the high-level goal is accomplished with the fulfillment of all the sub-goals. For instance, a high-level goal GetMedicalTreatment is fulfilled when its sub-goals GetMedicalConsult and GetMedicinePrescription have been fulfilled. Conversely, in the specialization structure, the fulfillment of a sub-goal implies the fulfillment of the high-level goal. For instance, the same hypothetical high-level goal GetMedicalTreatment is fulfilled when either one of the sub-goals GetHomeMedicalTreatment and GetHospitalizedMedicalTreatment is fulfilled. Figure 2 also shows the causal chain of goal satisfaction. An intention (of which a goal is its propositional content) causes an action (an instance of a Task) to be performed, i.e., since the agent is committed to the goal satisfaction, he acts accordingly to pursue its satisfaction. The action creates a situation that satisfies the goal. The use of situations to satisfy goals opens the possibility of using a Fuzzy mechanism to assess partial satisfaction (if necessary) of goals. Depending on the domain being specified using GSO, the domain specialists can define different goal satisfaction degrees.

In GSO, the ownership relation entitles the owner agent, i.e., a particular agent instantiating the specific service client type, to delegate the fulfillment of the goal to another agent. Moreover, by delegating a goal to an agent, the delegatee commits to the fulfillment of that goal. Therefore, the delegation relationship implies also a commitment between the delegator and the delegatee in relation to a goal. In GSO, this delegation relationship occurs when a service client delegates the fulfillment of a goal to a service provider. In the scope of this paper we are only considering the open delegation [5] of a goal. In this open delegation, a service client delegates the satisfaction of a goal to a service provider but does not prescribe any specific way of reaching this satisfaction. In other words, the service client only wants the goal satisfied without caring about how it is going to be satisfied. In contrast, in a close delegation the service provider should satisfy the service clients goal by means of a specific task. The relations of ownership, (close and open) delegations and satisfaction relations in GSO are also reflected in common goal-based requirements engineering languages such as i* and Tropos [6].

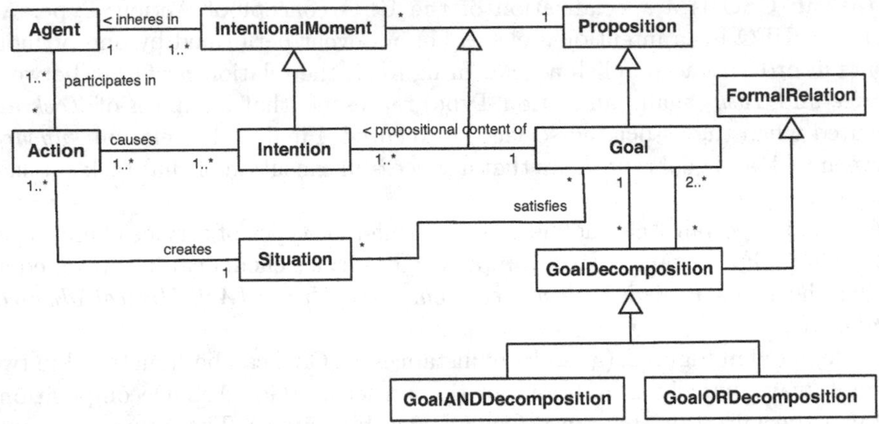

Fig. 2. Goal satisfaction and composition

3.3 Service

Although GSF aims at providing support for discovery and composition of computational services, at the ontological level we also consider services at the social level. This separation between social and computational services allows us to cope with situations where a computational service can be related to a social service and contribute to the fulfillment of a client goal. In GSO we define service as *a temporal entity related to the commitment (a service agreement) that a Service Provider, performs a task (a type of action) on behalf of a Service Client whose outcome satisfies a Service Client's goal.* This definition of service is based on the analysis of social services presented in [7].

Our definition encompasses some of the main characteristics of service as defined in the Marketing and Economics fields, namely, intangibility (as being a temporal entity) and the inseparability of production and consumption. As opposed to a product, when a service is delivered (the equivalent to the product's production) its outcome, which may satisfy the client's goal, is immediately perceived by the service client (the consumer). In [8], the authors state that the service's value "*is always uniquely and phenomenologically determined by the beneficiary*". In our framework this statement remains valid as the service client (the beneficiary) determined the service's value by the fulfillment of his goal.

In our definition two aspects can be considered, the service execution and the service agreement. Both have time-limited lifespan but represent different concepts. While the former represents that actual execution and consequent service provisioning, the later represents the validity of the service agreement. For example, the service execution of money withdraw from an ATM lasts as long as last the activities related to cashing out money from the teller. In this example, the agreement for the money withdraw service is valid for as long as the client has an account in his bank. This makes explicit that a service encompasses

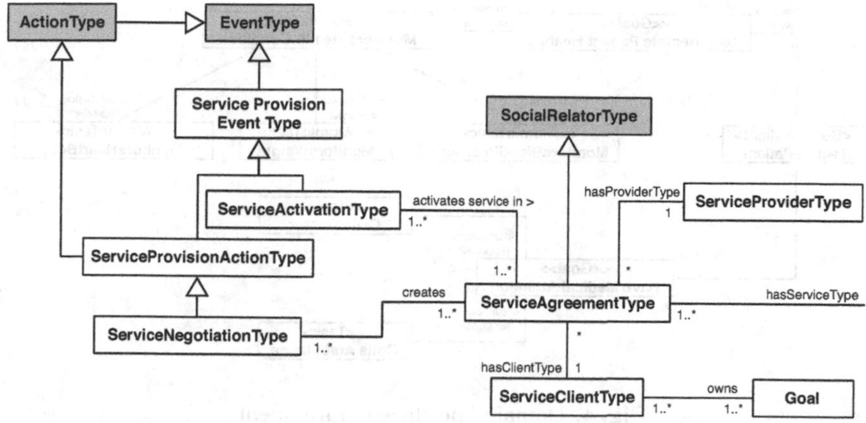

Fig. 3. Service negotiation and activation

a set of meta-commitments, e.g., commitments to commit to execute actions of a certain type [9].

Tying a service with a client's goal allows the analysis of the purpose of a service and its selection; namely, a service is selected because of its role on fulfilling a client's goal. Moreover, the relation between a goal and a service supports dynamic service discovery and selection by comparing situations that could satisfy a goal with the situations generated by services' outcomes. In other words, it is possible to discover services to fulfill a goal by verifying if the situation generated by the service's outcome is equivalent to a situation that can satisfy a goal.

Figure 3 depicts the relations between Service, Service Client Type and Service Provider Type. The *Service Provision Event Type* represents types of events that can participate in service provision such as *Service Negotiation Type* and *Service Activation Type*. When a service client discovers and selects a service, a negotiation takes place to determine the conditions and constraints for the service provisioning. A successful negotiation creates a service agreement type which is a social relator binding the service client and service provider and can be potentially composed of a set of commitments and claims, e.g., the commitment of providing the service under certain conditions and for an specified cost. This social relator (the *Service Agreement Type*) can be described in a contract (not depicted in the figure) which is a normative description [5].

4 Example Scenario

In this section we present an example scenario using GSF in the area of Home Health Care aiming at illustrating the feasibility and applicability of our approach. In this example we model the domain using GSO/GSM. The scenario is described as follows: "*John is a remote patient that receives health treatment at*

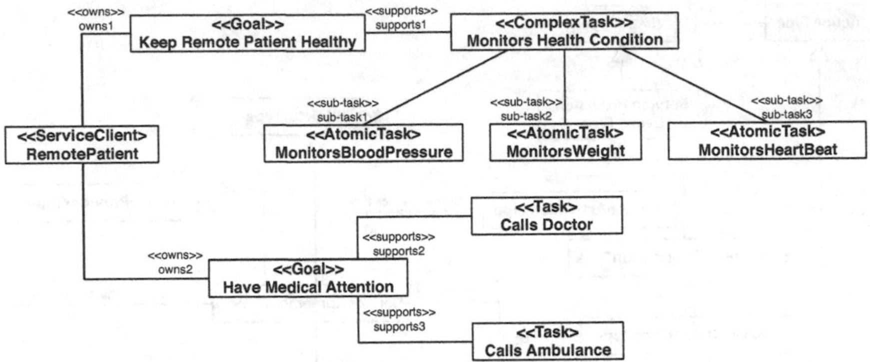

Fig. 4. Domain specification fragment

home. His house is equipped with several sensors that provide contextual information about his health condition such as weight, heart beat rate, blood pressure and glucose level. Moreover, movement sensors allow the determination of the householders' location and to assess whether their are in a responsive condition or not (e.g., asleep, fainted, etc). The main goal of John is to remain healthy. The house is equipped with the Context-Aware Service Platform, the Home Health domain has been specified and this domain specification is available to the platform. Several health-related services are available to the platform."

Figure 4 shows a fragment of the Home Health care domain specification. In this figure a *Remote Patient* which is a type of service client owns the two goals *Have Medical Attention* and *Keep Remote Patient Healthy*. The *Have Medical Attention* goal is supported by two tasks, namely, *Calls Doctor* and *Calls Ambulance*. Here we have an example of a goal being supported by two distinct tasks. The *Keep Healthy* goal is supported by the *Monitors Health Condition* complex task. This complex task is composed by the sub-tasks *Monitors Blood Pressure, Monitors Weight* and *Monitors Heart Beat.*

Figure 5 shows an UML object model of the instantiation of our illustrative domain specification. In this object model, *John* becomes a Remote Patient (a type of service client) when he pursues the fulfillments of his goals through services. Since *Keep Remote Patient Healthy* is a proposition, we have that *Keep John Healthy* represents a binding between an instance of *Remote Patient* and a generic proposition. However, for the sake of simplicity, we use a uniform representation for genuine instantiation and instance binding in a generic proposition.

Having John's goal, the GSF's Context-Aware Service Platform searches for instances of tasks that support John's goal *Keep John Healthy*. The supporting platform found that the complex task instance *Monitors Health Condition Inst* and its sub–classes *Monitors Weight Inst*, *Monitors Blood Pressure Inst* and *Monitors Heart Beat Inst* support John's goal. Having found the supporting tasks, the platform proceeds to search for services performing these tasks. In Figure 5 the platform found the services *Weight Monitoring Srv*, *Blood Pressure*

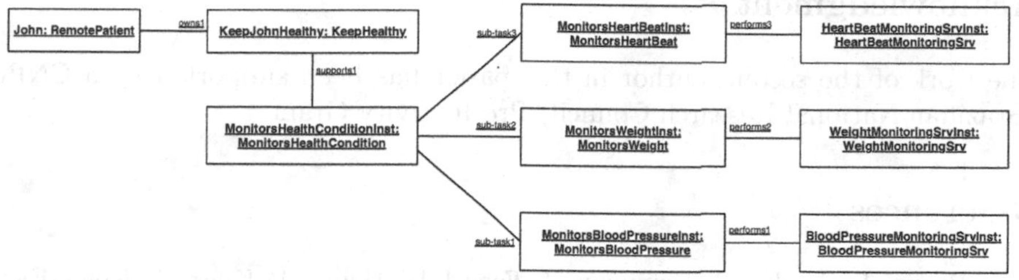

Fig. 5. John's instance model

Monitoring Srv and *Heart Beat Monitoring Srv* that perform the tasks *Monitors Weight Inst*, *Monitors Blood Pressure Inst* and *Monitors Heart Beat Inst*, respectively.

The Context-Aware Service Platform, acting on behalf of the service client negotiates a service agreement. In this example, this agreement stipulates the frequency of the monitoring activities and the threshold for emergency warnings in the case of abnormal health indicators' values, e.g., a blood pressure measurement above 200/160 or below 90/40.

5 Conclusions

This paper presented the main concepts comprising our characterization of a preliminary Goal-Based Service Ontology (GSO). This Ontology aims at providing the means for domain specialists to define domain ontologies. GSO is part of a framework (the Goal-Based Service Framework) for goal-based dynamic service discovery and composition. This framework is primarily target at application scenarios where the service clients are end-users without technological training in the scope of Pervasive and Service-Oriented Computing. For this purpose we propose the use of goal to express the service clients' requirements. In this manner, the service clients can express what they want to be accomplished by the services in a higher level of abstraction (by using goals).

Moreover, we presented and briefly discussed the ontological foundations of the main terms defined in this framework, i.e., goal, task, service client, service provider and service platform. This ontological foundation aims at providing an underlying conceptualization and at supporting the semantic definition of the terms used throughout our framework.

For the deployment of the framework, domain specialists should define their domain specifications (domain and task ontologies). Therefore, the framework is suitable for environments where the domain is clear and well known. Based on our experience in previous projects in the areas of Ambient Intelligence (AmI), Health Care and Mobile Pervasive Applications, we believe that these constitute examples of domains with suitable characteristics for our framework's deployment. A further investigation of the necessary characteristics and a more comprehensive list of suitable domains is in the scope of our future work.

Acknowledgment

The work of the second author in this paper has been supported by a CNPq (Brazilian National Research Council) Productivity Grant.

References

1. de Bruijn, J., Bussler, C., Domingue, J., Fensel, D., Hepp, M., Kifer, M., König-Ries, B., Kopecky, J., Lara, R., Oren, E., Polleres, A., Scicluna, J., Stollberg, M.: Web service modeling ontology (wsmo) (October 2006)
2. Martin, D., Burstein, M., Lassila, J.H.O., McDermott, D., McIlraith, S., Narayanan, S., Paolucci, M., Parsia, B., Payne, T., Sirin, E., Srinivasan, N., Sycara, K.: Owl-s: Semantic markup for web services (November 2004)
3. Guizzardi, G.: Ontological Foundations for Structural Conceptual Models. PhD thesis, University of Twente (2005)
4. Russel, S., Norvig, P.: Artificial Intelligence: A Modern Approach, 2nd edn. Prentice Hall, Englewood Cliffs (2002)
5. Guizzardi, G., Falbo, R., Guizzardi, R.S.S.: Grounding software domain ontologies in the unified foundational ontology (ufo): The case of the ode software process ontology. In: 1st Iberoamerican Workshop on Requirements Engineering and Software Environments (IDEAS 2008), Recife, Brazil (2008)
6. Bresciani, P., Giorgini, P., Giunchiglia, F., Mylopoulos, J., Perini, A.: Tropos: An agent-oriented software development methodology. Autonomous Agents and Multi-Agent Systems 8(3), 203–236 (2004)
7. Ferrario, R., Guarino, N.: Towards an ontological foundations for services science. In: Fensel, D., Traverso, P. (eds.) Proceedings of Future Internet Symposium 2008. Springer, Heidelberg (2008)
8. Lusch, R.F., Vargo, S.L.: The service-dominant logic of marketing: dialog, debate, and directions. M.E. Sharpe, Armonk (2006)
9. Silva Souza Guizzardi, R.: Agent-oriented Constructivist Knowledge Management. PhD thesis, University of Twente (February 2006)

ITGIM: An Intention-Driven Approach for Analyzing the IT Governance Requirements

Bruno Claudepierre[1] and Selmin Nurcan[1,2]

[1] Centre de Recherche en Informatique, Université Paris 1 - Panthéon Sorbonne
90 rue de Tolbiac - 75013 - Paris, France
[2] Institut d'Administration des Entreprises de Paris
20 rue Broca - 75005 - Paris, France

Abstract. Various research approaches on Information Technology (IT) management and IT engineering aim at understanding the new research area of IT governance. They identify the necessity to better formalize the domain of IT governance and to adapt it to a given organization. In this paper we aim at formalizing the user requirements regarding the IT governance and we propose an intentional model to capture these requirements. This allows us to formalize the requirements and the way-of-working of IT decision makers and IT stakeholders with regards to IT governance acivities.

1 Introduction

Various approaches in the literature identified the necessity of understanding and exploring the domain of IT governance. Most specifically Brown and Grant [BG05] argue the necessity to address a framework for IT governance and to adapt it to the particular situation of a given organization. Current researches [CN07] were exploring the adaptability and the enhancement of information systems engineering methods in order to better satisfy the software support provided to IT governance activities. Actually, at their creation, information systems do still not provide efficient support to IT governance activities. We identified the necessity to formalize also the requirements related to the IT governance in order to provide more complete methodological guidelines supporting information systems engineering. Various works proposed intention-driven approches in order to propagate the organizational change requirements over a technical system. An intention-driven model (MAP) has been used to define requirement engineering processes leading to the production of software specifications [RP01, Sal02, BN04]. This paper explores the cross-domain between IT management and IT engineering from the perspective of the requirement engineering.

The research question behind this work is the following: *how to formalize IT governance requirements to provide additional inputs to information system (re)engineering processes?*

We identified these requirements from the litterature related to the IT governance and organized them into an intentional model for IT governance.

C.A. Heuser and G. Pernul (Eds.): ER 2009 Workshops, LNCS 5833, pp. 275–285, 2009.

In the following, section 2 describes the context of IT governance and method engineering to determine the foundations of our approach. Section 3 presents the IT Governance Intentional Model (ITGIM).

2 Context of the Study

2.1 IT Governance

We define IT governance as an activity that aims at regulating and optimizing the IS management of an organization. It is generally performed under the responsability of the Chief Information Officer.

Strategic Level. The governance consists in distributing decision roles and responsibilities, and to organize the steering committee. We distinguish two types of strategic goals for the governance [Wir08], [CD98]: (i) the value creation where the executives are in line with the shareholders' requirements. In this context, the aim is to maximize the quotations of the shares; (ii) the value creation for the stakeholders which aims at improving the efficiency of the organization. IT alignment aims at providing the right IT support to business actors in order to improve business performance. Thus, governing IT with a strategic alignment perspective, generates value for stakeholders and the internal partnerships.

Luftman [LM04] concludes that alignment between the information system and the strategic goals of an organization is a crucial purpose for any CIO. Moreover [CB01] shows that alignment which is constructed between IT systems and business processes improves the organizational performance: an information system which is coherent with the business goals and processes generates value and improve the organizational performance. Various perspectives of alignment between external domains and internal domains have been identified by Henderson and Venkatraman [HV93]. These domains describe respectively : (i) the strategic level and (ii) the operational level for both IT and business purposes. Henderson and Venkatraman link the IT and business governance with the external domains.

Peter Weill [Wei04] analyzes the decisional aspects in the organization of information systems by comparing them with the traditional archetype of governance. He describes the possible organizations for decision making: a centralized decision process is thus compared with a monarchy and a collaborative decision process (between IT and business group) is related to a democracy. Decision processes are based on a typology of decisions and the study shows us that IT investment is the duty of business executives whereas CIOs are more concerned with decisions related to IT architecture or IT infrastructure. De Haes argues that "IT governance can be deployed using a mix of structures, processes and relational mechanisms" [DHVG05]; In which "structures involve the existence of responsible functions such as IT executives and a diversity of IT committees".

Supports the idea that a given organization for information systems is structured for decision making with a comity where decision roles and responsibility are defined and distributed.

Tactical Level. Process oriented approaches show an organizational view of IT projects. The information system life cycle is described by a set of IT processes where actors play roles (CIO, project owner). These approaches are sometimes joined with maturity evaluation frameworks. Indicators and metrics are specified to measure the maturity level of the IT processes.

The IT Governance Institute (ITGI) proposes also an evaluation framework for IT processes: Control Objectives for IT (COBIT) [ITG02]. More and more used, this framework proposes a set of processes organized by goals and linked to a corpus of metrics. This approach makes the assumption that a given IT process is enacted to achieve a measurable goal. Some researchers like Simonsson summarized a way of working of the COBIT system. For instance, IT Organization Modeling and Assessment Tool (ITOMAT) is a modeling and evaluating tool for the organization of the information system [SJ08]. Simonsson took into account the key concepts of COBIT, such as the description of processes and metrics.

The decision is the next step which succeeds to the organizational context analysis. ISO 9001 (2000) locates the decisional process over an analytical path which is composed of : (i) the measurement process, (ii) the management process and (iii) the support process. Izza [IVB07] describes the context of the organizations throughout a typology of processes. [SN07] proposes a framework which aims at supporting the analysis of the organizational context in order to integrate decisional aspects in the process engineering.

Operational Level. Governance systems should provide a support to decision making and to the process control based on measures. In the literature related to project management, we identified requirements for (i) activity and resource scheduling capabilities and (ii) risk management. During the development of an information system, risks are mainly related to the cost, the delay but also the conformity of the resulting system with user requirements.

Various authors work on risk management in the domains of project management, and process modeling and enactement [SLP07],[BZCC07]. The organizational change can be considered as an event which generates technical or financial risks. The negative impacts over projects and their cost must be minimized: [SLP07] proposed a conceptual model to integrate the risk management within the information system engineering processes. [BZCC07] proposed a set of indicators in order to measure the impact of change projects on the organization.

Indicators are the foundation of decision support systems. A well known method is the Balanced Score Card of Kaplan and Norton [KN96]: it allows managers to trace the evolutions of various parameters over projects. These authors proposed to organize the indicators into financial, customer, business process and learning perspectives. This cluster of indicators supports the risk tracking and evaluation over these perspectives.

2.2 Information Systems Engineering

Propagate the Change. Today information system engineering deals maintly with transformation and evolution of an existing information system rather than

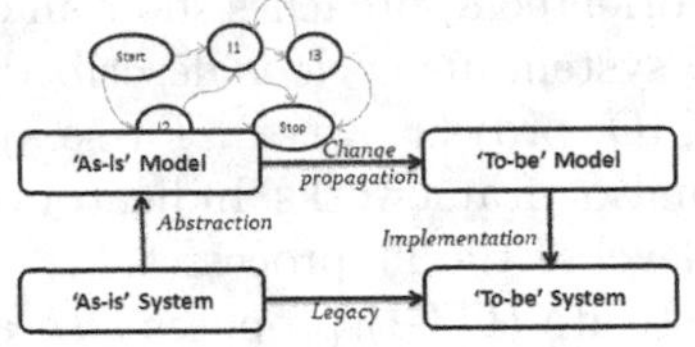

Fig. 1. Engineering process [Jac95]

creating new ones from scratch. Furthermore we base our reasoning on the requirement engineering vision of Mickael Jackson [Jac95] according to which the change process is the process of transforming the vision into a new model. Within the world in which the vision has to be realised, many habits (legacies) exist. Some are based on formally stated goals, policies, or competing visions. Others are just regularly observable phenomena for which no predefined structure or reasons are known a priori. The task is therefore twofold (see Fig. 1): (i) relevant habits must be analysed and the goals, policies and visions behind them must be made explicit and formal. This leads to the 'As-Is' model that defines the functionality of the current organisation. (ii) The new vision must be established by operating transformations on the 'As-Is' model, leading to the 'To-Be' model that defines the requirements for the envisioned organisation.

Modeling Requirements. The literature provides various formalisms for requirements modeling which are using different paradigms to specify user requirements for an information system under construction. According these approaches we are more concerned by the goal oriented ones because they allow us to capture the 'why' perspective which is essential both for IS engineering in an evolving environment and for IS governance. We can remind here KAOS which is a methodology for goal-driven requirement modeling where goals are structured into a refinement tree. Generic goals can be refined into a set of sub-goals. This hierarchy is described by a AND/OR graph [BDD+97]. The CREWS-L'écritoire [TVSBA00] approach is based on scenarii analysis for identifying goals. Another way of modeling requirements is provided by the well known I* framework [YM94]. In the following we use the intentional paradigm and more specifically the MAP approach [RPB99], [RP01]. The main advantages of this formalism with respect to the previously listed approaches are (i) the distinction between the concept of goal and the way of achieving it (ii) the formalization of a non-deterministic process and (iii) the refinement mechanism. A MAP is defined throughout a graph where a node represents a goal (or intention) and the way of reaching a goal is represented by an input edge (or strategy). We can note three main concepts : the intention, the strategy and the section. A section is a triplet *(source_intention, target_intention, strategy)* and can be performed only if the source intention is achieved. The MAP has been used in various domains for system engineering [RP01],[BN04], method engineering [RR01] and strategic alignement [TS07].

3 ITGIM: IT Governance Intentional Model

IT governance requirements are mainly focused on security, cost and risk management, compliance with laws and regulations, and the change definition. Thus, a CIO should simultaneously align IT with business strategy, manage risks and maintain the compliance of the information system with the laws regulating the business domain. The achievement of those intentions aims at generating value. The following subsections describe those intentions and the related strategies and summarize the resulting MAP fragments. We consider a MAP fragment as a section *(source_intention, target_intention, strategy)* as introduced above. IT governance requirements we will introduce below are based on our litterature survey and are not exhaustive.

3.1 Research Methodology and Construction Process

Our approach is a descriptive research that aims at formalizing IT governance goals and the ways to perform them. We use modeling notations based on the MAP model to show the result of a content analysis of textual documents related to IT governance.

Map models are built using the goal taxonomy of Prat [Pra97]. A map fragment is an extraction of a goal. Figure 2 shows the framework used to product MAP components based on text analysis. From the RE perspective, this research can be seen as a case study of RE formalisms and contributes to the validation, by example, of the MAP model.

Because of a lack of space we do not include details for the evaluation step.

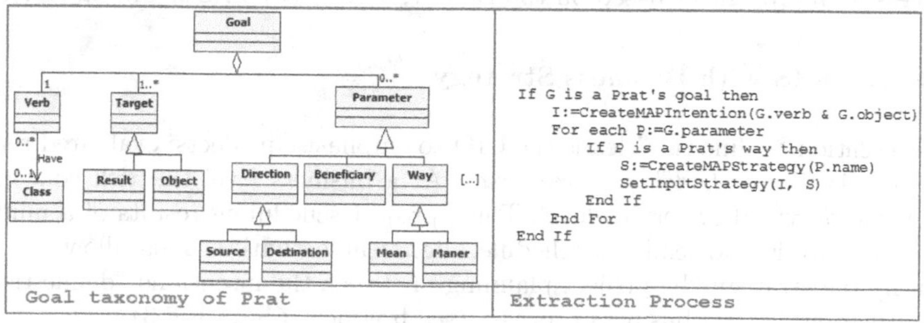

Fig. 2. Methodological framework

3.2 Model Risk

A main goal for IT governance is to steer information system projects and operations. This leadership takes place in an evolving and risky environment and CIOs have to take decisions under multiple interdependant constraints. We define the risk as an event which impacts negatively the assets of the organization and/or the completion of stakeholders' goals. According to Wikipedia, there are

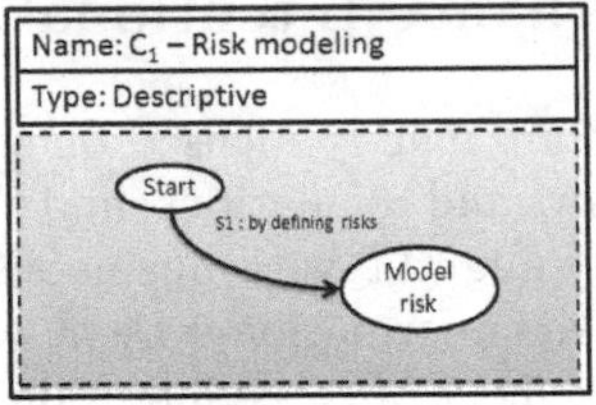

Fig. 3. Methodological fragments for risks

three ways to handle an identified risk: (i) to limit the occurrence of the event by using a prevention strategy; (ii) to accept the risk and to put it under control; (iii) and to categorically refuse it and to cancel projects which can potentialy generate this risk.

As mentioned in [HdBSLS02] by referencing the European project CORAS, the first step in risk management consists to define the risk and the related context. From an engineering perspective, defining the risk consists first in providing a representation of the risk as an artefact (for instance using a conceptual model) as well as a representation of the process which will deal with it. We represent "Model risk" in the IT governance Map as an intention which can be achieved by enacting the "by defining risk" strategy (see fragment C_1 on Fig.3). For instance, the main risks are about security failure on information assets or risks over project attributes as delay or cost. The standard ISO 27001 provides a framework for seting a management system for information integrity and safety. The standard is focused on a quality approach (ISO 9001) as a continuous process of improvement based on the Deming's cycle (Plan, Do, Check, Act).

3.3 Align IS with Business Strategy

The intention "define risk" leads the CIO to emphasise projects: evaluated risks are criteria like *cost, delay or resources*. At a business level the risk concerns the value provided or service level. Thus, project scheduling results of a multi-criteria analysis and leads to schedule alignment activities. This allows us to identify the strategy "by project planning" between the intentions "define risk" and "aligning IT and business process" (see fragment C_2 on Fig.3).

Strategic alignment is a "hard goal" – in I* perspective [YM94] – for CIOs: it consists in engineering an information system which is coherent with business strategies and goals. This activity become more and more important in the context of co-evolution, i.e. the simultaneous evolutions of information system, and the business structure and the orientations that it supports. In this area, Anne Etien [ES05] proposed various ways to propagate the change over layers of models specified for 'As-Is' and 'To-Be' states. This led us to draw the strategy "by modeling" between the intentions "Start" and "Align IT and business process" (see fragment C_3 on Fig.4). Thus aligning IS with business strategy aims at sustaining strategic activities. [Dey04] describes five categories of strategies for

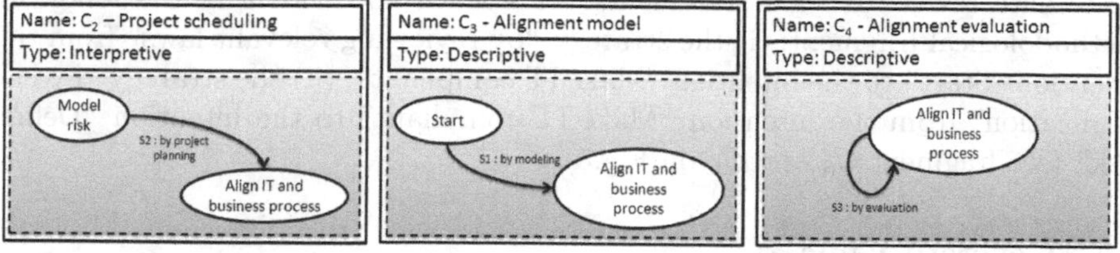

Fig. 4. Methodological fragments for alignment

organization, namely: 'innovation', 'market conquest', 'optimizing', 'improving' and 'differentiation'.

[Dam05] proposed a framework for evaluating the accuracy, visibility, delay of presentation and the security of information conveyed by the IS. This work addresses the control and regulation aspects, the need to comply with legislation, to manage the risk by applying a control framework like COSO or COBIT (see fragment C_8 on Fig.6). Finally, the strategic alignment is presented as a generator of performance and value. This supposes that the intention to align IS with business strategy is already active and the evaluation based on COSO or COBIT leads to improve the alignement. As a result, we mention the strategy "by evaluation" over the intention "Align IT and business process" (see fragment C_4 on Fig.4).

3.4 Comply with Laws

CIOs have to face challenges in maintaining their information systems also compliant with laws and regulations.

Scandals like Enron or Worldcom forced the governments to edit acts like Sarbanes-Oxley Act of 2002 (SOX) in USA or the *loi de régulation financière* in France. These laws gave some directives to public organizations in order to limit the risk generated by the manipulation of the information because of its lack of accessibility, quality or dificulty to agregate. For example if SOX is properly applied, the CEO must personally verify the balance sheet and income statement of the organization by signing it (SOX - Section 302). This forces the leader to be more vigilant on evaluation mechanisms and limits the risk of providing false financial results. The compliance can be satisfied only if IT owners have an

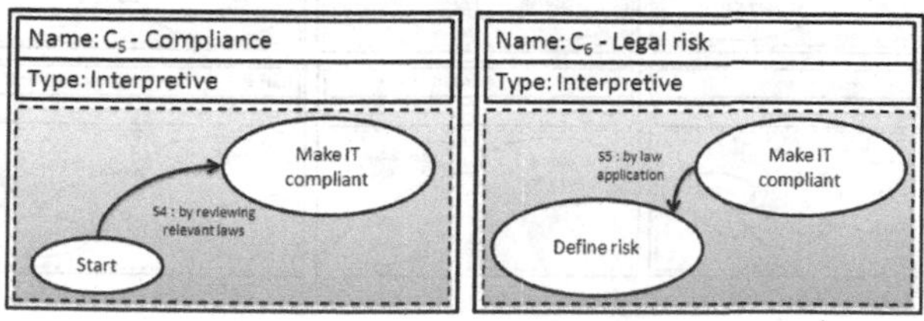

Fig. 5. Methodological fragments for compliance

up-to-date review of legal statements. This leads us to identify the following methodological fragment: (i) the strategy "by reviewing relevant laws" from the intention "Start" to the intention "Make IT compliant"; (ii) the strategy "by law application" from the intention "Make IT compliant" to the intention "Define risk" (see fragment C_5 and C_6 on Fig.5).

3.5 Generate Value

Value creation is the main goal of the corporate governance. We identified two types of value: the external value for which the purpose of the organization is to fulfill the expectations of investors and shareholders; and the internal value or partnership value for which the purpose of the organization is to develop synergies and improve internal performance of the organization [CD98]. The purpose of the value creation is propagated to the sub-systems of the corporate governance such as the IT governance, the business process governance or the production governance.

The compliance with laws is sometimes an obligation to access a market. For example international exchanges with U.S. are allowed if and only if partnership companies comply with SOX. To represent this fact, we introduce the strategy "by competitive advantage" from "Make IT compliant" to "Generate value" (see fragment C_7 on Fig.6).

The IT governance goals should fit with the value creation goal from the corporate governance. The success of the alignment between IT and business processes sustains the partnership value creation when IT provides an adequate and useful package of services to business actors. We define thus the intention "Generate value" that can be achieved by enacting the "by IT service proposal" strategy from "Align IT and business process" (see fragment C_9 on Fig.6). The IT governance process may stop (till the next instanciation) when value creation is complete. We add the strategy "by completeness" from the intention "Generate value" to the intention "Stop" (see fragment C_{10} on Fig.6). However we must also include the cases where value creation goal cannot be achieved and the IT governance process ends when: (i) the CIO will restrict the process to the

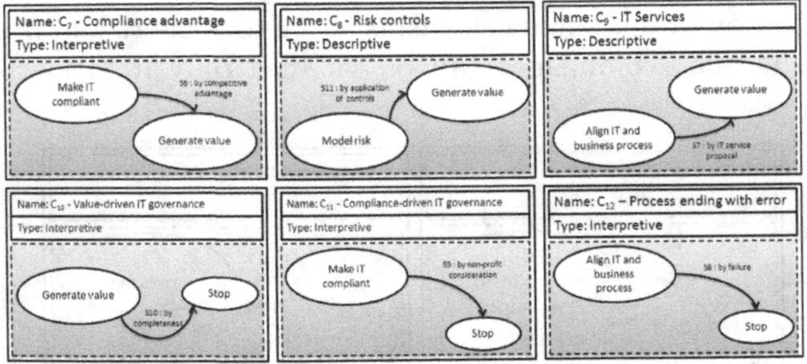

Fig. 6. Methodological fragments for value generation and IT governance process termination

mandatory application of the law or (ii) the alignment process between IT and business processes failed (see fragment C_{11} and C_{12} on Fig.6).

4 Conclusion

In this paper we proposed to analyze the area of IT governance and made a temptative to provide an intentional vision for IT governance processes (see Fig.7). The result shows us a process where requirements over risk management, business/IT alignment, value generation and compliance with laws and regulations are considered as centric aspects for IT governance purposes. The ITGIM model contains several paths to follow for performing IT governance.

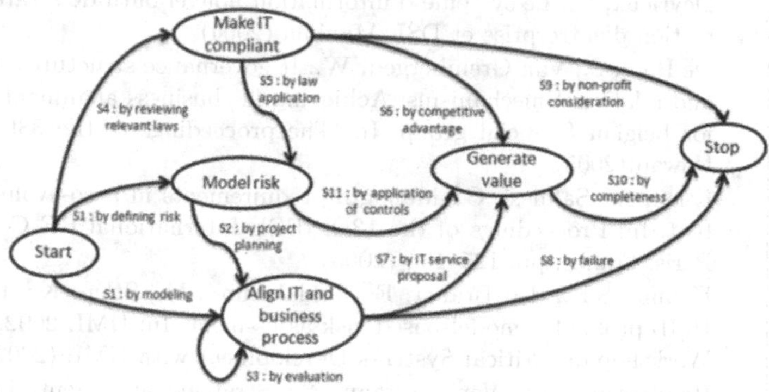

Fig. 7. An intentional view of IT governance processes

This work provides some inputs to the research efforts which aim at solving IT governance related problems. The IT governance requirements impact over IT engineering activities have not been explored yet. Our future work aims at providing methodological support for the IS engineering in order to build systems which integrate IT governance functionalities and capabilities.

References

[BDD+97] Bertrand, P., Darimont, R., Delor, E., Massonet, P., van Lamsweerde, A.: GRAIL/KAOS: An Environment for Goal-Driven Requirements Engineering. In: Proceedings of the 19th ICSE, Boston, Massachussetts, United States (1997)

[BG05] Brown, A.E., Grant, G.G.: Framing the frameworks: a review of IT governance research. Communications of the AIS (2005)

[BN04] Barrios, J., Nurcan, S.: Model driven architectures for enterprise information systems. In: Persson, A., Stirna, J. (eds.) CAiSE 2004. LNCS, vol. 3084, pp. 3–19. Springer, Heidelberg (2004)

[BZCC07] Ben Zaida, Y., Chapurlat, V., Crestani, D.: Construction et évaluation de projets de changement des entreprises manufacturières. In: 4ème workshop Ingénierie et gestion des processus, GDR I3, MACS (2007)

[CB01] Corteau, A.M., Bergeron, F.: An information technology trilogy: business strategy, technological deployment and organizational performance. Journal of Strategic IS (10), 77–99 (2001)

[CD98] Charreaux, G., Desbrière, P.: Gouvernance des entreprises: valeur partenariale contre valeur actionnariale. Finance Contrôle Stratégie (1), 57–88 (1998)

[CN07] Claudepierre, B., Nurcan, S.: A framework for analysing IT governance approaches. In: ICEIS, Funchal, Portugal, pp. 512–516 (2007)

[Dam05] Damianides, M.: Sarbanes–oxley and IT governance: New guidance on IT control and compliance. Information Systems Management (Winter, 2005)

[Dey04] Deyrieux, A.: Le système d'information nouvel outil de stratégie - Direction d'entrerprise et DSI. Maxima (2004)

[DHVG05] De Haes, S., Van Grembergen, W.: It governance structures, processes and relational mechanisms: Achieving IT/business alignment in a major belgian financial group. In: The proceedings of the 38th HICSS, Hawaii (2005)

[ES05] Etien, A., Salinesi, C.: Managing requirements in a co-evolution context. In: Proceedings of the 13th IEEE International RE Conference, Paris, France, pp. 125–134 (2005)

[HdBSLS02] Houmb, S.H., den Braber, F., Soldal Lund, M., Stlen, K.: Towards a UML profile for model-based risk assessment. In: UML 2002, Satellite Workshop on Critical Systems Development with UML (2002)

[HV93] Henderson, J.C., Venkatraman, N.: Strategic alignment: Leverating information technology for transforming organisations. IBM System Journal 32(1), 4–16 (1993)

[ITG02] ITGI. Les meilleures pratiques de gouvernance d'entreprise. ITGI (2002)

[IVB07] Izza, S., Vincent, L., Burlat, P.: Vers une typologie intégrée des processus d'entreprise. In: 4ème workshop Ingénierie et gestion des processus, GDR I3, GDR MACS (2007)

[Jac95] Jackson, M.: Software Requirements and Specifications - A Lexicon of Practice, Principles and Prejudices. Addison Wesley Press, Reading (1995)

[KN96] Kaplan, R., Norton, D.: Balanced Scorecard - Translating strategy into action. Harvard Business School Press, Boston (1996)

[LM04] Luftman, J., Maclean, E.R.: Key issues for IT executives. MIS Quarterly Executive (3), 89–104 (2004)

[Pra97] Prat, N.: Goal formalisation and classification for requirements engineering. In: Requirements Engineering: Foundation for Software Quality (REFSQ), Barcelona, Spain (1997)

[RP01] Rolland, C., Prakash, N.: Matching ERP system functionality to customer requirements. In: Proceedings of the 5th IEEE ISRE, pp. 66–75 (2001)

[RPB99] Rolland, C., Prakash, N., Benjamen, A.: A multi-model view of process modelling. Requirements Engineering (4), 169–187 (1999)

[RR01] Ralyté, J., Rolland, C.: An assembly process model for method engineering. In: Dittrich, K.R., Geppert, A., Norrie, M.C. (eds.) CAiSE 2001. LNCS, vol. 2068, p. 267. Springer, Heidelberg (2001)

[Sal02] Salinesi, C.: Cartographier le Système d'Information par les 'intentions' et 'stratégies' de métier. Colloque Urbanisme des Systèmes d'Information (2002)

[SJ08] Simonsson, M., Johnson, P.: The IT organization modeling and assessment tool: Correlating IT governance maturity with the effect of IT. In: Proceedings of the 41st HICSS (2008)

[SLP07] Sienou, A., Lamine, E., Pingaud, H.: Intégration de la gestion de processus et de la gestion des risques: vers un modèle conceptuel du risque processus. In: 4ème workshop Ingénierie et gestion des processus, GDR I3, MACS (2007)

[SN07] Saidani, O., Nurcan, S.: Prise en compte de l'aspect décisionnel dans l'ingénierie et la gestion des processus d'entreprise. In: 4ème workshop Ingénierie et gestion des processus, GDR I3, MACS (2007)

[TS07] Thevenet, L.-H., Salinesi, C.: Documenter l'alignement des objectif stratégiques de l'entrerprise et du SI avec la méthode INSTAL. In: 4ème workshop Ingénierie et gestion des processus, GDR I3, MACS (2007)

[TVSBA00] Tawbi, M., Velez, F., Souveyet, C., Ben Achour, C.: Evaluating the CREWS-L'écritoire requirements elicitation process. In: Fourth IEEE International Conference on RE, Schaumburg, Illinois, USA (2000)

[Wei04] Weill, P.: Don't just lead, govern: How top-performing firms govern IT. MIS Quarterly Executive 8(1), 1–17 (2004)

[Wir08] Wirtz, P.: Les meilleures pratiques de gouvernance d'entreprise. La Découverte (2008)

[YM94] Yu, E.S.K., Mylopoulos, J.: From e-r to a-r - modelling strategic actor relationships for business process reengineering. In: The proceedings of the 13th ICERA, Manchester (1994)

Adapting the i* Framework for Software Product Lines

Sandra António[1], João Araújo[1], and Carla Silva[2,*]

[1] CITI/FCT, Universidade Nova de Lisboa,
2829-516, Caparica, Portugal
{sia14381,ja}@di.fct.unl.pt
[2] Centro de Ciências Aplicadas e Educação, Universidade Federal da Paraíba,
58297-000, Rio Tinto, Brazil
ctaciana@ccae.ufpb.br

Abstract. Feature modeling is an important technique to capture commonalities and variabilities in a software product line (SPL). However, this kind of models shows a specific perspective, which is not sufficient to express all the characteristics and constraints of an SPL. Using a goal-oriented approach, such as i*, to complement (and help define) feature models would improve such models enhancing meaning and justification to features. Goal-oriented modelling provides a way to identify variabilities at an early phase of requirements, allowing alternative options to satisfy stakeholder's goals. The aim of this work is to benefit software product lines from the framework i*, a more expressive approach to requirements engineering of SPLs.

Keywords: i* Framework, Software Product Line, Requirements Engineering, Feature Model, Goal-Oriented Approach.

1 Introduction

Research in requirements for software product lines (SPL) has been exploring ways by which one can define a platform capable of serving as the basis for cost-effective derivation of products for individual users. Feature modeling is an important technique for capturing commonalities and variabilities in product lines. A feature may denote any functional or non-functional characteristic at the requirements, architecture, or any other abstraction level of software. However, feature models show a very specific perspective of a product line, so it is necessary to have an approach that shows other perspectives at the requirements level, and give them semantics to make an SPL more understandable.

The goal and agent-oriented paradigms have been used to develop complex systems and some approaches, such as the i* framework, have been developed to be used in requirements engineering. Organizational modeling with the i* framework offers social and intentional concepts, and stakeholder's desires are considered really important to develop systems that best meet their needs. Goal models provide a

* This work was partially performed at Centro de Informática – Universidade Federal de Pernambuco and partially supported by CAPES Proc. PNPD-0092088 research grant.

C.A. Heuser and G. Pernul (Eds.): ER 2009 Workshops, LNCS 5833, pp. 286–295, 2009.

natural way to identify variability at the early requirements phase, by allowing the capture of alternative ways by which stakeholders can achieve their goals. This characteristic of goal-oriented models can benefit the development of software product lines. However, the use of the i* framework to describe software product lines has not been explored sufficiently so far. Therefore, the purpose of this paper is to adapt the i* framework to develop software product lines in order to obtain a more expressive requirements engineering approach for such products.

This work is organized as follows. Section 2 gives an overview of the basic concepts of goal oriented requirements engineering and software product line engineering. Section 3 presents the IStarLPS approach. Section 4 illustrates IStarLPS using as an example the Media Shop system. Section 5 describes some related work. Finally, Section 6 summarizes our proposal and points out directions for future work.

2 Background

In this section, we briefly describe the i*framework, a goal oriented approach, and the feature model used in software product lines engineering. Both are the basis of our proposal.

2.1 Goal Oriented Requirements Engineering

Goal-Oriented Requirements Engineering (GORE) attempts to fill some gaps presented by other approaches [1]. According to Lamsweerde [10] a goal is what a system should reach with agents' cooperation in the software-to-be and in the environment.

The i* framework [2] is a goal-oriented approach widely used in early-phases of requirements engineering. It offers two models: Strategic Dependency (SD) and Strategic Rational (SR). With an SD model we can represent dependencies between actors and with an SR model we can show how an actor can satisfy their dependencies. An actor is capable to realize actions and there are two kinds of actors: a Depender (the depending actor) and a Dependee (the actor who is depended upon). The Dependum is the object around dependencies, which can be one of these four types: goal, task, resource or softgoal. Thus, an actor depends on others to achieve a goal, realize a task, obtain a resource or satisfy a softgoal. A softgoal differs from a goal because it has not an initially defined criterion to be achieved. SR models also have three types of links: Means-End, Task Decomposition and Contribution. These links are useful to refine dependencies links, showing how an actor satisfies their dependencies.

2.2 Software Product Line Engineering

Producing similar products separately would take longer than if they belonged to a product line. Hence, many companies decided to use Software Product Line Engineering (SPLE) to develop their products, based on the reuse of components. An SPL is a group of similar software products with commonalities and variabilities, and can be also called product family [3].

SPLE is recommended for large scale software production and to satisfy the needs of several clients. Besides, these software products are built using the same platform, which is a set of interfaces and sub-systems which represent a common structure for a group of products [4].

SPLE has two processes, domain engineering and application engineering. In the domain engineering process, the platform of a product line is defined, while in the application engineering process a specific product is configured [4]. To represent variabilities and commonalities of a product line, we can use feature models. A feature model is developed in the domain engineering process and configured in the application engineering process.

A feature is a property of a system relevant to some stakeholders and used to capture common and variable characteristics between products in the same family [5]. Features are organized in a hierarchical diagram such as a tree and the root represents a concept (e.g. software system). Each feature can be refined into sub-features. A feature can be classified as mandatory, optional or alternative. In this last case it may be an *Or* or *Xor* alternative. We can represent feature interaction through a requires link which is used when one feature requires another, but the opposite does not occur[5].

Czarnecki *et al.* [5] present a feature model with cardinality to remove ambiguous cases. The cardinality is an interval that shows the amount of times that a feature may be cloned, when an application is specified.

3 The IStarLPS Approach

Similarly to SPLE, the IStarLPS approach has two processes, domain and application engineering. The Domain engineering sub-process has four principal activities: a) SD model development, b) get features from the SD model, c) SR model development and d) feature model development. The application engineering sub-process has two activities: A) feature model configuration and B) i* models configuration, both for a specific application (or product).

The two main concepts in IStarLPS are goals and features. A feature represents a relevant characteristic of the system and allows a goal to be achieved. So, we can extract features from the description of a goal, because it contains the characteristics that are needed to satisfy the goal.

The concept of cardinality was added to represent variability in i* models. Cardinality is represented in the description of intentional elements. In i* models, cardinality helps mostly to identify optional and alternative cases. The cardinality used in i* models is similar to the one used in feature models. However, if we tag a goal with a cardinality (1..*), that means the property inside that goal (the simplest feature) can appear multiple times in an application, and the goal should be satisfied in all applications. Figure 1 illustrates some examples of intentional elements with cardinality and their respective obtained features.

Based on the analysis of the i * framework and feature models and their application to case studies, we defined hypotheses to produce a feature model from i* models. Each hypothesis is identified as **H**.

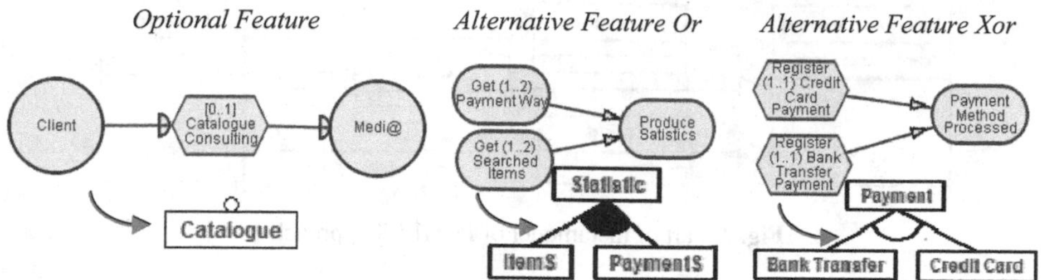

Fig. 1. Examples of intentional elements with cardinality

H.1. Features can be extracted from properties which describe an intentional element.

H.2. Features related with actions will probably have sub-features.

H.3. More than one feature can be extracted from the same intentional element.

H.4. If we obtain more than one feature from the same intentional element, probably those features will be related in feature model.

H.5. If a feature is obtained from an intentional element with cardinality [0..1] then, this feature will be represented as optional in the feature model.

H.6. If a feature is obtained from an intentional element with cardinality *(j..k), 0<j<k< max number of different features*, which is a mean in a means-end link, then that feature will be represented as an alternative Or in feature model.

H.7. In case a feature must be specified in a feature model but there is not enough information about it in the i* models, it is necessary another source to get that information.

H.8. If a feature is obtained from an intentional element with cardinality (1..1), which is a means in a means-end link then that feature will be represented as an alternative Xor in feature model.

H.9. If there are two features obtained from two intentional elements that are related through a means-end link, task-decomposition link or dependency link, then these features will be related in feature model.

H.10. The feature X requires feature Y when X needs Y and Y is independent of X.

When a feature is identified from the property of an intentional element, redundancy should be avoided and get the simplest feature, instead of generating more features than necessary. Although a feature can be obtained from all intentional element descriptions, that would not be necessary because some features can represent low level and do not bring new relevant information.

With these hypotheses we can obtain the features and information about their relationship and interactions, allowing feature model development in a systematic method.

The adaptation of the i* framework for SPLs to include cardinalities proved useful when deriving the respective feature model. This was contemplated when specifying the relations between the i* metamodel [2] with the feature model metamodel [5]. The resultant metamodel supports this adaptation which has been called the IStarLPS approach. The relevant concepts of the IStarLPS metamodel are represented in

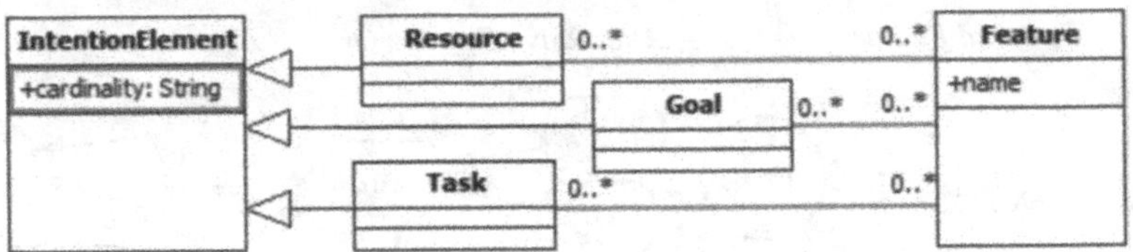

Fig. 2. Part of metamodel of IStarLPS approach

Figure 2, namely: the cardinality in the intentional element, the three associations between feature class and goal, task and resource classes. The associations relate a feature with the respective i* model elements. Softgoals were not analyzed so far because they normally are not contemplated in feature models. These models emphasize the functionalities.

4 Example: Media Shop

The IStarLPS approach has been applied to a product line of a Media Shop system to illustrate how it can be used. The Media Shop system represents a shop that sells media items through the Internet. First of all, the domain engineering process of the IStarLPS approach was applied.

According to the domain engineering process, initially an SD model must be developed. For that we define the actors, elicit the goals, tasks and resources dependencies. The SD model in Figure 3 represents part of the Media Shop system.

In Figure 3 there are three actors: Customer, Medi@ (Online Shop) and Media Shop. The Customer depends on Medi@ to obtain the "Catalogue" resource, to realize the "Place Order", "Keyword Search" tasks and to achieve the "Buy Media Items" goal.

On the other hand, Media Shop depends on Medi@ to achieve the "Statistics Provided" and "Process Internet Orders" goals. The "Catalogue" resource has cardinality [0..1] which means that depending on the application, a customer can obtain a Catalogue.

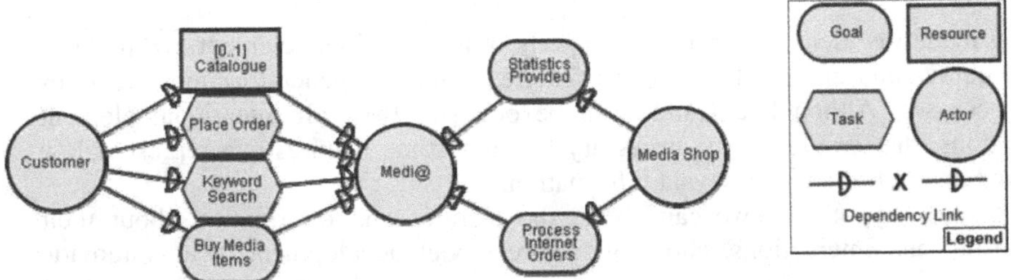

Fig. 3. Simplified SD model of Media Shop system

Next, Table 1 is built, where the **H.1** hypothesis is used to obtain features from intentional elements.

Table 1. Relationship between intentional elements and features

Intentional Element		Feature
Type	Name	
Objective	Buy Media Items	Media Items
Objective	Statistics Provided	Statistics
Objective	Process Internet Orders	Internet Orders
Task	Place Order	Order
Task	Keyword Search	Keyword
Resource	Catalogue	Catalogue

Afterwards, the SR model of the media shop system is developed. Figure 4 shows part of this model. From Figure 4 we can obtain the remaining system features, their relationships and interactions using hypotheses of the IStarLPS approach. In Table 2, the rest of features obtained through hypotheses **H.1** and **H.3** is presented.

Table 2. Obtaining features from intentional elements from SR model

Intentional Element		Feature
Type	Name	
Objective	Buy Media Items	Media Items
Objective	Confirm Payment	Payment
Objective	Get Bought Items	Item
Task	Manage Payment	Payment
Objective	Process Payment	Payment
Objective	Payment Method Processed	Payment Method
Task	Register Bank Transfer Payment	Bank Transfer Payment
Task	Register Credit Card Payment	Credit Card Payment
Objective	Get Payment Information	Payment Information
Objective	Get Used Payment Way	Payment Way
Objective	Get Customer Information	Customer Information
Objective	Get Customer Profile	Customer Profile
Task	Shopping Cart	Shopping Cart
Task	Add Item	Item
Task	Manage Internet Shop	Shop
Objective	Process Internet Orders	Orders
Objective	Get Searched Items	Item
Task	Produce Statistics	Statistics
Objective	Statistics Provided	Statistics
Objective	Item Searching Handled	Item
Task	Database Querying	-
Objective	Item Transaction	Item
Task	Get Item Detail	Item Detail
Objective	Item Selection	Item
Task	Choose Non-Available Items	Item
Task	Choose Available Items	Item
Task	Place Order	Order
Task	Catalogue Consulting	Catalogue
Resource	Catalogue	Catalogue
Task	Keyword Search	Keyword
Resource	Keyword	Keyword

Subsequently, the feature model using the information collected in previous activities is developed and the information about the relationships and interactions of features in the SR model is obtained.

Through **H.2** we know that features like "Order" or "Payment" probably will have sub-features.

Through **H.4** we know that features like "Media Item" and "Detail" will be related in the feature model.

Through **H.5** we obtain the optional feature "Catalogue", because this feature is originated from resource "[0..1] Catalogue".

Through **H.6** we get three different features from four different intentional elements with cardinality (1..k) linked to an objective by a means-end link, and this indicates the features obtained will be grouped by an Or alternative.

Through **H.7** we know that the "Detail" feature probably will have sub-features which are not specified in i* models.

Through **H.8** we get two features from two different intentional elements with cardinality (1..1) linked to an objective by means-end link, and this indicates the features obtained will be grouped by an XOR alternative.

Through **H.9** we know that features like "Statistic" and "Item" will be related in the feature model because they are related in the SR model through task-decomposition link.

Through **H.10** we know that features like "Catalogue" and "Item" will be related in a feature model, but that relation has the requires interaction, because they are related in the SR model through the "Catalogue Consulting" task. But although "Catalogue" needs "Item", "Item" does not need "Catalogue".

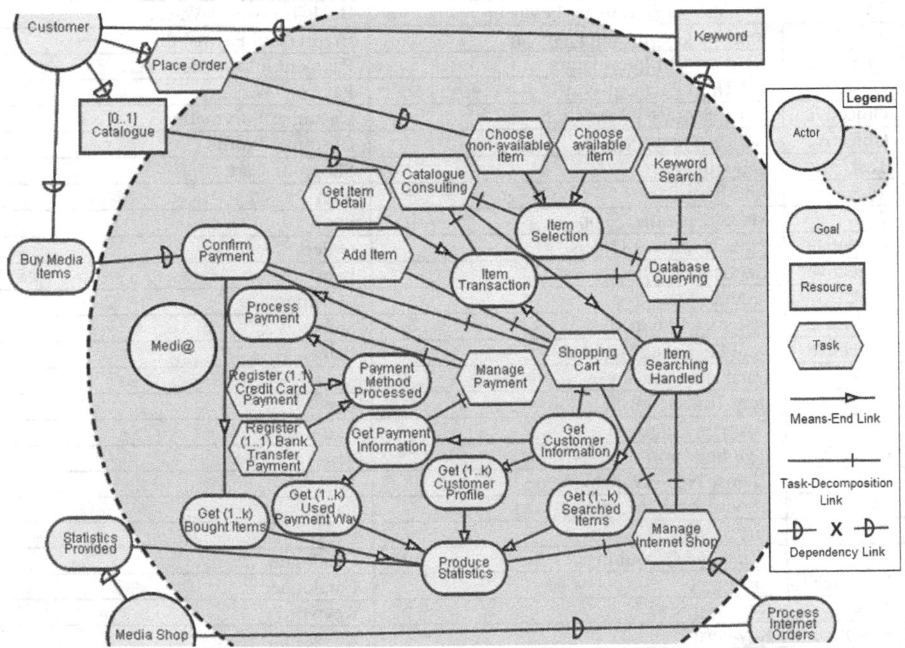

Fig. 4. Simplified SR model of Media Shop system

Figure 5 presents the final feature model.

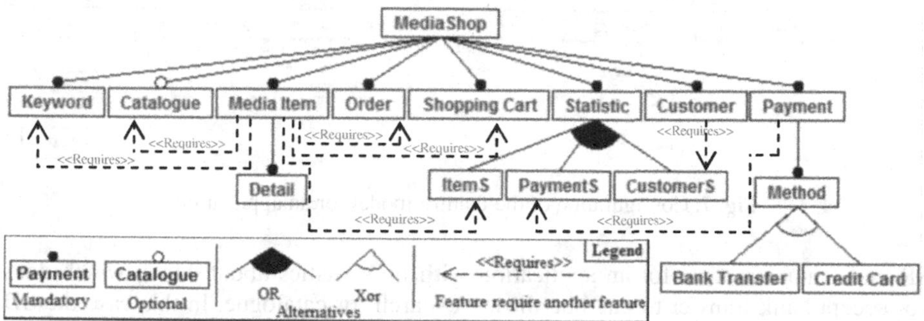

Fig. 5. Feature model

After the validation with the client of the SD, SR and feature models, the rest of the activities of the SPLE is performed. In the application engineering process the IStarLPS approach is applied where the configuration of the feature and i* models are realized. The feature model is configured as usual (selecting the desired features) and all i* models will show a configuration where all intentional elements that are not needed to the specific application are removed. Figures 6 and 7 show the configured

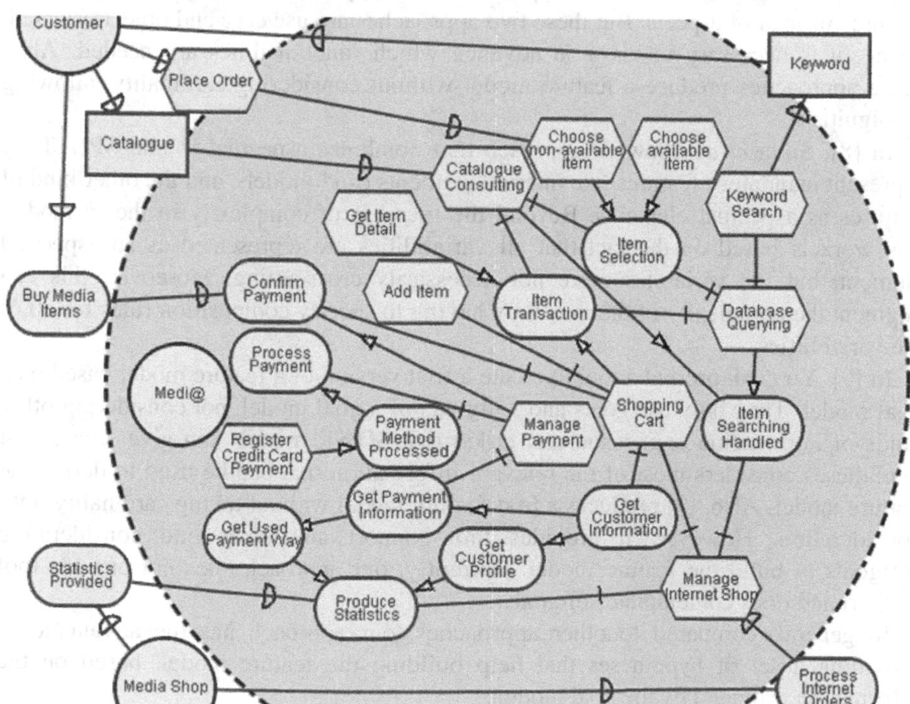

Fig. 6. Configuration of the SR model for an application

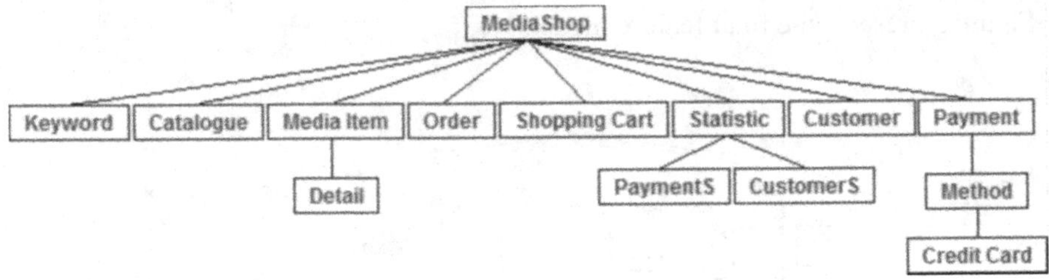

Fig. 7. Configuration of the feature model for an application

SR and feature models for an application with no statistics about items, which does not accept bank transfer to pay but allows a search for catalogue. In this case the SD model does not change.

5 Related Work

There are not many studies that relate SPL and GORE, but some approaches relate SPL and other approaches of requirements engineering [6, 7, 8, 9]. In [6] Gomaa presents a use case oriented approach which shows how to model SPL with use cases. In [7], Jayaraman *et al.* present a work that relates SPL and MATA (Modeling Aspects using a Transformation Approach). This approach uses MATA to represent features instead of aspects. But these two approaches are use case and object oriented, hence, it is necessary to know in advance which functionalities are needed. Also, those approaches produce a feature model without considering cardinality, allowing ambiguities.

In [8], Silva *et al.* show an approach that combines aspectual i* and SPL. They represent mandatory features like internal elements of i* models, and the other kind of features as aspectual elements. Beyond the increase of complexity in the i* model, this work is based on the fact that all variabilities are represented as an aspectual element, but the variabilities are not necessarily crosscutting. Moreover, this can augment the complexity of the model by having to specify composition rules for all of the variabilities.

In [9], Yu *et* al. present a tool to create a first version of a feature model based on a goal model. They only use goals and softgoals of a goal model, not considering other kinds of information (e.g. resources, tasks) that SD/SR models can give. Our set of hypotheses considers most of the relevant information that can be used to derive the feature model. Also, [9] produces a first feature model without taking cardinality into consideration. However, it provides tool support and takes into consideration softgoals to build the feature model. Currently, our approach does not provide tool support and does contemplate softgoals.

In general, compared to other approaches, our approach has the advantage of providing a set of hypotheses that help building the feature model, based on the information provided by the goal models.

6 Conclusions and Future Work

To conclude, this work improves requirements engineering of SPLE with a GORE approach. The IStarLPS approach is based on the i* framework, which helps to identify features, relationships and interactions between features to develop a feature model. To do this, we extended the i* framework by adding cardinality to the intentional elements, making it easier to derive a feature model from i* models.

A metamodel that relates the i* framework and feature model metamodels has been created to support the configuration of an individual product in the application engineering process.

As future work, it would be useful to analyze softgoals of i* models in order to help the development of feature models, probably integrating the Yu *et* al. approach. It would be an advantage to analyze how goals can be obtained from features, as a complement to this approach. Besides, the IStarLPS approach would benefit from more real case studies being applied to it.

References

1. Lapouchnian, A.: Goal-Oriented Requirements Engineering: An Overview of the Current Research. University of Toronto (2005)
2. Yu, E.: Modelling Strategic Relationships for Process Reengineering. Dept. of Computer Science. University of Toronto (1995)
3. Clements, P., Northrop, L.: Software Product Lines: Practices and Patterns. Addison-Wesley, Boston (2007)
4. Pohl, K., Böckle, G., Van Der Linder, F.: Software Product Line Engineering Foundations, Principles, and Techniques. Springer, Heidelberg (2005)
5. Czarnecki, K., Eisenecker, S.: Formalizing cardinality-based feature models and their specialization. Software Process: Improvement and Practice 10(1), 7–29 (2005)
6. Gomaa, H.: Designing Software Product Lines with UML: From Use Cases to Pattern-based Software Architectures. Addison-Wesley, Reading (2004)
7. Jayaraman, P., Whittle, J., Elkhodary, A., Gomaa, H.: Model Composition in Product Lines and Feature Interaction Detection Using Critical Pair Analysis. LNCS, p. 151. Springer, Heidelberg (2007)
8. Silva, C., Alencar, F., Araújo, J., Moreira, A., Castro, J.: Tailoring an Aspectual Goal-Oriented Approach to Model Features. In: 20th Inter. Conf. on Software Engineering and Knowledge Engineering San Francisco Bay, USA (2008)
9. Yu, Y., Leite, J., Lapouchnian, A., Mylopoulos, J.: Configuring features with stakeholder goals. In: Proc. of the 2008 ACM Symposium on Applied computing. ACM, Brazil (2008)
10. Lamsweerde, A.: Goal-Oriented Requirements Engineering: A Guided Tour. In: RE 2001 – 5th IEEE Inter. Symposium on Requirements Engineering, Toronto, Canada (2001)

Preface to SeCoGIS 2009

Claudia Bauzer Medeiros[1] and Esteban Zimányi[2]

[1] University of Campinas, Brazil
[2] Université Libre de Bruxelles, Belgium

Recent advances in information technologies have increased the production, collection, and diffusion of geographical data, thus favoring the design and development of geographic information systems (GIS). Nowadays, GIS are emerging as a common information infrastructure, which penetrate into more and more aspects of our society. This has given rise to new methodological and data engineering challenges in order to accommodate new users' requirements for new applications. Conceptual and semantic modeling are ideal candidates to contribute to the development of the next generation of GIS solutions. They allow to elicitate and capture user requirements as well as the semantics of a wide range of applications.

The SeCoGIS workshop brings together researchers, developers, users, and practitioners carrying out research and development in geographic information systems. It aims at stimulate discussions on the integration of conceptual modeling and semantics into current geographic information systems, and how this will benefit end users. The workshop provides a forum for original research contributions and practical experiences of conceptual modeling and semantic web technologies for GIS, fostering interdisciplinary discussions in all aspects of these two fields, and highlights future trends in this area. The workshop is organized in a way to highly stimulate interaction amongst the participants. The co-location with the Entity-Relationship Conference allows cross-fertilization and mutual interactions between the ER and GIS research communities.

The call for papers attracted 18 papers, which were submitted by authors from 12 countries and 3 continents, clearly illustrating the international nature of the domain. The program committee, consisting of 42 researchers, conducted three to four reviews of each paper and selected 6 papers for presentation and discussion at the workshop. The workshop was organised in two sessions of 3 papers each, devoted, respectively, to foundational issues and semantical issues. Additionally, Stefano Spaccapietra, from EPFL, Switzerland, presented a keynote talk concerning practical and theoretical issues on trajectories.

Many people helped in putting together this workshop. First of all the Steering Committee was in charge of the initial conception and reality of the workshop. The Program Committee carefully reviewed the papers under a very tight schedule. We thank Philippe Rigaux for providing the MyReview program, used during the reviewing process. We hope that you find the program and presentations beneficial and enjoyable and that during the workshop you had many opportunities to meet colleagues and practitioners.

C.A. Heuser and G. Pernul (Eds.): ER 2009 Workshops, LNCS 5833, p. 296, 2009.

A New Point Access Method Based on Wavelet Trees*

Nieves R. Brisaboa[1], Miguel R. Luaces[1], Gonzalo Navarro[2], and Diego Seco[1]

[1] Database Laboratory, University of A Coruña
Campus de Elviña, 15071, A Coruña, Spain
{brisaboa,luaces,dseco}@udc.es

[2] Center for Web Research, Department of Computer Science, University of Chile
Blanco Encalada 2120, Santiago, Chile
gnavarro@dcc.uchile.cl

Abstract. The development of index structures that allow efficient retrieval of spatial objects has been a topic of interest in the last decades. Most of these structures have been designed for secondary memory. However, in the last years the price of memory has decreased drastically. Nowadays it is feasible to place complete spatial indexes in main memory.

In this paper we focus in a subcategory of spatial indexes named Point Access Methods. These indexes are designed to solve the problem of indexing points. We present a new index structure designed for two dimensions and main memory that keeps a good trade-off between the space needed to store the index and its search efficiency. Our structure is based on a *wavelet tree*, which was originally designed to represent sequences, but has been successfully used as an index in areas like information retrieval or image compression.

Keywords: spatial index, point access methods, wavelet tree.

1 Introduction

Recent improvements in hardware have made the implementation of Geographic Information Systems (GIS) affordable for many organizations. An outstanding feature of this kind of systems is that huge amounts of spatial data have to be stored and processed. Therefore, a topic of interest in this research area has been the development of spatial indexing methods, which allow efficient access to these data. Many different spatial index structures have been proposed along the years. These structures can be broadly classified into Point Access Methods (PAMs) and Spatial Access Methods (SAMs) [1]. PAMs are used to improve the access time in collections of spatial points. SAMs are more general and are used

* This work has been partially supported by "Ministerio de Educación y Ciencia" (PGE y FEDER) ref. TIN2006-15071-C03-03, by "Xunta de Galicia" ref. 2006/4 and ref. 08SIN009CT, and by Fondecyt Grant 1-080019, Chile.

C.A. Heuser and G. Pernul (Eds.): ER 2009 Workshops, LNCS 5833, pp. 297–306, 2009.

to improve the access time in collections of geographic objects (e.g. points, lines, polygons, etc.).

Most of these spatial index methods have been designed for secondary memory. This is mainly due to historical reasons. A few years ago, the main memory was small and very expensive. Thus, the development of spatial index structures for main memory was unimaginable. However, in the last years the price of memory has decreased drastically and nowadays it is feasible to place complete spatial indexes in main memory. Hence, new requirements in the design of spatial access methods must be considered in order to develop structures suitable for main memory. In this paper we present a new PAM for two dimensions that stores both the index and the collection of points in a compact structure. This structure reaches a good trade-off between the space needed and its search efficiency. This makes it suitable for main memory.

In the last years, the idea of storing both the data and the index in a compact form has been widely used in the design of index structures in several research fields. These structures are known as *self-indexes*. In [2], a new approach for document indexing using wavelet trees is presented. A wavelet tree [3] is a self-index organized as a binary tree, originally designed to represent and index a sequence. Here we adapt this structure to the special characteristics of spatial data.

2 Related Work

Many different SAMs and PAMs have been proposed along the years. A good survey of these structures can be found in [1]. The main goal of these structures is to improve the performance in the retrieval of geographic objects that satisfy a search query. A common kind of search query that must be solved by both categories of methods is the *region query*. This operation defines a query window (i.e. a rectangular region in the geographic space) and it returns all the geographic objects that overlap that region.

One of the most popular spatial access methods and a paradigmatic example is the R-tree [4]. The R-tree is a balanced tree derived from the B-tree that splits the space into hierarchically nested, possibly overlapping, MBRs (minimum bounding rectangles). The number of children of each internal node varies between a minimum and a maximum. The tree is kept balanced by splitting overflowing nodes and merging underflowing nodes. MBRs are associated with the leaf nodes, and each internal node stores the MBR that contains all the nodes in its subtree. The decomposition of the space provided by an R-tree is adaptive (dependent on the rectangles stored) and overlapping (nodes in the tree may represent overlapping regions). Several variations of the original R-tree have been proposed to improve its efficiency (e.g. the R+-tree or the R*-tree) and to take into account some specific problems (e.g. the STR R-tree for static data). Most of these proposals have been summarized in [5].

The K-d-tree [6] is a d-dimensional data structure and one of the most prominent PAMs. When this structure is used to index a collection of points, it is also known as Point K-d-tree. The K-d-tree is a binary search tree that represents a recursive subdivision of the space based on the value of just one coordinate at each level of the tree. Many variations of this structure differ in the manner in which they partition the space. In our experiments we use a static approach, proposed in [6], that assumes that all the data points are known a priori. In this variation, the partition lines must pass through the data points and the partition axis changes cyclically in a fixed order.

As we noted before, the wavelet tree [3] is a compact structure used in other fields to store and index data in a compressed way. For instance, in [2] a wavelet tree is used to index and retrieve documents and in [7] it is used to index images. It is known to be efficiently implementable [8]. The basic tool used in the wavelet tree is the bit-vector *rank* operation: given a bit vector $B[1, n]$, the query $rank(B, i) = rank_1(B, i)$ returns the number of bits set to 1 in the prefix $B[1, i]$ of B. Symmetrically, $rank_0(B, i) = i - rank_1(B, i)$. The dual query to $rank_1$ is $select_1(B, j)$. It returns the position of the j-th bit set to 1 in B. The definition of $select_0(B, j)$ is analogous. For example, given a bitmap $B = 1000110$, $rank_1(B, 5) = 2$, and $select_0(B, 4) = 7$. Both rank and select operations can be implemented in constant time and using little additional space on top of B [9,10,11].

3 Spatial Indexing Using Wavelet Trees

3.1 Index Construction

Given a set of N points $P = P_1 \dots P_N$, each point consisting of two coordinates (e.g. latitude and longitude) that define its position in the geographic space with regard to a spatial reference system, we can assume that these points can be distributed in an $N \times N$ matrix with only one point in each row and column. This is not a strong restriction because if two points have the same coordinate we can order them arbitrarily and assign them consecutive rows or columns in the matrix. It is important to note that the matrix is only used to keep the relative positions of the points. Neither the distances nor the proportions are kept in it. This is a very important characteristic because it allows us to construct the matrix for any set of points, even if there are points with duplicate coordinates in the set. The translation from the geographic space to a matrix is illustrated with an example in the left part of Figure 1.

The wavelet tree is a compact structure that can be used to store this matrix with little storage cost. Given an $N \times N$ representative matrix, a wavelet tree with $\lceil \log_2 N \rceil$ levels and N bits per level can be built to store the permutation from the order of the points in one dimension (e.g. longitude) to their order in the other (e.g. latitude). Let $X = P_{X_1} \dots P_{X_N}$ and $Y = P_{Y_1} \dots P_{Y_N}$ be the permutations where the points are ordered by their longitudes and latitudes, respectively. For example, in Figure 1 we can name the points from left to right (i.e. P_i is the i-th point counting from the left). Therefore, the first permutation can be written as $X = P_1 P_2 \dots P_{16}$ and the second as $Y = P_2 P_{13} P_{11} \dots P_1 P_5$.

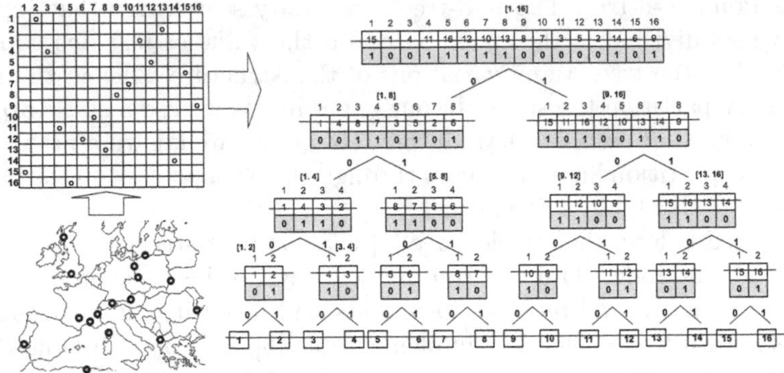

Fig. 1. Wavelet tree construction. Only the greyed data are stored.

The point P_1, for instance, is the first one in the order of the longitudes and it is the second to last in the order of the latitudes.

The root of the wavelet tree is a bitmap $B = b_1 \ldots b_N$ with the same length of the set of points (i.e. N positions). Each position i represents the i-th point assuming them ordered in the first dimension (e.g. longitude). In the example, $P_{X_1} = P_1$, $P_{X_2} = P_2$, etc. Then, $b_i = 0$ if $P_{Xi} \in P_{Y_1} \ldots P_{Y_{N/2}}$, and $b_i = 1$ if $P_{X_i} \in P_{Y_{N/2+1}} \cdots P_{Y_N}$. The sequence of the points given a 1 in this vector are processed in the right child of the node, and those marked 0 are processed in the left child of the node. In this way, each node indexes half the symbols indexed by its parent node. This process is repeated recursively in each node until the leaf nodes where the sequence of indexed symbols corresponds to the permutation in the second dimension (e.g. latitude). The right part of Figure 1 shows the wavelet tree that represents the matrix on the left. Each position in each node of the wavelet tree has been annotated with the order of the corresponding point in the second permutation (these orders are crossed out because in fact they are not stored in the wavelet tree).

3.2 Solving Queries

Obtaining the order of a point in a dimension knowing its order in the other dimension is quite simple. If we know the order of a point in the first dimension (in our example, the longitude) we can go down the wavelet tree to obtain its order in the second dimension (in our example, the latitude). The value at a certain position and the *rank* operation are used to go down in the wavelet tree. The bit b_i in the bitmap of a node defines whether the corresponding point is indexed by either the left ($b_i = 0$) or right ($b_i = 1$) branch of this node. In addition, $rank_{b_i}(B, i)$ gives us the position of that point in the bitmap of the child node. This process is repeated until a leaf node is reached, which gives us the position of the point in the other permutation. As an example, in the wavelet tree of Figure 1, the point in column 6 is at row 12. To obtain this result we first retrieve the bit at position 6 of the root node. That bit is set to 1. Then, we obtain $rank_1(B, 6) = 4$. Both results indicate that we have to repeat the operation in position 4 of the right node. If we repeat this process until a leaf node is reached, we obtain the result 12 (i.e. the order in the second dimension of the element at the sixth position in the first dimensions).

On the other hand, if we know the order of a point in the second dimension (in our example, the latitude) we can go up the wavelet tree to obtain its order in the first dimension (in our example, the longitude). The value in the label of the branch that gives access to the node and the *select* operation are used to go up the wavelet tree. As our structure is a perfect binary tree, it is very easy to know at each level of the tree whether the current node is a left or right child of its parent. In the example, the point in row 13 is at column 8. To obtain this result we first calculate $select_0(B, 1) = 1$. We have made a $select_0$ because the position 13 in the leaf level is stored in a left node (i.e. a branch labeled with a 0) and the position 1 because 13 is the first position of its node. In the next level, we have to calculate $select_0(B, 1) = 3$ and then $select_1(B, 3) = 6$ (1 because it is a right child and 3 because this is the position computed in the previous step). Finally we reach the root, where we obtain the result $select_1(B, 6) = 8$.

We can also use the wavelet tree to solve *region query* operations. However, for this purpose we need three auxiliary structures: two arrays with the coordinates ordered in each dimension and the point identifiers ordered in the same order as one of the other arrays. The arrays of ordered coordinates are used to translate spatial queries to ranges of valid rows and columns in the wavelet tree. Once the query has been translated, the range of columns (longitudes) is the range of valid positions in the root node of the wavelet tree. We can go down through the structure using the algorithm that we have sketched before. Nevertheless, the performance of that algorithm can be easily improved taking into account that consecutive points in a parent node remain consecutive in the corresponding child node. Hence, only two *rank* operations (one for the first position of the range and one for the last one) have to be calculated. Furthermore, the range of valid rows (latitudes) obtained in the translation of the query can be used to prune the search tree. Each node in the wavelet tree contains points in a certain range. If this range does not intersect with the range of rows, the algorithm does not continue in that branch.

Figure 2 shows the wavelet tree of the example with the auxiliary structures. In the figure, two arrays with the point identifiers are shown (IDs(X) and IDs(Y)). The structure needs only one of them, and the decision of which one to employ has pros and cons, as we see later. The figure shows an example of a region query q = <(27.53, 15.75), (30.71, 19)>. The translation of this query to the representative matrix defines the range of valid columns [6, 10] and the range of valid rows [9, 14]. The algorithm to solve the query begins with the traversal of the wavelet tree. As we noted before, only the first and the last positions of a chunk (i.e. several consecutive positions) are relevant to decide the chunks of interest in the next level. Therefore, in the first step the algorithm has to calculate $rank_0(B, 6-1)+1 = 3$, $rank_0(B, 10) = 4$, $rank_1(B, 6-1)+1 = 4$ and $rank_1(B, 10) = 6$. Actually, only two of them have to be computed because

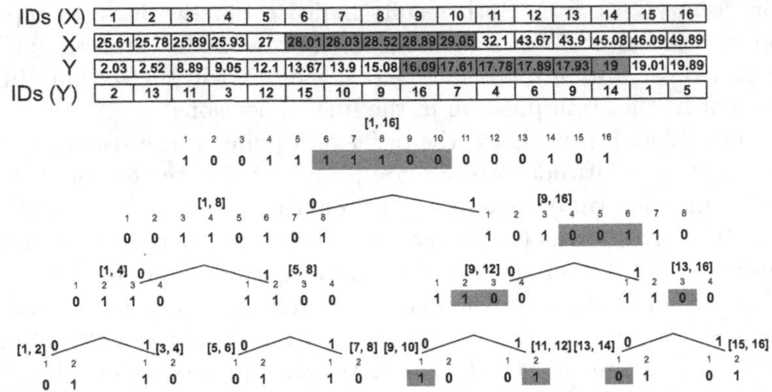

Fig. 2. Query solution using the wavelet tree

$rank_0(B, i) + rank_1(B, i) = i$. Thus, the valid chunks in the second level are [3, 4] in the left node and [4, 6] in the right one. However, solutions to the query cannot be in the left node because it covers the range of rows [1, 8], which does not intersect with the range of rows in the query ([9, 14]). Hence, this branch is discarded. The algorithm repeats this process until the leaf level is reached.

Once the leaf nodes are reached, the way the algorithm continues depends on the order selected for the point identifiers. If the point identifiers are ordered in the same way of the second array of coordinates (in our example, latitudes), the positions of the leaf nodes can be directly translated to the positions in the array of identifiers. Hence, this version of the algorithm is simpler and, as we will see in the next section, it is more efficient too. On the other hand, if the point identifiers are ordered in the same way of the root node (in our example, longitudes), the algorithm is more complex because when the algorithm discovers that the latitude of a point is valid for the query, the algorithm has to go up the wavelet tree again to obtain its identifier. Since the validity of a latitude can be discovered at any level of the tree, and therefore the algorithm does not always have to reach the leaf nodes, this ordering of the identifiers could improve the performance of the solution. However, as we will see in the next section, its performance is worse than that of the previous version.

4 Experiments

We compare the efficiency of our structure with respect to other spatial index structures, considering first the space requirements and then their efficiency to solve region queries. The results show that our structure achieves a good trade-off between the required space and its time efficiency.

We compare four spatial index structures that run in main memory. The first two are the variants of our index structure presented in Section 3. In the first one, called DPW-tree (*down point wavelet tree)*, the identifiers of the points are stored ordered following the permutation of the leaf nodes and it is only necessary to descend the wavelet tree to obtain the identifiers of the points that fulfil the query. In the second one, called UPW-tree (*up point wavelet tree*), the identifiers of the points are stored ordered following the permutation of the root node and once that a point is known to belong to the query result it is necessary to ascend the tree to retrieve its identifier. The third index structure is a classical R-tree adapted to run in main memory [12]. Although this index structure is not a point access method but a spatial access method, and therefore it is not optimized for point indexing, it is nonetheless the most used index structure in the geographic information systems that are developed nowadays. We use two variations of the original structure: the R*-tree [13] and the static construction of the STR R-tree [14]. We count its space assuming a contiguous layout in memory. Finally, the fourth index structure is a K-d-tree that represents the point access methods. The K-d-tree variant that we have selected [15] is probably the most efficient one because it is optimized for scenarios where the set of points to be indexed is known *a priori*.

To build the query sets, we implemented an algorithm that generates query windows of a given size. This algorithm is based on the one used on the evaluation of the R*-tree in [13]. The query windows generated by the algorithm are distributed uniformly in the space. Furthermore, the size of the window sides is adjusted so that the ratio between the horizontal and vertical extensions varies uniformly between 0.25 and 2.25.

4.1 Space Comparison

Both variants of our structure need to store the coordinates of the N points (two arrays of N 8-byte floating-point numbers), the identifiers (an array of N 4-byte integer numbers) and the wavelet tree. The wavelet tree is a very compact structure that needs only $N \times \lceil \log_2 N \rceil$ bits (1 bit per point per level, that is, N bits per level, and there are $\lceil \log_2 N \rceil$ levels). Moreover, in order to perform *rank* and *select* operations in constant time, some auxiliary structures are needed that use an additional space of around 37.5% of the wavelet tree size [10]. Therefore, the complete structure requires $20 \times N + (N \times \lceil \log_2 N \rceil \times 1.375)/8$ bytes.

The space needed by an R-tree over a collection of N points can be estimated considering an average arity (M). The leaves store the point identifiers. A static structure can store the leaves contiguously without spare space. Thus the leaves amount to $4 \times N$ bytes, and with the table storing the coordinates of the points, we add up to $20 \times N$ bytes. Each leaf costs a MBR and a pointer at its parent, which requires 36 bytes. Over all the levels, there are $N/(M-1)$ nodes, so the total R-tree space is $20 \times N + 36 \times N/(M-1)$. The best performance of the STR R-tree is achieved with an effective M value of 30.

Finally, a K-d-tree that indexes N points has height $h = \lceil \log_2 N \rceil$ and $2^h - 1 + (N \bmod 2^{\lfloor \log_2 N \rfloor})$ nodes, where each node needs 16 bytes (a floating point number and two pointers). Just like the R-tree, we must also consider the $20 \times N$ bytes of the table of points.

To finalize the space comparison, we show the space per point necessary in each spatial index structure. First, both variants of our structure need the same space: 23.69 *bytes/point*. In the same way, both variants of the R-tree need 21.24 *bytes/point*. Finally, the K-d-tree needs 36.00 *bytes/point*. The main conclusion that we can extract from these results is that our structure needs less space than the K-d-tree and more than the R-tree.

4.2 Time Comparison

To perform the time comparison we take into account the two variables that can affect the tests: the selectivity of the queries and the size of the test collections. The query selectivity depends directly on the size of the query windows used. In our tests, we created windows of four different sizes that represent 0.01%, 0.1%, 1% and 10% of the area of the space where the points are represented. We use four synthetic collections with 2^{19}, 2^{20}, 2^{23} and 2^{24} points uniformly distributed in the space. Figure 4 shows four graphs where one can appreciate the influence of these variables in the time needed to solve the queries.

We have also experimented with non-uniform spaces. Figure 4 shows the results with two synthetic collections and two real collections. Both synthetic collections have one million points each, the first one with a Zipf distribution (world size $= 1000 \times 1000$, $\rho = 1$) and the second one with a Gauss distribution (world size $= 1000 \times 1000$, mean $= 500$, sigma $= 200$). The two real collections have 123,593 postal addresses from New York, Philadelphia and Boston (NE dataset available at http://www.rtreeportal.org) and 2,693,569 populated places distributed all over the world (available at http://www.geonames.org).

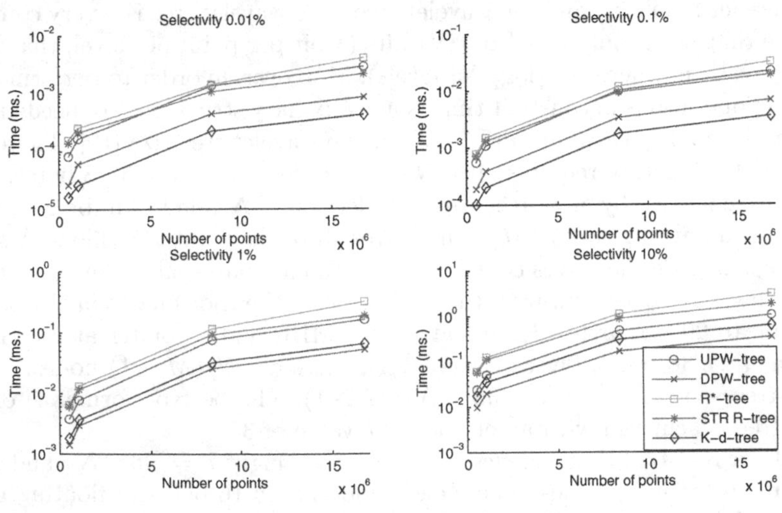

Fig. 3. Time comparison. Note the logscale.

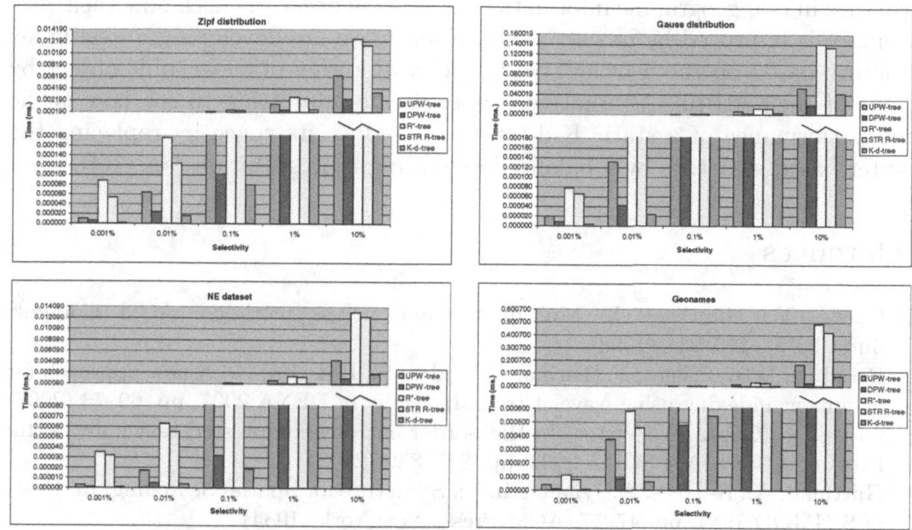

Fig. 4. Time comparison (other collections)

The main conclusion that can be extracted from these results is that our structure is competitive with respect to query time efficiency. The K-d-tree is generally the most efficient structure, but the DPW-tree is always close, and becomes better for low selectivities. On the other hand, the K-d-tree requires significantly more space. Both the R*-tree and the STR R-tree uses less space than the DPW-tree but they are not competitive in time as a point access method. We have included them in the experiments because the R-tree is the paradigmatic example of spatial index structures, and it must be taken into account because it is widely used nowadays. Finally, regarding the two variants of our structure, the DPW-tree (the version that only needs to descend the wavelet tree) is more efficient than the UPW-tree (the version that requires ascending in the wavelet tree).

5 Conclusions and Future Work

We have presented a new point access method based on the *wavelet tree*, a compact structure widely used in other areas such as information retrieval. Our spatial index structure is designed for two dimensions and for main memory, and keeps a good trade-off between the space needed to store the index and its search efficiency. This is an important advantage, as main-memory spatial indexes are becoming popular.

We are currently working on several research lines. First, we are working on allowing the insertion or removal of points once the structure has been constructed. Second, we plan to design algorithms to solve other kinds of queries such as k-nearest neighbor queries or spatial joins. We are also integrating this structure in real geographic information systems in order to check how their performance is improved by our structure. Finally, we are developing a new index structure based on the wavelet tree to index any type of geographic object by means of their MBRs. Alternatively, it could be interesting to see how is the time performance of a static K-d-tree if we reduce its space by replacing the pointer-based structure by a balanced extending representation (see [16]).

References

1. Gaede, V., Günther, O.: Multidimensional access methods. ACM Comput. Surv. 30(2), 170–231 (1998)
2. Brisaboa, N.R., Cillero, Y., Fariña, A., Ladra, S., Pedreira, O.: A new approach for document indexing using wavelet trees. In: Proc. of DEXA 2007, pp. 69–73 (2007)
3. Grossi, R., Gupta, A., Vitter, J.: High-order entropy-compressed text indexes. In: Proc. of ACM-SIAM SODA 2003, pp. 841–850 (2003)
4. Guttman, A.: R-Trees: A Dynamic Index Structure for Spatial Searching. In: Proc. of SIGMOD 1984, pp. 47–57. ACM Press, New York (1984)
5. Manolopoulos, Y., Nanopoulos, A., Papadopoulos, A.N., Theodoridis, Y.: R-Trees: Theory and Applications. Springer-Verlag New York, Inc., Heidelberg (2005)
6. Bentley, J.L.: Multidimensional binary search trees used for associative searching. Commun. ACM 18(9), 509–517 (1975)
7. Mäkinen, V., Navarro, G.: On self-indexing images - image compression with added value. In: Proc. of DCC 2008, pp. 422–431. IEEE Computer Society, Los Alamitos (2008)
8. Claude, F., Navarro, G.: Practical rank/select queries over arbitrary sequences. In: Amir, A., Turpin, A., Moffat, A. (eds.) SPIRE 2008. LNCS, vol. 5280, pp. 176–187. Springer, Heidelberg (2008)
9. Munro, I.: Tables. In: Chandru, V., Vinay, V. (eds.) FSTTCS 1996. LNCS, vol. 1180, pp. 37–42. Springer, Heidelberg (1996)
10. González, R., Grabowski, S., Mäkinen, V., Navarro, G.: Practical implementation of rank and select queries. In: Proc. of 4th WEA (Poster), pp. 27–38 (2005)
11. Okanohara, D., Sadakane, K.: Practical entropy-compressed rank/select dictionary. In: Proc. of 9th ALENEX (2007)
12. Hadjieleftheriou, M.: Spatial index library, `http://research.att.com/~marioh/spatialindex/` (retrieved March 2009)
13. Beckmann, N., Kriegel, H.P., Schneider, R., Seeger, B.: The R*-tree: an efficient and robust access method for points and rectangles. SIGMOD Rec. 19(2), 322–331 (1990)
14. Leutenegger, S., Lopez, M., Edgington, J.: Str: A simple and efficient algorithm for r-tree packing. In: Proc. of ICDE 1997, pp. 497–506 (1997)
15. Tagliasacchi, A.: Kd-tree for matlab, `http://www.mathworks.com/matlabcentral/fileexchange/21512` (retrieved March 2009)
16. Navarro, G., Mäkinen, V.: Compressed full-text indexes. ACM Comput. Surv. 39(1) (2007)

A Reference System for Topological Relations between Compound Spatial Objects

Max J. Egenhofer

National Center for Geographic Information and Analysis and Department of Spatial Information Science and Engineering,
Department of Computer Science, University of Maine Boardman Hall,
Orono, ME 04469-5711, USA
max@spatial.maine.edu

Abstract. A current trend in the development of spatial-relation ontologies is to capture more and more details about the geometries of the spatial objects that are related, primarily by topological relations. In an effort to bridge between often disparate approaches, a reference system for topological relations between compound spatial objects is introduced. Its framework comprises the base relations' conceptual neighborhood graphs, which, when nested, provide a means for visually analyzing the completeness and consistency of the set of derived relations.

1 Introduction

Spatial-relation ontologies are becoming increasingly more refined as domain-specific requirements call for spatial relations over detailed spatial data types. While the relations for the conventional spatial data types—points and simple lines and regions—have reached a level of maturity up to standardization [15], the quest for topological relations over other spatial data types in ongoing. Examples include the relations that involve regions with broad boundaries [1,2,4], concave regions [3], extended points [18,21], regions with a hole [23], objects with potential separations and holes [22], lines with uncertainties [20], directed lines [17], and intervals with gaps [6]. Despite such a wide variety, the spectrum has certainly not yet been exhausted, as the thirst for capturing and analyzing more details of the semantics of particular spatial configurations appears to be unsatisfiable. Rather than devoting a new methodology to each and every new specialized set of spatial relations, time is ripe for a framework within which the set of relations for arbitrarily complex geometries can be derived consistently, while also providing a path to those relations' conceptual neighborhoods for similarity assessments and their compositions for qualitative spatial reasoning.

Two different approaches compete for the derivation of the consistent and complete sets of such relations:

C.A. Heuser and G. Pernul (Eds.): ER 2009 Workshops, LNCS 5833, pp. 307–316, 2009.

- The use of the vanilla 9-intersection [9] for arbitrarily complex objects [19,22] yields a coarse topological-relation model that favors, for multi-part objects, non-empty intersections (as they overwrite empty values) and, therefore, is often insufficient to capture critical differences between distinct topological relations.
- The mere Cartesian product of relations among components [1,12,23] models relations with compound objects, however, without capturing explicitly how the components are related.

This paper provides a framework for deriving the sets of topological relations involving compound spatial objects in a canonical way. This framework also serves as a *reference system* for the abstract space of binary topological relations as it addresses locations in this space and perform analyses about neighborhood and remoteness. Since it offers quantitative measures to compare qualitative relations this framework essentially acts as a coordinate system for topological relations. Since the references are symbolic rather than numeric, it resembles more a semantic reference system [16] than a traditional Cartesian coordinate system.

A framework for spatial relations is at the core of modeling geospatial semantics as it offers an opportunity to depart from loosely coupled specifications of specialized spatial relations to a coherent way of dealing with topological relations for arbitrarily complex spatial objects. It enables not only the enumeration of realizable relations and how to distinguish them, but also leads directly to methods for assessing the relations' similarity and deriving the qualitative inferences over such relations. As a new property, not previously discussed in the literature, the framework offers a natural way for incrementally generalizing topological relations by focusing on the objects' most relevant features, yielding coarser topological relations, up to the level of the generic relations between points, lines, and regions.

The remainder of the paper is structured as follows: Section 2 briefly reviews the most frequently used model for tailored topological relations. Section 3 introduces the formation of compound spatial objects and the formalization of their topological relations. Section 4 develops *nested neighborhood graphs* as reference systems for compound-object relations and Section 5 analyzes their visual properties. The paper closes with conclusions and a discussion of future work (Section 6).

2 Topological Relations and Conceptual Neighborhood Graphs

The 9-intersection models the binary topological relation between two spatial objects on the two objects' interiors (denoted as A° and B°), boundaries (∂A and ∂B), and exteriors (A^- and B^-) [9]. For two spatial regions—objects that are homeomorphic to 2-disks, that is, each with a continuous boundary, no holes, no spikes, and no cuts—an embedding in $\mathbb{R}^2$ yields eight relations, each with different combinations of empty and non-empty values applying (Figure 1). This set of relations is jointly exhaustive and pairwise disjoint so that any combination of two regions in $\mathbb{R}^2$ exhibits exactly one of the spatial relations.

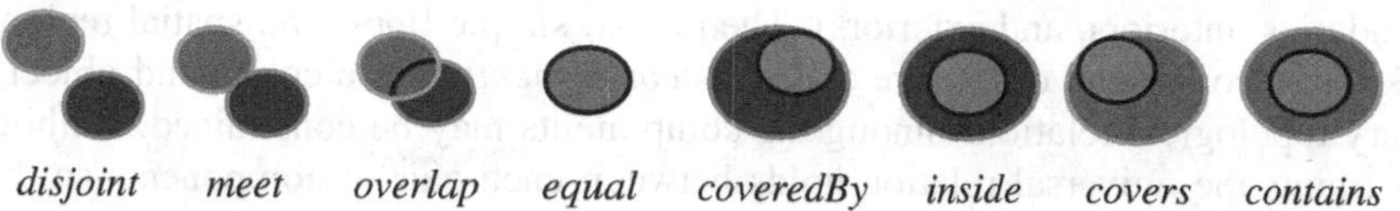

Fig. 1. The eight topological relations between two regions in $\mathbb{R}^2$ with their 9-intersection matrices and the relations' labels

For two simple lines (with exactly two endpoints and no self-intersections), the 9-intersection identifies 33 different relations in $\mathbb{R}^2$, while for the relation between a spatial region and a simple line 19 different matrices have valid interpretations in $\mathbb{R}^2$ [9]. The relations of a domain can be arranged according to their conceptual neighborhoods, grouping the most similar relations. These conceptual neighborhood graphs [7,10] expose high regularities and also serve as a framework for identifying individual relations or their disjunctions (Figure 2).

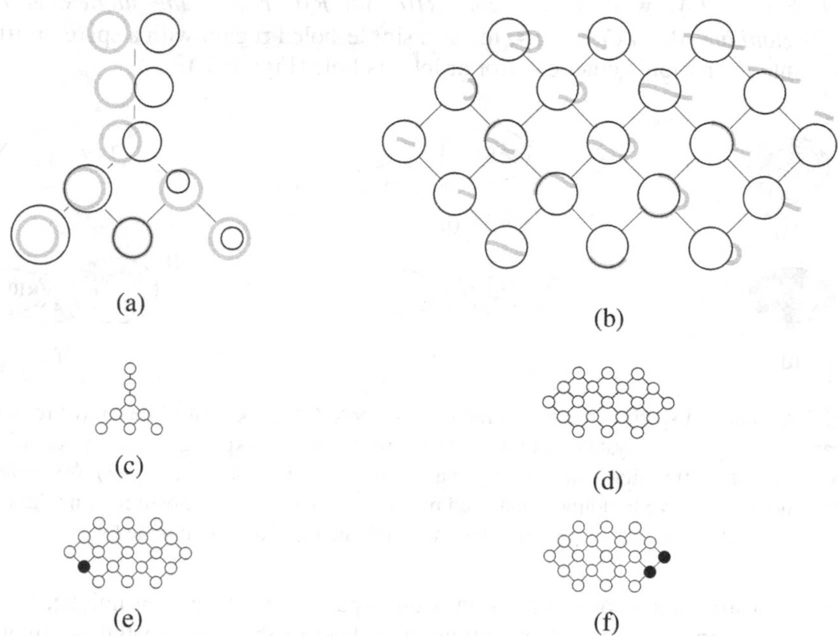

Fig. 2. The conceptual neighborhoods of (a) the eight region-region relations and (b) the 19 region-line relations with their respective conceptual neighborhood graphs (c and d) and selecting relations the neighborhood graph (e) *regionLine_singleCovers* and (f) *regionLinedisjoint* xor *regionLine_singleMeet*

3 Compound Spatial Objects

Compound spatial objects result from geometric combinations—union and difference—of basic spatial object types or their topological parts (i.e., their boundaries, interiors, and exteriors). The points, simple lines, and spatial regions that make up a compound object are called its *components.* For a compound object some binary topological relations among the components may be constrained. Without any constraints the universal relation holds between each pair of components so that the compound spatial object would become a mere collection of components.

For the regions *R1* and *R2*, with the constraint *R1 contains R2*, $R1 \setminus R2^{\circ}$ forms a region with a single hole (Figure 3a). On the other hand, if *R3 disjoint R4* then $R3 \cup R4$ forms a region with separations (Figure 3b). Compound objects also result from combinations over different object types, for instance over regions and lines, such as $R5 \cup L1$—when combined with the constraint *R5 regionLine_singleMeet L1*—yields a simple region with a spike (Figure 3c), whereas $R6 \setminus L2$, with *R6 regionLine_singleCovers L2*, creates a region with a cut (Figure 3d). More complex compound objects may result from conjunctions or disjunctions of relation specifications. For example, $R7 \cup R8 \cup L3$, with the constraints *R7 disjoint R8*, *R7 regionLineRelation_singleMeet L3*, and *R8 regionLine_singleMeet L3* creates a pair of regions that are connected by a linear conduit (Figure 3e), while $(R9 \setminus R10^{\circ}) \cup L4$, with *R9 contains R10* and *R10 regionLine_singleMeet L4* or *R10 regionLine_singleCovers L4*, forms a single-holed region with a spike protruding either into the region's outer exterior or into its hole (Figure 3f).

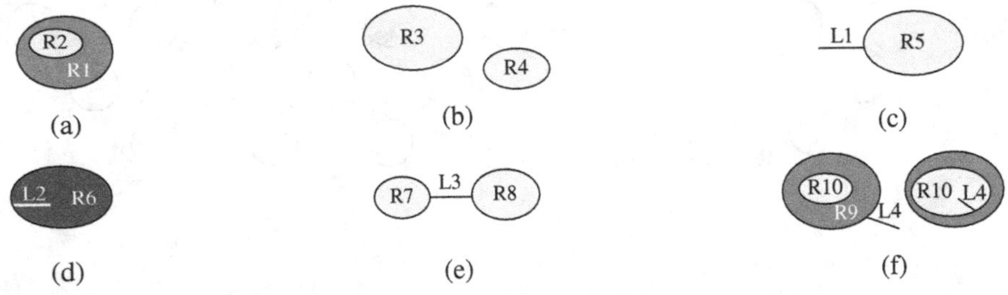

Fig. 3. Compound spatial objects: (a) region with hole (e.g., lake with island), (b) region with separations (e.g., water system of two lakes), (c) region with spike (e.g., water system of a river feeding a lake), (d) region with cut (e.g., navigable area of a lake with a pier), (e) separations with conduit (e.g., two buildings connected by an underpass), and (f) holed region with internal or external spike (e.g., water system of lake with island and river feeding the lake)

The construction steps of such compound spatial objects are not unique, however, as two or more sequences of operations may lead to the same spatial configuration. The determination of when two compound-object specifications yield the same configuration is outside the scope of this paper, since a single specification is sufficient for capturing compound-object relations.

The detailed topological relation of a simple spatial object with respect to a compound spatial object is captured through the topological relations between the simple spatial object and each component. The set of relevant component relations of a compound object with *n* components comprises the *n*(*n*–1)/2 relations above the main diagonal in a spatial scene description [11], yielding a detailed specification of the topological relation with the compound object. This set of constituent relations also includes the topological constraints that govern the formation of the compound spatial object. The approach generalizes to specifying the detailed topological relations between configurations of two compound objects (with n and m respective components) with $(n+m)(n+m-1)/2$ relations.

4 Nested Neighborhood Graphs for Compound-Object Relations

While the Cartesian product of the component relations specifies a compound-object relation, it does not enforce sufficiently any dependencies that may hold among the different relations and, therefore, may easily lead to specifications of compound-object relations that cannot be realized in a particular embedding space.

To *derive* the set of relations that are feasible between a simple and a compound object we introduce *nested neighborhood graphs* as the reference system for compound-object relations. A nested neighborhood graph starts with the possible relations between a simple spatial object and one of the components of the compound spatial object and then iteratively adds details about the relations with respect to the components. The choice of the initial component of the compound object leads to the *principal relation*, while the relations due to the consideration of further components are called *refining relations*. This reference framework for topological relations of compound spatial objects has an origin, an orientation, a granularity, and a scale.

The *origin* is given by the choice of the principal relation, to which all refining relations are attached. Different choices of a principal relation for the same configuration leads to different specifications, much like selecting a different location as the origin for a Cartesian coordinate system.

The *orientation* is given by the sequence in which the refining relations are considered. Again different sequences of refining relations, while maintaining the same origin through the same principal relation, lead to different specifications, in analogy to selecting azimuth as the grounding for the orientation of a Cartesian coordinate system. The vector $principalRel \oplus refiningRel \oplus \ldots \oplus refiningRel$ captures a compound relation's origin and orientation.

The *granularity* of the topological-relation reference system refers to the addressable units to describe a compound-object relation. This compound object's components and their topological relations with respect to the reference object steer this property. For a region component the granularity with respect to a region reference the granularity is eight, while it is 19 with respect to a line reference, and three with respect to a point reference [9]. For a simple-line component, the granularity is three with respect to a line reference, 19 with respect to a region reference, and 33 with another simple line. The most basic case is with respect to a point with granularities of two (for a reference point), three (for a reference line), and three (for a reference region).

The total granularity G of the topological relation with a compound spatial object C_l with r_l region component, l_l linear components, and p_l point-like components, therefore, depends on the type of reference object, that is, a point (Eqn. 1a), line (Eqn. 1b), or region (Eqn. 1c).

$$G_P(C_1) = 2p_1 + 3l_1 + 3r_1 \quad (1a)$$

$$G_L(C_1) = 3p_1 + 33l_1 + 8r_1 \quad (1b)$$

$$G_R(C_1) = 3p_1 + 19l_1 + 8r_1 \quad (1c)$$

The granularity of the topological relation between two compound objects, C_1 and C_2, results then from the Cartesian product of the components of C_1 and C_2 (Eqn. 2).

$$G(C_1, C_2) = p_2 \cdot G_P(C_1) + l_2 \cdot G_L(C_1) + r_2 \cdot G_R(C_1) \quad (2)$$

The *scale* of the topological-relation reference system is given by the lengths of the legs that connect relations in the neighborhood graph, yielding the longest shortest path between any pair of relations in the graph. For mere topological transformations (e.g., translation, scaling, rotating) typically an equal length of 1 is assumed for each leg. With the creation of complex spatial objects (i.e., through union or set difference), configurations may arise that may require unequal leg lengths, for instance to account for the addition or elimination of holes and spikes in order to account for the balance with the topological transformations.

Granularity is independent of the origin and the orientation (i.e., it can be changed without affecting either, and vice versa). Likewise origin and orientation are independent of each other as the chosen principal relation and the refining relations are mutually exclusive. Finally, although the choices of origin and orientation typically create different nested neighborhood graphs, origin and orientation do not affect the scale of the topological-relation reference system.

5 Visualization of Nested Neighborhood Graphs

The depiction of nested neighborhood graphs offers new insights about the stepwise refinement of a compound spatial relation. It builds on the neighborhood graphs for the base set of topological relations and iteratively addresses those relations that can be realized for a component. This process leads to a tree in which the neighborhood graph of the principal relation is the root. Each node stands for a feasible relation, where only feasible relations may be parents for further nodes (i.e., the corresponding relations offer further refinements). Feasible relations are depicted as filled nodes in the nested graph, while infeasible relations remain empty. This visual reference system preserves the relations' neighborhood and, therefore, organizes them according to their highest similarity.

Assume the relation between a region *R1* and a region with a spike (modeled as $R2 \cup L$ such that *R2 regionLine_singleMeet L*) starts with the principal relation between *R1* and *R2*, so that the relation between *R1* and *L* is the refining relation. The origin of the nested neighborhood graph is then the frame of the region-region

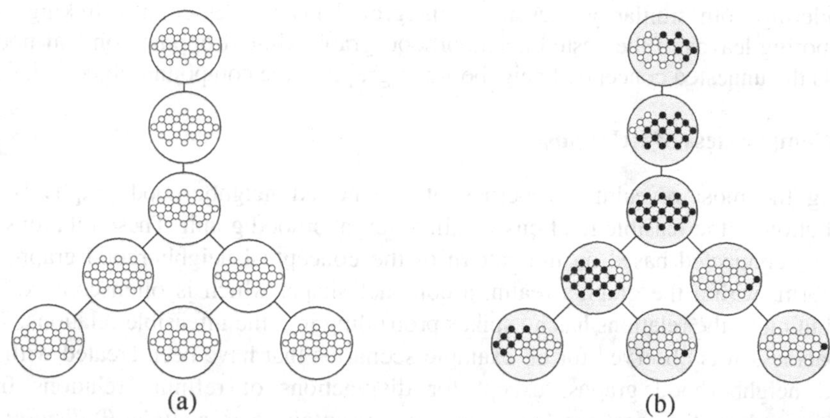

Fig. 4. For the relation between a region and a spiked region (Fig. 3c) its nested neighborhood graph (a) featuring the refining region-line relations inside the principal region-region relations and (b) with the feasible relations highlighted

relations' conceptual neighborhood graph, while the region-line's neighborhood graph captures the refining relations that are nested inside each node of the region-region graph (Figure 4a). Since all eight region-region relations are feasible as principal relation between the two regions, all eight top-level nodes are selected (i.e., highlighted), while for all but one of the principal relations only a subset of the refining region-line relations is feasible (Figure 4b). For instance, if *R1* and *R2* are *equal*, then only one relation is possible between *R1* and *L* so that only one node—the one for *regionLine_singleMeet*—is filled in the graph inside the top-level relation *equal*. On the other hand if *R1 covers R2*, then two refining region-line relations are feasible—*regionLine_Disjoint* and *regionLine_singleMeet*.

5.1 Axes of Similarity

The nested neighborhood graphs integrate two similarity axes:

- The similarity among the refining relations with the same parent. Such relations typically form a connected subset in the neighborhood graph of the refining relations [13]. For example, Figure 4b shows for the principal relation *R1 covers R2* two refining relations *R2 regionLine_singleMeet L* and *R2 regionLine_disjoint L*, which are connected in the neighborhood implying the neighborhood of $covers \oplus regionLine\_singleMeet$ and $covers \oplus regionLine\_disjoint$.
- The similarity between the same refining relations whose parents are neighbors. For instance, in Figure 4b the refining relation *R2 regionLine_singleMeet L* for *R1 equal R2* has no neighbors at the same level, but both neighbors of *equal*—*coveredBy* and *covers*—have *regionLine_singleMeet* as refining relation as well; therefore, $equal \oplus regionLine\_singleMeet$ is a neighbor of both $coveredBy \oplus regionLine\_singleMeet$ and $covers \oplus regionLine\_singleMeet$.

Considering both similarity axes in an integrated fashion, essentially linking only neighboring leaves of the nested neighborhood graph while dropping non-leaf nodes, reveals the unnested conceptual neighborhood graph of the compound-object relation.

5.2 Compactness of Relations

Among the most revealing properties of the nested neighborhood graphs is the distribution of the feasible relations within a neighborhood graph. These relations are not only connected based on the pattern of the conceptual neighborhood graph, but they form, within the graph's realm, a compact shape, which is often convex. The distribution of the relations has no spikes protruding into the infeasible relations. This compactness was observed for all example scenarios that have been created with the nested neighborhood graphs, except for disjunctions of refining relations from opposite ends of the spectrum (e.g., *contains* ⊕ *contains* xor *contains* ⊕ *disjoint* for $(R1 \setminus R2) \cup R3$ with *R3 inside R2* xor *R3 disjoint R1*), which lead to two disconnected clusters of compact relations. Disjunctions with neighboring constraints (e.g., *R3 inside R2* xor *R3 coveredBy R1*) or any other constructs have led to a single, connected set of compact relations in each set of refining relations.

The compactness serves as a useful consistency checker for the completeness of a set of derived relations, as it should not have any holes or dents. A good initial approximation for identifying the set of feasible relations for a complex compound object is the identification of some extreme cases (i.e., configurations that are driven by the peripheral relations in a neighborhood graph, such as *disjoint*, *inside*, and *contains* for region-region relations) and their mapping onto the nested neighborhood graph. For instance, to determine for $R1 \setminus R2^{\circ}$ with $R1$ contains $R2$ (region with hole, Figure 3a) the feasible relations with $R1$ overlaps $R3$, the two extreme cases *overlap* ⊕ *disjoint* and *overlap* ⊕ *contains* are a good start (Figure 5a). When visualized in the nested neighborhood graph the two relations are disconnected. The completion to compactness adds the three relations *overlap* ⊕ *meet*, *overlap* ⊕ *overlap*, and *overlap* ⊕ *covers* (Figure 5b). Since neither *overlap* ⊕ *inside* nor *overlap* ⊕ *equal* are feasible for a region with a hole, this set of feasible relations wrt. *overlap* is complete [12]. The potential refinement *overlap* ⊕ *coveredBy* is eliminated, because without *overlap* ⊕ *equal* or *overlap* ⊕ *inside* it would create a non-compact subset of the neighborhood graph (Figure 5c).

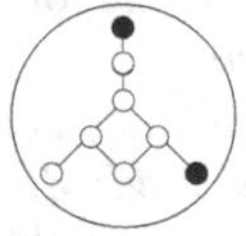
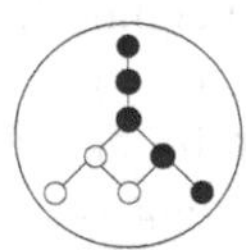
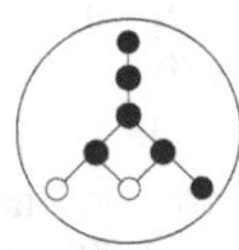

Fig. 5. Derivation of feasible relations for a region that overlaps a region with a hole: (a) the segment of the nested neighborhood graph with the two extreme cases disjoint and contains; (b) the completion to compactness; and (c) a non-compact shape if adding the infeasible relation *overlap* ⊕ *coveredBy*

6 Conclusions and Future Work

Nested neighborhood graphs were introduced as reference systems for topological relations with compound spatial objects. They support the identification of consistent and complete sets of feasible relations by incrementally building relations from the base set of topological relations between simple regions, lines, and points. The nesting of refining relations leads to a natural way of coarsening relations by moving up towards the graph's root. The nested neighborhood graphs also lend themselves to the conceptual neighborhood graphs over the compound spatial objects, and have the potential to derive compositions of compound-object relations directly from the base relations' compositions.

While the visual approach to presenting nested neighborhood graphs supports the confirmation of connectedness and compactness among feasible relations, their depictions for nesting of more than two or three sets of relations becomes graphically and visually challenging. Such an approach clearly needs tailored methods for panning and zooming so that the most relevant parts and connections are highlighted.

A number of further issues will need to be addressed in the future, such as the impact of the difference between A, B, and C neighborhoods [7,13] on nested neighborhood graphs, for what kinds of configurations the nested neighborhood graphs reveal multiple clusters or feasible relations, how to transform a nested neighborhood graph into one with either a different origin (i.e., a principal relation) or with a different orientation (i.e., sequence of refining relations), and a comparative analysis with Galton's Dominance Diagrams [14].

Acknowledgment

This work was partially supported by the National Geospatial-Intelligence Agency under grant number NMA201-01-1-2003.

References

1. Bejaoui, L., Pinet, F., Schneider, M., Bédard, Y.: An Adverbial Approach for the Formal Specification of Topological Constraints Involving Regions with Broad Boundaries. In: Li, Q., Spaccapietra, S., Yu, E., Olivé, A. (eds.) ER 2008. LNCS, vol. 5231, pp. 383–396. Springer, Heidelberg (2008)
2. Clementini, E., Di Felice, P.: An Algebraic Model for Spatial Objects with Indeterminate Boundaries. In: Burrough, P., Frank, A. (eds.) Geographic Objects with Indeterminate Boundaries, pp. 155–170. Taylor & Francis, Bristol (1996)
3. Cohn, A., Bennett, B., Gooday, J., Gotts, N.: Qualitative Spatial Representation and Reasoning with the Region Connection Calculus. GeoInformatica 1(3), 1–44 (1997)
4. Cohn, A., Gotts, N.: The 'Egg-Yolk' Representation of Regions with Indeterminate Boundaries. In: Burrough, P., Frank, A. (eds.) Geographic Objects with Indeterminate Boundaries, pp. 171–187. Taylor & Francis, Bristol (1996)
5. Egenhofer, M.: A Model for Detailed Binary Topological Relationships. Geomatica 47(3&4), 261–273 (1993)

6. Egenhofer, M.: Temporal Relations of Intervals with a Gap. In: Goranko, V., Wang, X.S. (eds.) 14th International Symposium on Temporal Representation and Reasoning, Alicante, Spain, pp. 169–174. IEEE Computer Society, Los Alamitos (2007)
7. Egenhofer, M., Al-Taha, K.: Reasoning about Gradual Changes of Topological Relationships. In: Frank, A.U., Campari, I., Formentini, U. (eds.) GIS 1992. LNCS, vol. 639, pp. 196–219. Springer, Heidelberg (1992)
8. Egenhofer, M., Franzosa, R.: Point-Set Topological Relations. International Journal of Geographical Information Systems 5(2), 161–174 (1991)
9. Egenhofer, M., Herring, J.: Categorizing Binary Topological Relationships Between Regions, Lines, and Points in Geographic Databases, Department of Surveying Engineering, University of Maine, Orono, ME (1991)
10. Egenhofer, M., Mark, D.: Modeling Conceptual Neighborhoods of Topological Line-Region Relations. International Journal of Geographical Information Systems 9(5), 555–565 (1995)
11. Egenhofer, M., Sharma, J.: Assessing the Consistency of Complete and Incomplete Topological Information. Geographical Systems 1(1), 47–68 (1993)
12. Egenhofer, M., Vasardani, M.: Spatial Reasoning with a Hole. In: Winter, S., Duckham, M., Kulik, L., Kuipers, B. (eds.) COSIT 2007. LNCS, vol. 4736, pp. 303–320. Springer, Heidelberg (2007)
13. Freksa, C.: Temporal Reasoning based on Semi-Intervals. Artificial Intelligence 54(1), 199–227 (1992)
14. Galton, A.: Dominance Diagrams: A Tool for Qualitative Reasoning About Continuous Systems. Fundamenta Informaticae 46(1-2), 55–70 (2001)
15. ISO, Geographic Information—Simple Feature Access—Part 1: Common Architecture (2004)
16. Kuhn, W.: Semantic Reference Systems. International Journal of Geographical Information Science 17(5), 405–409 (2003)
17. Kurata, Y., Egenhofer, M.: The Head-Body-Tail Intersection for Spatial Relations Between Directed Line Segments. In: Raubal, M., Miller, H.J., Frank, A.U., Goodchild, M.F. (eds.) GIScience 2006. LNCS, vol. 4197, pp. 269–286. Springer, Heidelberg (2006)
18. Lee, B., Flewelling, D.: Spatial Organicism: Relations between a Region and a Spatially Extended Point. In: Egenhofer, M., Freksa, C., Miller, H. (eds.) Extended Abstracts and Poster Summaries GIScience 2004, Adelphi, MD, pp. 144–147 (2004)
19. Li, S.: A Complete Classification of Topological Relations Using the 9-Intersection Method. International Journal of Geographic Information Science 20(6), 589–610 (2006)
20. Reis, R., Egenhofer, M., Matos, J.: Conceptual Neighborhoods of Topological Relations between Lines. In: Ruas, A., Gold, C. (eds.) The 13th International Symposium on Spatial Data Handling (SDH 2008), Montpellier, France, pp. 557–574. Springer, Heidelberg (2008)
21. Santos, M., Moreira, A.: Topological Spatial Relations between a Spatially Extended Point and a Line for Predicting Movement in Space. In: Wachowicz, M., Bodum, L. (eds.) 10th AGILE International Conference on Geographic Information Science, Aalborg, Denmark (2007)
22. Schneider, M., Behr, T.: Topological Relationships between Complex Spatial Objects. ACM Transactions on Database Systems 31(1), 39–81 (2006)
23. Vasardani, M., Egenhofer, M.: Single-Holed Regions: Their Relations and Inferences. In: Cova, T.J., Miller, H.J., Beard, K., Frank, A.U., Goodchild, M.F. (eds.) GIScience 2008. LNCS, vol. 5266, pp. 337–353. Springer, Heidelberg (2008)

A Model for Geographic Knowledge Extraction on Web Documents

Cláudio E.C. Campelo and Cláudio de Souza Baptista

Computer Science Department
University of Campina Grande
{campelo,baptista}@dsc.ufcg.edu.br

Abstract. There is an increasing interest on doing research in the field of information retrieval which aims to incorporate new dimensions, apart from text based retrieval, to the Web search engines. Geographical Information Retrieval (GIR) aims to index Web resources using a geographic context. The process of identifying the geographic context starts with the detection of different types of geographic references associated to the documents, as for example, the occurrence of place names. This paper presents a model for detecting geographic references in Web documents based on a set of heuristics. Moreover, new concepts and methods for disambiguation of many places with the same name are addressed. Finally, a prototype was built, called GeoSEn which aimed to validate the effectiveness of the proposed model.

Keywords: Geographic Knowledge, Spatial Information Extraction, Web Information Retrieval.

1 Introduction

The aim of GIR is to provide approaches for crawling, indexing and retrieving information using the spatial dimension. There are situations in which a unique search using a GIR based search engine means several searches on traditional search engines. Furthermore, there are searches using a geographic enabled search engine which may not be expressed at all using traditional search engines. For instance, the query "retrieve pages about football from Brazil's neighbors". Hence, one can express neither neighboring nor distance and other topological operators in traditional search engines. They are only possible using a geographic enabled search engine.

Recent experiments have demonstrated that a considerable amount of Web pages have references to terms which may be derived from geographic places, as for instance, placenames, telephone numbers, zipcode, and so on [1], [2], [3]. McCurley [1] assures that approximately 8,5% of web pages have a telephone number, 4,5% have a zip code, and 9,5% contain one of the two. Silva et al [4] state that there is average occurrence of 2.2 references per document to some of the 308 Portuguese cities, in a total of 3,775,611 pages analyzed. Obviously, when taking into account other placenames apart from cities these statistics increase a lot.

C.A. Heuser and G. Pernul (Eds.): ER 2009 Workshops, LNCS 5833, pp. 317–326, 2009.

Research in this new field may be categorized into crawling spatial-related documents, modeling the geographic scope of a document, indexing these documents using textual and spatial features, and the building of spatially-enabled searching and ranking. The crawling process involves the detection of geographic references (which may happen in the document content, in the URL, in whois data, and so on); the solving of ambiguity problems (e.g. two places sharing the same name, or places with people or things names); and the conversion of the valid references into spatial footprints.

We have developed a new geographical search engine called GeoSEn - Geographic Search Engine. This paper focuses on the GeoSEn geographic references detection model. The proposed model is based on a set of heuristics to recognize the references and assignment of confidence values to them.

The reminder of this paper is structured as follows. Section 2 addresses related work. Section 3 presents an overview of the GeoSEn architecture. Section 4 focuses on geographic references detection. Section 5 discusses the results obtained from experimental validation. Finally, section 6 concludes the paper and highlights further work to be undertaken.

2 Related Work

In the process of disambiguation proposed by Martins et al [5], all detected references in a given document have an associated weight. At the end of the process, only the reference which has the highest weight is selected to represent the geographic scope of the document. Otherwise, if all assigned weights were lesser than a given threshold no reference is chosen. The values assigned to the entities may be propagated to others through ontological relationships among them, by applying inference methods to probability graphs.

Li et al [6] describe a method for disambiguation, known as Toponym Resolution. In their approach, it is assigned a probability value to each possible location in an ambiguous place name. The initial value is given according to the information returned from TGN (www.getty.edu/TGNServlet), as for example, whether the place is in a capital or in an inhabited place. After that, other heuristics are used to increment the initial values: (i) the occurrence of places spatially related to the references of which are near in the document; (ii) population statistics; (iii) geographic terms (e.g. "country") close to the analyzed reference.

Volz et al [7] propose a method for disambiguation based on an ontology. Initially, the candidate references are associated to a given weight, which is computed based on the properties of the reference obtained from a gazetteer, similarly to Li et al [6]. After that, there is an analysis of the textually neighbors terms, using a distance between –5 and +5 terms, which seeks words that may define a place (e.g. city, country, river, mountain) and correlated geographic references. At the end, just one reference is selected, the one which has the highest value for the multiplication of the weight by the number of occurrences.

Other proposed methods take into account the contextual information of the references through the investigation of neighbor terms [8], [9]. Indeed previous and next terms are analyzed using a distance between 2 and 5 of the analyzed reference. In some research works the disambiguation process also considers that usually the locations which are geographically near also have their references near in the document [8], [9], [10].

Markowetz et al [11] propose that terms which represent city names should be grouped into classes, as for example, strong terms (words used exclusively to report cities); and weak terms (words which may refer to other things apart from cities). The idea is to firstly detect the strong terms, and then to detect the weak ones. There are other classes of terms such as killer terms, validator terms, and general killer terms. One drawback of the proposed approach is that the classification is done manually.

3 The GeoSEn Architecture

Fig. 1 presents the GeoSEn architecture. The GeoSEn was developed as an extension of the Apache Nutch Framework - http://lucene.apache.org/nutch, adding to it the ability to manipulate and retrieve geographic information.

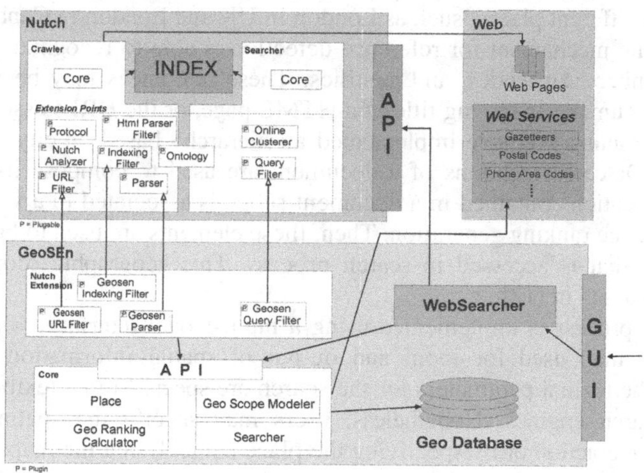

Fig. 1. The GeoSEn architecture

Nutch is based on a plugin oriented architecture, which offers many extension points. In Fig. 1, there is an area identified by *Extension Points* which addresses such extension points. We have chosen some of these points to extend in GeoSEn. Hence we have implemented the following plugins:

- *Geosen Parser*, which is responsible for detecting geographic terms during the parsing process. These terms will be used later to analyze document's geographic scope;
- *Geosen Indexing Filter*, which adds to the index the information on geographic scope;

- *Geosen Query Filter*, which enables to query the index taking into account the information about the geographic scope added by the *Geosen Indexing Filter*;
- *Geosen URL Filter*, which offers different ways to detect geographic references and spatial relationships at the URLs accessed during the crawling process.

These plugins work as a bridge to the core of the GeoSEn prototype the functions of which are available through an API. This core also contains auxiliary functions such as the geographic scope, and the relevance ranking. There are also a database server (*Geo DataBase*) to store the geographic data, and W*eb Services*, which provide additional geographic data to the system.

The process of detection of geographic references is based on a set of heuristics. Each detected reference is associated to a confidence factor, the value of which is composed of an assignment of a weight to each related heuristic. This varies according to the reference type, the type of the place and the position in the document that contains the reference was found. The references in which the confidence factors do not reach a given threshold are rejected. This threshold was set to 0.5, that is to say that the selected references have more than 50% probability of being valid. This value is also used during the desambiguity process. For instance, a given name which is used to locate different places, such as London in UK and London in Canada.

Currently the mechanism for reference detection is able to recognize placenames, telephone numbers, zip codes, and gentilics. These references may be found in the body of the document, in the tag title of a HTML page, or the URLs associated to the captured documents. We have implemented an hierarchy based on city, state, region and country. Detected locations of a document are used to compose its geographic scope. Each location contained in a document scope is associated to a value which is used for relevance ranking generation. Then, these elements are used to build a spatio-textual index, that is accessed in search process. This geographic scope modeling process is discussed in [12].

The search process is implemented using a multi-modal interface, which contains an interactive map used for input and output of spatial information during user interaction. The textual parameters for the search are specified in a textbox area as in traditional search engines. Nonetheless, users may specify the spatial dimension through map interaction or by specifying the place name. Using the map interaction it is possible to select pre-defined geometries, such as countries, states and cities; or to draw a rectangle in the map to set the area of interest. After specifying the spatial dimension, users may use spatial operators such as inside, outside, distant from, etc.

4 Geographic References Detection

The mechanism of detecting geographic references is part of the crawling process. It aims to identify and extract geographic information from the documents retrieved with the web crawler. Examples of such information include place names, postal code, and telephone code.

Once this information is obtained, it is started the process of converting it into geographic places recognized by the system. For instance, a telephone code may be associated into a geographic area (e.g. a country or state). The geographic references may be detected in the body, title and URL of the document.

The detected references are filtered by a process of disambiguation. Then, they are used in the modeling of the geographic scope of the document, aiming to verify the set of places which will be associated to the documents, with their respective relevance values, which will be used in the searching process.

4.1 Geographic Terms Confidence

A confidence rate is assigned to the detected geographic terms. This rate is defined as:

Confidence Rate (CR): a measure which is assigned to a parser detected geographic reference which represents the probability of this reference to be a valid place. CR is the main factor used in the disambiguation. It is a value between 0 and 1. There is a threshold, the value of which is 0.5, in which the terms with CR lesser than this value are ignored. Hence, the detected references with low probability values are ignored. This happens because there are places with either people or thing names. When there is more than one place name for the detected place, as for instance London in the UK and London in Canada, the CR is computed for each place and then the one with highest value is selected. In this case a reference for each place is created, and then the one which has the greater confidence factor is chosen. In this case, the place has been selected probably due to the occurrence of some related place in the same text, such as a city of the same country. This is explained in the section 4.4.

The process of geographic reference detection analyses several features related to the candidate references, aiming to assign the value of CR. Nonetheless, the CR value comes from the values of the entities known as confidence factors, which are associated to the analyzed reference and are defined as follows.

Confidence Factor (CF): a measure associated to each analyzed feature. Each CF has a weight in the CR computation.

The CFs used may vary according to the local in which the term appears in the text (e.g., page title, body of text, URL); to the reference type (e.g. place name, zip code, telephone code area); and to the place type (e.g. city, state, country). The CFs used by GeoSEn are:

- CF_{ST}: it analyzes the occurrence of special terms associated to geographic references;
- CF_{TS}: it considers probabilities computed from textual searches;
- CF_{CROSS}: it analyzes the occurrence of cross references; and
- CF_{FMT}: it evaluates the syntax used to describe the geographic references.

The value of each CF is obtained from the values of their confidence modifiers – CM, which are defined as:

Confidence Modifier (CM): it is part of the CF computation. It is used when a given reference is related many times to a CF feature of the same type.

Each CF has one or many CMs. Hence, for a factor related to a feature X, which is identified by CF_X, its n modifiers are identified as CM_{X1}, CM_{X2}, ... , CM_{Xn}. Therefore, a reference may be associated to one or more confidence factors, while each factor may be associated to one or many modifiers.

4.2 Recognition of Special Terms

One of the features analyzed during the evaluation of the candidate geographic reference is the occurrence of special terms (ST - *Special Terms*), which are defined as:

Special Term (ST): a term that may increase the confidence that there is a geographic reference in a given document.

Examples of STs include "in" (e.g. "in Paris); "city" (e.g. "city of London"); "Zip" (e.g. "CT2 7NZ"). Hence, the confidence factor CF_{ST} of a geographic reference is modified according to the quantity and type of STs related to it. Thus each special term related to a same reference represents a modifier of type CM_{ST}, the values of these modifiers are combined to compose the value of CF_{ST}. The main attributes of a ST are:

- *Term:* the special term;
- *Type of geographic reference:* types of geographic references (e.g. place names, zip, telephone code);
- *Type of place:* types of place (e.g. city, state, country);
- *Minimum distance (D_{MIN}):* minimum distance to the associated reference, which can be positive or negative. For instance, in the expression "in the city of Rio de Janeiro" the terms "in the" and "city" are apart from the reference "Rio de Janeiro" -3 and -2, respectively.
- *Maximum distance (D_{MAX}):* maximum distance to the associated reference which can be positive or negative.
- *Maximum confidence rate:* this is the maximum confidence rate added to the reference through the special term.

A value of a CM_{ST} is calculated so that as the special term is closer to its associated reference, greater will be the influence to the value of the modifier. For example, for a given ST, if the minimum distance is equal to 2 and the maximum distance is equal to 4, there are three possibilities for the ST to add confidence to the reference. If the distance is 2, the maximum confidence is added to the reference. If the distance is 3, 2/3 of the maximum confidence is added to the reference; is the distance is 4, 1/3 is added to the maximum confidence.

Each geographic reference may be associated to one or more special terms. If there is no special term associated to the reference, then $CF_{ST} = 0$. As the maximum value for CF_{ST} is 1, any result greater than this maximum value is set to 1. The occurrence of only a special term is enough to the value of CF_{ST} be greater than half of its maximum value. Each occurrence of a special term contributes with its value of CM_{ST} (multiplied by a weight value - w) to the value of the factor CF_{ST}. This weight w is used to determine the number of special terms that can contribute to the value of CF_{ST}. For example, let us suppose a reference, which was associated to three STs, with values of CM_{ST} equal to 0.8, 1 and 0.5, respectively. Hence, the computed value for the CF_{ST} of this reference would be $0.5 + (w \times 0.8) + (w \times 1) + (w \times 0.5)$.

The GeoSEn detector of special term contains thirty four terms stored. The attributes of which and the value of w were chosen empirically based on experimental evaluation using several documents.

4.3 Confidence from Textual Search

In order to solve the ambiguity problem for place names which are also name of things or people, it was proposed a mechanism which is able to extract features and to associate probability rates based on the result of the textual search (TS – *Textual Search*) for a place name. During the execution of this process, each place name stored in the system was analyzed and the resultant value is stored in the database. Thus, during the parsing process the values previously computed are retrieved and they represent the CM_{TS} of the analyzed reference. This process is defined as:

Confidence Extraction of Textual Search (TS): it is a process which aims to associate a place name to the probability of a term to refer to a location, based on the analysis of the result provided by a textual search engine using this term.

By querying the textual search engine using a given name, each item of the result set contains a snipet (text fragment) with the term. Then, these fragments are analyzed to discover whether the term is a geographic place. For instance, the fragment: "...Campina Grande is the second largest city in terms of population in the Brazilian state of Paraíba...", extracted from one of the results to a query on "campina grande", contains some keywords (e.g. "city", "population") that may indicate that this fragment is describing something about the city. Hence, the values of CM_{TS} are obtained based on the amount of keywords verified in the result set. After that, these values are normalized, so that they are in the interval between 0 and 1. Once that search engines show the results according to document relevance, it is possible to analyze only the first ten documents in the result set, which is the case of GeoSEn.

4.4 Cross References

Another feature evaluated in GeoSEn is the cross reference, which is defined as:

Cross Reference: geographic reference found in the document which has topological spatial relationships in relation to other reference. Let $L(R_X)$ be the place referenced by the geographic reference R_X. Then, for an analyzed reference R_A, a reference R_B is considered a cross reference from R_A is any of the following is true:

- $L(R_B)$ contains $L(R_A)$;
- $L(R_B)$ is contained into $L(R_A)$;
- $L(R_A)$ and $L(R_B)$ are in the same hierarchic level and exists a place L_X, so that L_X is in a ancestral level of $L(R_A)$ and $L(R_B)$ and L_X contains $L(R_A)$ and $L(R_B)$.

Each cross reference is represented by a modifier of type CM_{CROSS}. The value of this modifier varies according to the distance (in number of words in the text), among the references analyzed (*D*) and the hierarchic distance between the places referenced by them (*N*). Therefore, the value of CM_{CROSS} rises proportionally to the dropping of the result of the multiplication $N \bullet D$. The final cross relevance confidence value (CF_{CROSS}) is obtained analogously to CF_{ST}, where each CM_{CROSS} contributes to compose its value.

We have observed that the identification of cross references plays an important role in solving ambiguities, mainly when there is more than a place for the same name. For example, let us consider a document in which it was detected a reference to the city named *Atalaia*. However, in *Brazil* there are two cities with this name located in the Northeast and South regions. In this case, the system will create internally two references, in which only one will be selected at the end of the process. Nonetheless, let us consider that the mentioned document also refers the Brazilian state of Alagoas which contains the Atalaia city in the Northeast region. Hence, the CM_{CROSS} value computed to the reference *Atalaia / Alagoas* will be greater than the one computed for *Atalaia / South*. This difference is considered in the disambiguation.

4.5 Reference Format

Some types of geographic references may be identified in the text using different written styles. Different confidence rates are associated to each one of the styles. For example, a Brazilian postal code may be represented by strings “58.109-000” or “58109000”. This confidence factor is identified by CF_{FMT} and its modifiers by CM_{FMT}.

The recognition of references of the type place name, uses two specific modifiers. The first one evaluates if a detected place name is written in uppercase in the initial characters. This is a simple technique, but it is worthy for disambiguation, once place names usually are written following this style (exception for the words taken as *Place Stop Words,* which are the words of the place names that are ignored). The possible values for this modifier are either 0 or 1. If any of the terms does not have the initial character in uppercase, then the modifier is set to zero, otherwise the modifier is set to 1, if the previous uppercase rule is obeyed.

The second CM_{FMT} for place names is used to measure how abbreviated is the detected place. The less is abbreviated, more confident is the reference. The computation of this confidence factor is done by dividing the number of terms abbreviated by the total number of terms that belong to the place name (including the *Place Stop Words*). It assumes values between 0 and 1. An example of the importance of detecting abbreviated place names, one may observe the result set from Google search engine for some abbreviations of the Brazilian city of *Cabo de Santo Agostinho*: for the expression "cabo de s agostinho" there were 749 occurrences; and for "cabo de santo a", approximately 453 occurrences. Another example is that there are 38,100 occurrences for the query "R. de Janeiro", which may represent either the city or state of *Rio de Janeiro*.

The CF_{FMT} value is computed using the average of its modifiers.

4.6 Computing the Confidence Value

Once the values of the many confidence factors are obtained, the final confidence value (CR) is computed for each reference detected in the parsing process. Then, the references which have not achieved the minimum value are eliminated, and the more confident ones are selected.

The CFs used (and their importance) vary according to the place of the reference detection, the type of detected reference and the type of referenced place. For example, there is no sense in verify uppercase in postal code references. Then, the CR value is computed by the sum of each CF associated to the reference, weighting them accordingly to their importance. At the end of this process, all references that have not been eliminated are used in the composition of the geographic scope of the document.

The mechanism of scope generation associates to each local a value of relevance. In the computation of this value it is considered as one of the variables the CR value, which is computed during the step of detecting geographic references.

5 Experiments

The mechanism of detecting geographic references implemented in GeoSEn was executed using fifty documents previously selected, aiming to quantify the effectiveness of the proposed system. From this analysis, we have obtained the following results:

- 71% of the valid references were correctly detected;
- 92% of the invalid references were correctly ignored;
- Among the references that were correctly detected and there was ambiguity, 84% processed correctly the disambiguation;
- Among the references ignored incorrectly (23%), 65% had their places detected by other references in the same document;

The results have shown that the proposed mechanism has a good effectiveness, mainly when regarding the elimination of the invalid references and the solving of ambiguity problems. Notice that the majority of the valid references which were initially discarded were detected through other references in the same document (item d).

6 Conclusion and Future Work

Research on GIR is still in its infancy and there is a long way to investigate new techniques. We have noticed, among related work, a lack of more deep models, which, for example, may eliminate ambiguities in the process of detecting geographic references and the set of geographic scope in a way more complete and flexible.

In this paper, we presented a model for detecting geographic references in Web documents, based on a set of heuristics. In the proposed approach we introduced the concepts of confidence factor and confidence modifier, which aim to measure the probability of a detected geographic reference be a valid reference and be associated to a correct place, even when there exists ambiguity.

As further work we plan to introduce new confidence factors, as for example, a factor related to demographic data in places. Moreover, we plan to improve the proposed techniques, as for instance, the extension of the recognition of special terms to support expressions and not only isolated terms. Similar methods to confidence assignment may use other sources such as dictionaries and encyclopedias.

We believe that by applying these new proposals we can improve the system. Finally the use of other heuristics already proposed in related works is also desirable.

References

1. McCurley, K.S.: Geospatial mapping and navigation of the web. In: Proceedings of the WWW 2001, pp. 221–229. ACM, Hong Kong (2001)
2. Buyukkokten, O., Cho, J., Garcia-Molina, H., Gravano, L., Shivakumar, N.: Exploiting geographic location information of web pages. In: ACM SIGMOD, Workshop on the Web and Databases. ACM, Philadelphia (1999)
3. Ding, J., Gravano, L., Shivakumar, N.: Computing geographic scopes of web resource. In: The International Conference on VLDB, pp. 545–556. Morgan Kaufman, Cairo (2000)
4. Silva, M.J., Martins, B., Chaves, M.S., Afonso, A.P., Cardoso, N.: Adding geographic scopes to web resources. Computers, Environment and Urban Systems 30(4), 378–399 (2006)
5. Martins, B., Chaves, M., Silva, M.J.: Assigning geographical scopes to web pages. In: Losada, D.E., Fernández-Luna, J.M. (eds.) ECIR 2005. LNCS, vol. 3408, pp. 564–567. Springer, Heidelberg (2005)
6. Li, Y., Moffat, A., Stokes, N., Cavedon, L.: Exploring probabilistic toponym resolution for geographical information retrieval. In: Proceedings of the 3rd ACM Workshop On Geographical Information Retrieval, pp. 17–22. ACM, Seattle (2006)
7. Volz, R., Kleb, J., Mueller, W.: Towards ontology-based disambiguation of geographical identifiers. In: Proceedings of The WWW 2007, CEUR-WS.org, Banff (2007)
8. Rauch, E., Bukatin, M., Baker, K.: A confidence-based framework for disambiguating geographic terms. In: Proceedings of the HLT-NAACL Workshop on Analysis of Geographic References, pp. 50–54. ACL, Morristown (2003)
9. Amitay, E., Har'El, N., Silvan, R., Soffer, A.: Web-a-where: Geotagging web content. In: Proceedings of SIGIR, Workshop on Geographical Information Retrieval, pp. 273–280. ACM, Sheffield (2004)
10. Zong, W., Wu, D., Sun, A., Lim, E., Goh, D.: On assigning place names to geography related web pages. In: Proceedings of JCDL, pp. 354–362. ACM, Denver (2005)
11. Markowetz, A., Chen, Y.Y., Suel, T., Long, X., Seeger, B.: Design and implementation of a geographic search engine. In: WebDB, Baltimore, pp. 19–24 (2005)
12. Campelo, C.E.C., Baptista, C.S.: Geographic Scope Modeling for Web Documents. In: Proceedings of The 5th International Workshop on Geographic Information Retrieval (GIR 2008), pp. 11–18. ACM, Napa Valley (2008)

A Semantic Approach to Describe Geospatial Resources

Sidney Roberto de Sousa

Institute of Computing, University of Campinas, 13083-970, Campinas, SP, Brazil
sidney@lis.ic.unicamp.br

Abstract. Geographic information systems (GIS) are increasingly using geospatial data from the Web to produce geographic information. One big challenge is to find the relevant data, which often is based on keywords or even file names. However, these approaches lack semantics. Thus, it is necessary to provide mechanisms to prepare data to help retrieval of semantically relevant data. This paper proposes an approach to attack this problem. This approach is based on semantic annotations that use geographic metadata and ontologies to describe heterogeneous geospatial data. Semantic annotations are RDF/XML files that rely on a FGDC metadata schema, filled with appropriate ontology terms, and stored in a XML database. The proposal is illustrated by a case study of semantic annotations of agricultural resources, using domain ontologies.

1 Introduction

The Web became an immense repository of geospatial data in different geographic formats like remote sensing images, maps, sensor data temporal series, textual data files, among others [1, 2]. The retrieval of these data requires special attention due the geographic distribution of the sources and the heterogeneity of the data. Geographic metadata standards and geospatial information portals were created as an initiative to attack this problem. In these portals, users can create their own queries using keywords and metadata fields from some metadata schema such as ISO 19115 and FGDC Metadata. These metadata fields are often filled with natural language text, which can cause ambiguities, while keywords can restrict the result of the queries if different terminology is used or if terms are homonymous [3].

One solution to overcome these problems is the use of domain ontologies - as can be seen in [4] - to identify and associate common concepts. Ontologies are frequently used to explain knowledge about some domain of interest. In the geographic domain, an ontology must have terms and concepts about useful issues to describe geospatial resources, for instance, spatial references, time periods, geographic formats details, and other kinds of meta-information that may improve the retrieval of geospatial information.

C.A. Heuser and G. Pernul (Eds.): ER 2009 Workshops, LNCS 5833, pp. 327–336, 2009.

The World Wide Web Consortium (W3C) proposed the Resource Description Framework (RDF) to describe resources available in the Web as an initiative for providing semantic interoperability. RDF identifies resources using their URIs and describes them using statements. A statement is a triple <*subject, predicate, object*>. From the geospatial point of view, a subject is a geospatial resource, a predicate is a metadata field of this resource, and an object is the value filling the metadata field. Applying this model in a way so ontologies could be included, the object can be an ontology term that semantically associates the metadata field content to some appropriate concept.

Based on this approach, this paper discusses the use of semantic annotations to describe geospatial data, extending the work of [2] to cover implementation aspects. This work defines a semantic annotation as a set of RDF triples, where each triple is basically composed of a FGDC metadata schema, where each metadata field is filled with appropriate terms from domain ontologies. The annotations are stored in an XML database, where they can be retrieved using XQuery and XPath statements.

The rest of this paper is organized as follows. Section 2 describes the approach for semantic annotation of geospatial resources presented in this paper. Section 3 explains how annotations are stored in an XML database. Section 4 shows how the presented approach is applied in a case study of semantic annotations of agricultural resources. Section 5 describes related work. Finally, section 6 contains conclusions and ongoing work.

2 An Approach for Representing Semantic Annotations

A semantic annotation of a geospatial resource must provide semantic descriptions about geographic characteristics of this resource. Such characteristics are structurally organized using geographic metadata standards. The role of ontologies in this scenario is to enhance the annotations, providing appropriate terminology. This section describes the representation of the semantic annotations in RDF/XML format, detailing geographic metadata schema, and ontology concepts.

2.1 Geographic Metadata Schema

Metadata can be considered as data about other data. Their principal role is to add important information to a resource so that ambiguities can be avoided and the retrieval of the resource can be done in an easier way. Absence of metadata may lead to unreliability and re-work when it comes to interoperability among distinct systems, hampering data exchange and integration [5]. Geographic metadata describe geospatial resources, enhancing them with useful information such as reference system used, producer identification, and location information.

Use of geographic metadata is strongly disseminated by geographic catalogs, such as GeoNetwork[1], which use geographic metadata standards. ISO 19115 is a

[1] http://sourceforge.net/projects/geonetwork Accessed in March 30th, 2009.

proprietary standard of geographic metadata, developed by the ISO Committee. It has a UML based structure, where each metadata element is defined in context of a class and is characterized by a *name*, *definition*, *obligation*, *multiplicity*, *data type*, and a *domain*. This standard has a minimal set of elements which is defined for the most important information needed to describe some resource, called *core data*. It is possible to extend this set of elements to serve special needs [6].

The Federal Geographic Data Committee Metadata (FGDC Metadata) is an open standard which defines some particularities needed to catalog and publish geographic meta-information. It provides knowledge about the kind of the resource, indicating whether it meets the users expectation, and where/how to find it. Use of a specific section or element is either mandatory or optional [7].

2.2 Using Ontology Terms

In geographic catalogs, metadata fields are filled with natural language text, which most times can lead to ambiguities or bad understanding. Despite the structure and semantics that metadata can provide, the content of the fields may not be able to avoid this and other kinds of problems [3]. The use of ontology terms guarantees unique meaning, associating metadata fields to concepts that semantically represent their content. Ontologies also provide a hierarchical structure that helps to understand their concepts. Figure 1 shows the solution for the example seen, using terms of NASA SWEET Numerics ontology[2]. It indicates that the *Graph* term is a 2D distribution.

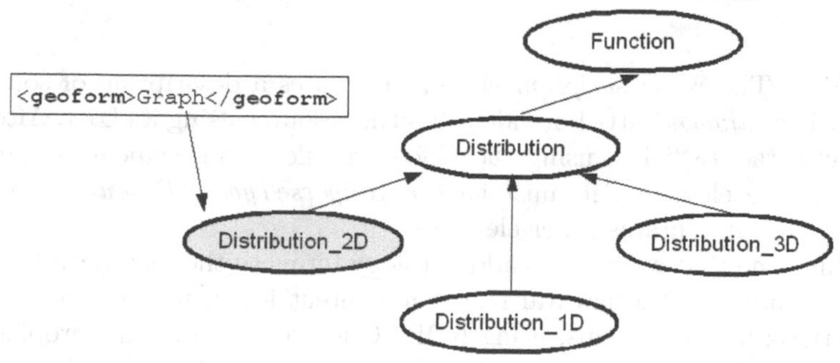

Fig. 1. Use of an ontology term to represent a metadata field

2.3 Representation in RDF

Once a metadata schema is chosen, it is possible to use RDF to semantically describe a resource. Figure 2 illustrates a possible representation in RDF/XML (without the use of ontology terms) of a graph that shows the evolution of some phenomenon with time, as measured per seasons. It uses metadata fields

[2] http://sweet.jpl.nasa.gov/ontology/ Accessed in March 31st, 2009.

```
<rdf:RDF xmlns:rdf="http://www.w3.org/1999/02/22-rdf-syntax-ns#"
       xmlns:fgdc="http://www.fgdc.gov/metadata/fgdc-std-001-1998.xsd#">

   <rdf:Description
    rdf:about="http://www.lis.ic.unicamp.br/efarms/NDVI_graphs/graph01.jpg">
      <fgdc:citeinfo rdf:parseType="Resource">
         <fgdc:origin>eFarms</fgdc:origin>
         <fgdc:pubdate>20080526</fgdc:pubdate>
         <fgdc:title>NDVI Graph</fgdc:title>
         <fgdc:edition>Digital image version</fgdc:edition>
         <fgdc:geoform>Graph</fgdc:geoform>
         <fgdc:serinfo rdf:parseType="Resource">
            <fgdc:sername>NDVI graphs set</fgdc:sername>
            <fgdc:issue>NDVI calculus of rural areas</fgdc:issue>
         </fgdc:serinfo>
         <fgdc:pubinfo rdf:parseType="Resource">
            <fgdc:pubplace>Campinas - SP</fgdc:pubplace>
            <fgdc:publish>LIS, IC-UNICAMP</fgdc:publish>
         </fgdc:pubinfo>
      </fgdc:citeinfo>
   </rdf:Description>
</rdf:RDF>
```

Fig. 2. Representation in RDF of metadata for a graph, using fields from the FGDC metadata standard

```
<fgdc:origin rdf:parseType="Resource">
     <rdfs:comment>eFarms</rdfs:comment>
     <rdf:type rdf:resource="http://sweet.jpl.nasa.gov/1.1/data.owl#Project"/>
</fgdc:origin>
```

Fig. 3. Adding an ontology term to *fgdc:origin* element

from FGDC. The *rdf:Description* element indicates a description of some Web resource. The *rdf:about* attribute identifies the resource using its URI. After this, come the metadata fields, using the following rule: if an element is composed of one or more elements, it must have a *rdf:parseType="Resource"* attribute indicating that it contains other elements.

Now, imagine that we want to add ontology terms to the metadata fields, but we want to preserve the natural language content for future use in a publication interface: how to do this, using RDF? One way to solve this problem is to keep the natural language text as a human readable description of the metadata field's content, using the property *rdfs:comment* from RDF Schema (RDFS), an extension to RDF for defining application-specific classes and properties[3]. In addition, we can specify that the content of the metadata field is an instance of an ontology class (the ontology term), using the property *rdf:type*. Figure 3 shows this solution. In this example, the field *origin* contains a human readable description that says that the resource was originated by "eFarms" and a reference to the class *Project* that specifies that the originator of the resource is an instance of this class. Thus, we want to say that "the resource was originated by a project called eFarms".

[3] http://www.w3.org/TR/rdf-schema/ Accessed in June 23rd, 2009.

3 Storing RDF Annotations

RDF can be represented by more human-readable languages like Notation3[4] (N3) or by more structured languages like RDF/XML, which is the most used one. An essential characteristic of a good quality geographic metadata standard is that it should be XML compatible. Both FGDC Metadata and ISO 19115 have this feature, as well as metadata standards from other domains such as Dublin Core [8] and e-GMS [9]. These facts lead towards the use of XML databases to store RDF/XML.

An XML database is a data persistence software that allows storage of data in XML format, generally mapping these data from XML to some storage format, which can be a relational database or even other XML documents [10]. Queries over a XML database are generally executed using XPath or XQuery statements. It is possible to retrieve RDF/XML data using XQuery, once this language was designed to query XML data not just from XML files, but anything that is structured in XML.

Both XPath and XQuery allow retrieval of full XML-based documents or subtrees of these, using their DOM trees[5]. If we know the schema of an annotation of interest, we can retrieve the full annotation or parts of these. For instance, if someone wanted to know who originated the NDVI graph of the previous example, he could retrieve this information using an XPath statement (*/rdf:RDF/rdf:Description/fgdc:citeinfo/fgdc:origin*).

Another solution for storing and querying RDF is to use some framework for these purposes, like Sesame [11] and Jena [12]. These frameworks play the role of a layer that manage persistent storage of RDF in files or relational databases and provide queries over RDF in SPARQL or in other specific languages. Moreover, such frameworks provide reading and writing of RDF in different notation languages.

4 A Case Study: Semantic Annotation of Agricultural Resources

We propose an architecture for semantic annotation of agricultural geospatial data to illustrate the approach of this paper, taking as example a NDVI graph. Normalized Difference Vegetation Index (NDVI) is a numerical indicator used to analyze whether some region of interest has live green vegetation or not. Using this index, it is possible to verify some aspects like density of vegetation or crops in some area of interest. A NDVI graph is a 2D distribution containing return values of the NDVI function in a certain time period, where the y axis is the NDVI index and the x axis is a date, thus characterizing a time series. This kind of annotation can be useful for activities like crop management and monitoring [2].

[4] http://www.w3.org/DesignIssues/Notation3.html Accessed in June 23rd, 2009.

[5] The XML DOM (Document Object Model) defines a standard way for accessing and manipulating documents compatible to XML, presenting them as a tree structure where elements, attributes, and text are nodes.

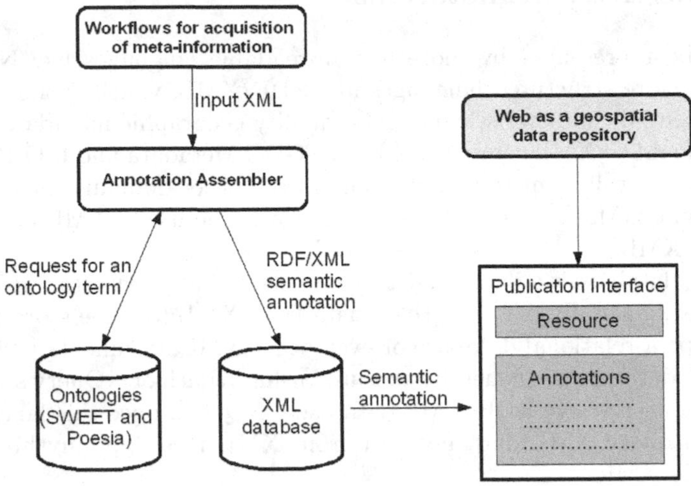

Fig. 4. Proposed architecture

Figure 4 illustrates the proposed architecture. It is composed by the following modules:

- ***Workflows for acquisition of meta-information:*** Acquisition of meta-information depends on the kind of resource. Here, the acquisition of such information is performed by specific workflows described in [2], where a specific workflow is activated for each kind of geospatial data. For a NDVI graph, crop identification is done comparing the curve of the graph to other existing curves where crops were already identified. A more detailed explanation about this work is given in section 5. The meta-information is organized in a set of FGDC metadata, encapsulated in a simple XML file, and submitted to the annotation assembler;
- ***Annotation assembler:*** This module receives the meta-information submitted by a specific workflow, which contains a specific set of FGDC metadata fields filled with natural language text. An ontology term is associated to each metadata field using the approach presented in section 2.3. The choice of the ontology term is done by a mechanism that queries the ontology base for URI of terms;
- ***Ontology base:*** The ontology base is composed by geospatial and agricultural ontologies. NASA SWEET ontologies provide terms about issues in various domains like geography, physics, chemistry, among others. Poesia Agricultural Zoning ontology [13] provides terms about crops and Brazilian locations. Some of these ontologies were extended to attend specific needs like terms about crop production;

Metadata element	Description	Short name	Metadata Schema	Obligation	Ontology used
Citation Information	Reference to be used for the data set	citeinfo	FGDC	Yes	Nasa SWEET
Indirect Spatial Reference	Means which locations are referenced	indspref	FGDC	Yes	POESIA Agricultural Zoning
Horizontal Coordinate System Definition	System which linear/angular quantities are measured and assigned to the position that a point occupies	horizsys	FGDC	Yes	Nasa SWEET
Time Period Information	Time period for which the data set corresponds	timeperd	FGDC	Yes	Nasa SWEET
Digital Transfer Information	Description of the form of the data to be distributed	digtinfo	FGDC	Yes	Nasa SWEET
Crop Identification	Information about identification of crops	cropid	Agricultural Extension	No	POESIA Agricultural Zoning
Soil Identification	Information about identification of soils	soilid	Agricultural Extension	No	POESIA Agricultural Zoning
Productivity Identification	Information about productivity issues	productivity	Agricultural Extension	No	POESIA Agricultural Zoning

Fig. 5. Composition of a semantic annotation of a NDVI graph

- ***XML database:*** After an annotation in RDF/XML is created, it is stored in a XML database from where it can be retrieved using XQuery statements;
- ***Publication interface:*** A Web interface where agricultural researchers can see the Web resource and its semantic annotation.

Figure 5 shows a table that explains the contents of the semantic annotation of a NDVI graph. In order to cover agricultural needs, a agriculture metadata set was created containing elements about crops, soil, and productivity issues (Agricultural Extension).The first column shows the metadata elements used to describe a NDVI graph. Each FGDC element shown in the table is composed by other specific elements, which were abstracted in the table. For details about the acquisition of agricultural meta-information, see [2]. The second column shows a brief description of each element. The third column shows the short name of each element, defined in their respective XML Schemas. The fourth column shows the metadata schema to which each element belongs. The fifth column specifies whether the presence of the element is mandatory or not. Finally, the last column shows the ontologies used to describe each metadata element.

5 Related Work

One of the aims of this work is to provide implementation support to the work of [2], which proposes a framework for semantic annotation of agricultural resources. In that work, each different geospatial resource has a specific workflow for acquisition of spatial and crops meta-information, linkage to ontologies terms, and production and publication of semantic annotations. Our architecture provides the infrastructure needed to associate semantics to annotations, via the linkage module.

There are several research initiatives related to the work reported in this paper. One such trend concerns semantic interoperability in GIS, dealing with problems in data exchange and retrieval. There are some efforts to provide interoperability among metadata standards, as can be seen in [5, 7]. Use of ontologies to deal with interoperability problems in the geospatial domain is discussed in [1, 4, 14, 15].

Another area is representation of information. RDF is being widely used for representing geographic meta-information. In [16], RDF is used to define a catalog of geographic resources from various Web sites. Córcoles and Gonzáles [17] propose an approach for providing queries over spatial XML resources with different schemas using a unique interface, where the resources are integrated using RDF.

Due to the conventional use of XML to represent meta-information, some works have used XML databases to store metadata. In [18], a XML database is used to store metadata in a prototype of a digital library system, which provides queries over metadata from art pieces. The use of XML databases for the management of metadata in the MPEG-7[6] format is discussed in [19], where a survey concerning XML database solutions for this issue was done. A schema-independent XML database used to store metadata about scientific resources is presented in [20].

6 Conclusions and Ongoing Work

Geographic distribution and heterogeneity are issues that hamper the retrieval of geospatial data. Geographic metadata standards were created to solve these problems, but filling metadata fields with natural language text can cause ambiguities. To attack this problem, this paper discussed an approach based on RDF, geographic metadata and ontologies to describe geospatial resources, bringing together Semantic Web and geographic standards technologies. Moreover, it discussed the storage of semantic annotations in XML databases, considering the RDF/XML notation.

Based on this approach, a mechanism is being implemented that chooses and ranks appropriate ontology terms to the metadata fields. At the moment, the choice of terms is done over specific ontologies (Nasa SWEET and Poesia Agricultural Zoning), but the mechanism is intended to be ontology-independent, so that it can choose appropriate ontologies and hence appropriate terms to fill the fields. Once an annotation in RDF is created, the mechanism stores it in a XML database. However, it is intended to use a RDF framework for storing and querying the semantic annotations and so make a comparison about the two approaches.

Acknowledgments

The authors thank FAPESP-Microsoft Research Virtual Institute (eFarms project), CNPq and CAPES for the financial support for this work.

[6] A standard for the description of multimedia content.

References

[1] Fonseca, F., Rodriguez, A.: From Geo-Pragmatics to Derivation Ontologies: new Directions for the GeoSpatial Semantic Web. Transactions in GIS 11(3), 313–316 (2007)

[2] Macário, C.G.N., Medeiros, C.B.: A Framework for Semantic Annotation of Geospatial Data for Agriculture. Int. J. Metadata, Semantics and Ontology - Special Issue on "Agricultural Metadata and Semantics" (2008) (accepted for publication)

[3] Klien, E., Lutz, M.: The Role of Spatial Relations in Automating the Semantic Annotation of Geodata. In: Cohn, A.G., Mark, D.M. (eds.) COSIT 2005. LNCS, vol. 3693, pp. 133–148. Springer, Heidelberg (2005)

[4] Klien, E., Einspanier, U., Lutz, M., Hübner, S.: An Architecture for Ontology-Based Discovery and Retrieval of Geographic Information. In: Proceedings of the 7th Conference on Geographic Information Science (AGILE 2004), Heraklion, Greece, pp. 179–188 (2004)

[5] Nogueras-Iso, J., Zarazaga-Soria, F.J., Lacasta, J., Bejar, R., Muro-Medrano, P.R.: Metadata Standard Interoperability: Application in the Geographic Information Domain. Computers, environment and urban systems 28(6), 611–634 (2003)

[6] Karschnick, O., Kruse, F., Topker, S., Riegel, T., Eichler, M., Behrens, S.: The UDK and ISO 19115 Standard. In: Proceedings of the 17th International Conference Informatics for Environmental Protection EnviroInfo. (2003)

[7] Chandler, A., Foley, D.: Mapping and Converting Essential Federal Geographic Data Committee (FGDC) Metadata into MARC21 and Dublin Core: Towards an Alternative to the FGDC Clearinghouse. D-Lib Magazine 6 (2000)

[8] Weibel, S., Kunze, J., Lagose, C., Wolf, M.: RFC2413: Dublin Core Metadata for Resource Discovery (1998)

[9] Alasem, A.: An Overview of e-Government Metadata Standards and Initiatives based on Dublin Core. Electronic Journal of e-Government 7, 1–10 (2009)

[10] XML:DB Initiative: Frequently Asked Questions About XML:DB, `http://xmldb-org.sourceforge.net/faqs.html` (accessed in April 4th, 2009)

[11] Broekstra, J., Kampman, A., van Harmelen, F.: Sesame: A Generic Architecture for Storing and Querying RDF and RDF Schema, pp. 54–68. Springer, Heidelberg (2002)

[12] Wilkinson, K., Sayers, C., Kuno, H., Reynolds, D.: Efficient RDF Storage and Retrieval in Jena2. Exploiting Hyperlinks 349, 35–43 (2003)

[13] Fileto, R., Liu, L., Pu, C., Assad, E.D., Medeiros, C.B.: POESIA: An ontological workflow approach for composing Web services in agriculture. The VLDB Journal 12(4), 352–367 (2003)

[14] Visser, U., Stuckenschmidt, H., Schuster, G., Vögele, T.: Ontologies for geographic information processing. Comput. Geosci. 28(1), 103–117 (2002)

[15] Fonseca, F.T., Egenhofer, M.J.: Ontology-driven geographic information systems. In: GIS 1999: Proceedings of the 7th ACM international symposium on Advances in geographic information systems, pp. 14–19. ACM, New York (1999)

[16] Córcoles, J.E., González, P., López-Jaquero, V.: Integration of Spatial XML Documents with RDF. In: Cueva Lovelle, J.M., Rodríguez, B.M.G., Gayo, J.E.L., del Ruiz, M.P.P., Aguilar, L.J. (eds.) ICWE 2003. LNCS, vol. 2722, pp. 407–410. Springer, Heidelberg (2003)

[17] Córcoles, J.E., González, P.: Using RDF to Query Spatial XML. In: Koch, N., Fraternali, P., Wirsing, M. (eds.) ICWE 2004. LNCS, vol. 3140, pp. 316–329. Springer, Heidelberg (2004)

[18] Baru, C., Chu, V., Gupta, A., Ludäscher, B., Marciano, R., Papakonstantinou, Y., Velikhov, P.: XML-based information mediation for digital libraries. In: DL 1999: Proceedings of the fourth ACM conference on Digital libraries, pp. 214–215. ACM, New York (1999)

[19] Westermann, U., Klas, W.: An analysis of XML database solutions for the management of MPEG-7 media descriptions. ACM Comput. Surv. 35(4), 331–373 (2003)

[20] Jones, M.B., Berkley, C., Bojilova, J., Schildhauer, M.: Managing Scientific Metadata. IEEE Internet Computing 5(5), 59–68 (2001)

An Ontology-Based Framework for Geographic Data Integration

Vânia M.P. Vidal[1], Eveline R. Sacramento[1], José Antonio Fernandes de Macêdo[1,2], and Marco Antonio Casanova[3]

[1] Universidade Federal do Ceará, Department of Computing, Brazil
{eveline,vvidal,jose.macedo}@lia.ufc.br
[2] EPFL - Ecole Polytechnique Fédérale, Database Laboratory, Switzerland
jose.macedo@epfl.ch
[3] Department of Informatics – Pontifical Catholic University of Rio de Janeiro
Rio de Janeiro, RJ – Brazil
casanova@inf.puc-rio.br

Abstract. Ontologies have been extensively used to model domain-specific knowledge. Recent research has applied ontologies to enhance the discovery and retrieval of geographic data in Spatial Data Infrastructures (SDIs). However, in those approaches it is assumed that all the data required for answering a query can be obtained from a single data source. In this work, we propose an ontology-based framework for the integration of geographic data. In our approach, a query posed on a domain ontology is rewritten into sub-queries submitted over multiples data sources, and the query result is obtained by the proper combination of data resulting from these sub-queries. We illustrate how our framework allows the combination of data from different sources, thus overcoming some limitations of other ontology-based approaches. Our approach is illustrated by an example from the domain of aeronautical flights.

Keywords: data integration, schema mappings, geographic information retrieval, query processing, Web Feature Service, ontologies.

1 Introduction

Spatial Data Infrastructures (SDIs) provide access, reuse and integration of geographic information (GI) from multiple sources. Service providers currently offer access to geospatial data and expose basic processing functionality using Web services technology [1, 2, 3, 6]. This strategy not only offers a standardized, flexible and transparent way to publish underlying data but it also hides details of data access and retrieval from the application. In OGC-compliant SDIs, geospatial data are served via Web Feature Services (WFS). Each WFS offers a feature type schema (FTS), which is the XML schema of the feature type exported by the service. Users can query and update data sources through an FTS. The specifications provided by the Open Geospatial Consortium (OGC) enable syntactic interoperability and cataloguing of GI.

C.A. Heuser and G. Pernul (Eds.): ER 2009 Workshops, LNCS 5833, pp. 337–346, 2009.

In any data sharing architecture, including SDIs, reconciling semantic heterogeneity is a key issue. No matter whether the query is issued or whether the data is shared, the semantic differences between data sources need to be reconciled. Typically, semantic mappings are used to define how translate data from one data source into another, preserving the semantics of the data or, alternatively, to rewrite a query posed on one source into a query on another source. However, the specification of these mappings is labor intensive and error prone, representing over half of the effort spent in a typical data integration scenario. Moreover, the problem of semantic heterogeneity is exacerbated when dealing with semi-structured data due to its flexibility in adding new attributes and, consequently, generating more schema variations. One possible approach to overcome this problem is the explicitation of knowledge by means of ontologies [5]. In this sense, the idea is to use ontologies to describe terms of the domain and the data WFSs services.

Current research [2,3] in the geospatial context have proposed the use of DL ontologies for enhancing discovery and retrieval of geographic information. The framework proposed in [3] adopts a hybrid ontology approach [5], where each feature type schema offered via WFS is described by specific application concepts that are built using properties and classes from a shared vocabulary. The shared vocabulary is represented by a domain ontology that contains basic terms (the primitives) of a domain. These terms are combined to describe the semantic of feature types in separate application ontologies. It is assumed that all actors within a domain share a common understanding of the concepts contained in the domain ontology. In this framework, the requester formulates a query using terms from the domain ontology. Reasoning services are used to determine whether existing application concepts (describing feature types) are a match for the query concepts. When an appropriate feature type is discovered, the query can be used to generate a request to retrieve data from its WFS. The translation of the query into the actual WFS query, which is formulated in terms of the feature type schema, is based on so-called registration mappings [3], which map the structure of the schema to ontology concepts. The restriction of this approach it that a data source is discovered only if it contains all the information required for answering the user's question.

In [4], it is proposed a methodology that uses rules for both the discovery of data sources and, based on the discovered data, answering queries in SDIs. Their approach allows inferences that use relationships between individuals and the combination of data from different sources. Query answering is realized in three steps: First, schema mapping and domain rules are used in the discovery of appropriate data sources that can answer a specific user query. Then, the knowledge base has to be populated with data of the relevant data sources. Finally, using domain rules, new knowledge is inferred to answer the user query. The major drawback of this approach is that a large amount of data that is materialized in the knowledge base may not be relevant to the user query.

In this paper, we propose a framework that deals with the situation where data from several data sources have to be combined in order to answer a given question. In our approach, a query formulated in terms of a domain ontology is rewritten into sub-queries submitted over multiples data sources, and the query results are obtained by the proper combination of data resulting from these sub-queries. Our approach takes advantage of DL reasoning to discard sub-queries that are not consistent.

The remainder of the paper is structured as follows. Section 2 describes our framework for integration of geographic data. Section 3 describes the proposed approach with the help of an example. Section 4 presents the conclusions and directions for future research.

2 A Framework for Geographic Data Integration

Figure 1 describes the main components of the proposed framework. The mediated schema is represented by a domain ontology (DO), which provides a conceptual representation of the application domain (a global shared vocabulary). Each feature type schema, offered via a WFS, is described by an application ontology (AO) whose vocabulary is restricted to be a subset of the vocabulary of DO. The Global Ontology consists of the union of the application ontologies, and a set of axioms that define inter-ontology properties. The *mediated mapping* defines the concepts and properties of the domain ontology in terms of the vocabularies of the global ontology, whereas the *local mappings* define the classes and properties of the application ontologies in terms of the elements of its feature type schemas.

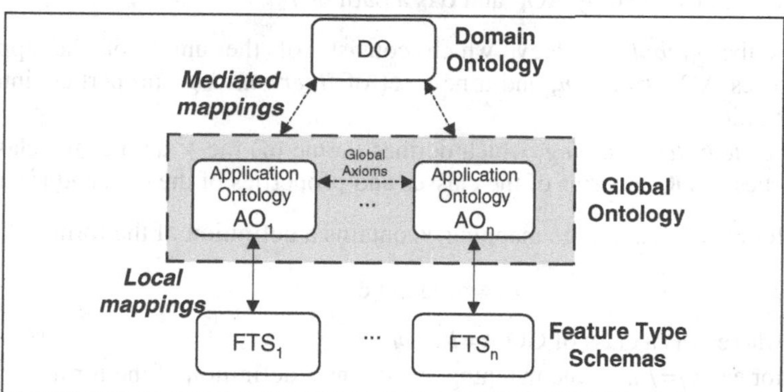

Fig. 1. Ontology-based Architecture for discovery and retrieval of geographic information

In our approach, the global ontology plays a key role in order to deal with data integration. Application ontologies help breaking the query answering problem into two sub-problems, as discussed in Section 4. They are also a notational convenience to divide the definition of the mappings into two stages: the definition of the mediated mapping and the definition of the local mappings.

In order to represent ontologies and mappings, we adopt a family of logics called Description Logics (DL) [7,8]. The following definition formally introduces the notion of mediated environment.

Definition 2.1: (***Mediated Environment***). A *mediated environment* is a 6-tuple ME = (DO, FTS_k, AO_k, γ_k, GO, γ), $k=1,...,n$, where

- DO is a *domain ontology*, which represents the mediated schema. We assume that the classes and properties in DO are $C_1,...,C_u$ and $P_1,...,P_v$.

- for each $k=1,...,n$,
  - o FTS_k is a feature type schema
  - o AO_k is an *application ontology*, which describes exactly the feature type FTS_k. The vocabulary of AO_k is a subset of the vocabulary of DO. We adopt namespace prefixes to distinguish the occurrence of a symbol in the DO vocabulary from the occurrence of the same symbol in the vocabulary of AO_k. We assume that:
    - o the classes and properties in DO are $C_1,...,C_u$ and $P_1,...,P_v$. So, for each class C_i (or property P_j) in the vocabulary of DO, we denote the occurrence of C_i (or P_j) in the vocabulary of AO_k by $AO_k:C_i$ (or $AO_k:P_j$)
    - o (*Domain Disjointness Assumption*) for any interpretation ξ_i and ξ_j for the alphabets of AO_i, AO_j,, ξ_i and ξ_j have disjoint domains, for each $i, j \in [1,k]$, with $i \neq j$
  - o γ_k is a set of correspondence assertions, called a *local mapping*, each one of the form $A \equiv T_k/\delta$, where A is a class or property of AO_k, T_k is the feature type schema described by AO_k, and δ is a path of T_k
- GO is the *global ontology*, which consists of the union of the application ontologies AO_k, $k=1,...,n$, and a new set of *inter-ontology* properties, introduced by definition.
- γ is the *mediated mapping*, which defines (some of) the γ defines the classes and properties of DO in terms of the classes and properties of the GO, and is such that:

  1. for each $i=1,...,u$, the mapping γ contains a definition of the form

     $$C_i \equiv c_1 \sqcup ... \sqcup c_n \tag{1}$$

     where c_k is a class of GO, $k=1,...,m$.
  2. for each $j=1,...,v$, the mapping γ contains a definition of the form

     $$P_j \equiv p_1 \sqcup ... \sqcup p_m \tag{2}$$

     where p_k is a property of GO, $k=1,...,m$.

In following, we explain in more detail our mediated environment through a data integration example, adapted from [4].

Feature Type Schemas

In this example, we assume that the user provides feature type schemas. We consider three data sources. The first data source is based on the Digital Aeronautical Flight Information File (DAFIF), the second is based on the Aeronautical Information Exchange Model (AIXM) and the third data source concerns Aircraft Database (AIRFRAMES). The DAFIF and AIXM data sources provide information about airports and their runways. Particularly, the DAFIF data source has airports with runway length less than 5,000 meters. The AIRFRAMES data source provides information about aircrafts. Figure 2 shows the feature type schemas exported by these data sources via Web Feature Services.

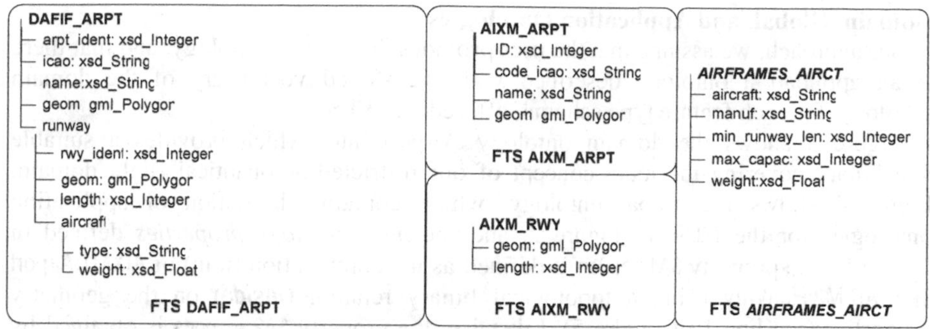

Fig. 2. Feature Type Schemas

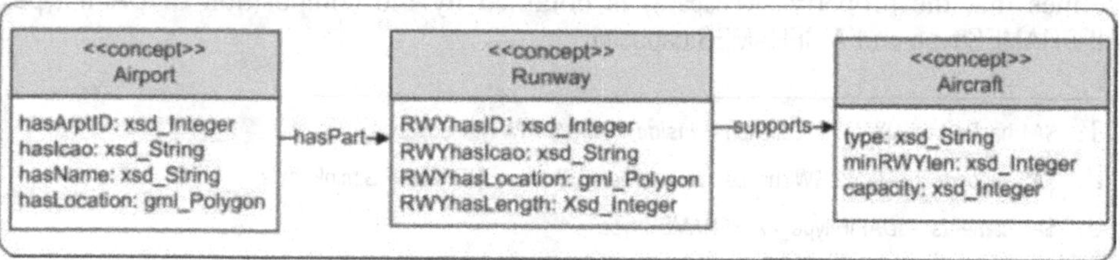

Fig. 3. Domain Ontology AirportOnto

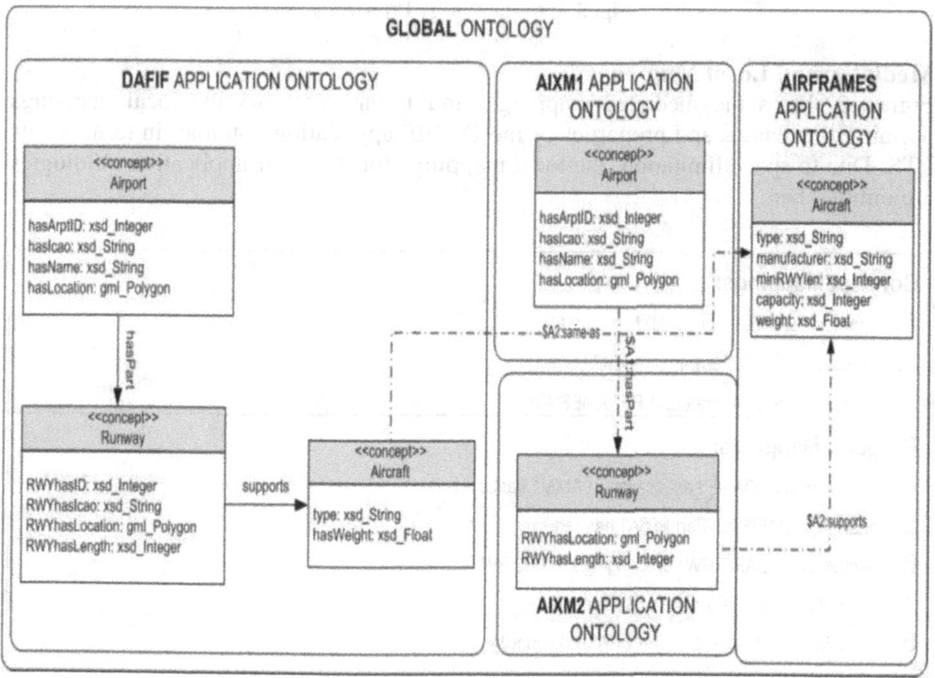

Fig. 4. Application Ontologies and Global Ontology

Domain, Global, and Application Ontologies

In our approach, we assume that the user provides the domain ontology, and that there is an application ontology described with the shared vocabulary of the domain ontology, for each feature type schema, offered via WFS.

Figure 3 shows the domain ontology AirportOnto, which provides a suitable vocabulary covering the main concept of our restricted aeronautical flight domain. Figure 4 shows the global ontology, which contains the union of application ontologies for the FTSs in Figure 2, and the *inter-ontology properties* defined in Figure 5. The property $A1:hasPart is defined as the combination (join) of AIXM1:Airport and AIXM2:RunWay using a topological binary relation (***inside***) on the geometry properties (see line 1 of Figure 5). Likewise, the property $A2:supports is obtained by the combination of AIXM2:RunWay and AIRFRAMES:Aircraft using the binary relation ***greater_than*** on the lengths properties (see line2 of Figure 5). The axiom in line 4 defines that the property $A4:capacity is obtained by the composition of DAFIF:type, AIRFRAMES:type and AIRFRAMES:capacity.

1. $A1:hasPart ≡ AIXM1:hasLocation ∘ ***inside*** ∘ AIXM2:RWYhasLocation⁻
2. $A2:supports ≡ AIXM2:RWYhasLength ∘ ***greater_than*** ∘ AIRFRAMES:minRWYlen⁻
3. $A3:same-as ≡ DAFIF:type ∘ AIRFRAMES:type⁻
4. $A4:capacity ≡ $A3:same-as ∘ AIRFRAMES:capacity

Fig. 5. Inter-Ontology Properties

Mediated and Local Mappings

Figure 6 shows the mediated mappings, and Figure 7 shows the local mappings defining the classes and properties of the DAFIF application ontology in terms of its FTS. Due to space limitation, the local mappings for the other application ontologies are omitted here.

Concept Mappings:

1. Airport ≡ DAFIF:Airport ⊔ AIXM1:Airport
2. Runway ≡ DAFIF:Runway ⊔ AIXM2:Runway
3. Aircraft ≡ DAFIF:Aircraft ⊔ AIRFRAMES:Aircraf

Property Mappings:

4. hasLocation ≡ DAFIF:hasLocation ⊔ AIXM1:hasLocation
5. hasPart ≡ DAFIF:hasPart ⊔ $A1:hasPart
6. hasLength ≡ DAFIF:RWYhasLength ⊔ AIXM2:RWYhasLength
7. supports ≡ DAFIF:supports ⊔ $A2:supports
8. capacity ≡ $A4:capacity ⊔ AIRFRAMES:capacity

Fig. 6. Mediated Mappings

Concept Mappings:

DAFIF:Airport ≡ DAFIF_ARPT

DAFIF:Runway ≡ DAFIF_ARPT/runway

DAFIF:Aircraft ≡ DAFIF_ARPT/runway/aircraft

Property Mappings:

DAFIF:hasArptID ≡ DAFIF_ARPT / arpt_ident

DAFIF:hasIcao ≡ DAFIF_ARPT / icao

DAFIF:hasName ≡ DAFIF_ARPT / name

DAFIF:hasLocation ≡ DAFIF_ARPT / geom

DAFIF:hasPart ≡ DAFIF_ARPT / runway

DAFIF:RWYhasId ≡ DAFIF_ARPT / runway / rwy_ident

DAFIF:RWYhasIcao ≡ DAFIF_ARPT / runway / icao

DAFIF:RWYhasLocation ≡ DAFIF_ARPT / runway / geom

DAFIF:RWYhasLength ≡ DAFIF_ARPT / runway / length

DAFIF:supports ≡ DAFIF_ARPT / runway / aircraft

DAFIF:type ≡ DAFIF_ARPT / runway / aircraft / type

DAFIF:hasWeight ≡ DAFIF_ARPT / runway / aircraft / weight

Fig. 7. Local Mappings from the feature type schema DAFIF_ARPT to its Application Ontology

As already mentioned, the DAFIF data source has airports with runway length less than 5,000 meters. Translating this constraint to the vocabulary of the DAFIF application ontology, we have the following constraint (expressed in DL):

$$\text{DAFIF:Airport} \sqsubseteq \exists\text{DAFIF:hasPart.}(\exists\text{DAFIF:RWYhasLength.}\{<5000\})$$

3 Query Processing

We propose a two-step strategy for answering a query Q posed on the domain ontology, summarized as follows:

1. The user's query is decomposed into a set of elementary sub-queries over the application ontologies that are relevant to the query. Each such (elementary) sub-query aims at extracting data from a single application ontology. The result of this step is a query expressed as unions and joins over elementary sub-queries. This step is performed using the mediated mappings in conjunction with reasoning.
2. Sub-queries resulting from the previous step are rewritten in terms of FTSs with the help of the local mappings. Hence, we obtain a *global execution plan*, which is a combination of WFS queries using joins, unions and other (possibly spatial) operations.

It is worth noting that each step mentioned above can take advantage of ontological reasoning tasks (e.g. subsumption and classification) in order to facilitate the rewriting process.

1. Rewrite the query Q using the mediated mappings:

Q'≡(DAFIF:Airport ⊔ AIXM1:Airport) ⊓ e^1

where e^1= ∃(DAFIF:hasPart ⊔ $A1:hasPart) .
(∃(DAFIF:RWYhasLength ⊔ AIXM2:RWYhasLength).{>13000} ⊓
(∃(DAFIF:supports ⊔ $A2:supports).
(∃ ($A4:capacity ⊔ AIRFRAMES:capacity.{>300})))

2. Apply the Distributive Property over Union:

Q'≡Q_1⊔Q_2

where Q_1≡(DAFIF:Airport ⊓e^1)
Q_2≡(AIXM1:Airport⊓e^1)

3. Simplify e^1 by using domain disjointness axiom:

Q_1'≡ DAFIF:Airport⊓e_1^1
where e_1^1= ∃ DAFIF:hasPart.(∃DAFIF:¨RWYhasLength.{>13000} ⊓
∃ (DAFIF:supports.(∃ $A4:capacity.{>300}))
Q_2'≡ AIXM1:Airport⊓e_2^1
where e_2^1= ∃$A1:hasPart.(∃AIXM2:RWYhasLength.{>13000} ⊓
∃ ($A2:supports.(∃ AIRFRAMES:capacity.{>300}))

4. Replace the inter-ontology properties in $e_1^1$and e_2^1:

Q_1'≡ DAFIF:Airport⊓e_1^2
where e_1^2= ∃ DAFIF:hasPart.(∃DAFIF:RWYhasLength.{>13000} ⊓
∃(DAFIF:supports.
(∃ ((DAFIF:type∘ AIRFRAMES:type⁻∘ AIRFRAMES: capacity).{>300}))

Q_2'≡ AIXM1:Airport⊓e_2^2

where e_2^2= ∃(AIXM1:hasLocation ∘***inside***∘AIXM2:RWYhasLocation⁻).
(∃AIXM2:RWYhasLength.{>13000} ⊓
∃ (AIXM2:RWYhasLength∘***greater_than***∘AIRFRAMES:minRWYlen⁻).
(∃ (AIRFRAMES: capacity).{>300}))

5. Apply property of role composition to decompose e_1^2 and e_2^2 into sub-queries where each sub-query is applied over a single application ontology.

Q_1'≡ DAFIF:Airport ⊓ e_1^3
where e_1^3= ∃ DAFIF:hasPart.(∃DAFIF:RWYhasLength.{>13000} ⊓
∃(DAFIF:supports.(∃ DAFIF:type. Q_3)
where Q_3 ≡∃(AIRFRAMES: type⁻∘ AIRFRAMES: capacity).{>300}))

Q_2'≡ AIXM1:Airport ⊓ e_2^3

Fig. 8. Query Answering Example

Where $e_2^3 \equiv \exists$(AIXM1:hasLocation $\circ$ ***inside***). $Q_4 \sqcap$

$\exists$ ((AIXM2:RWYhasLength $\circ$ ***greater_than***).Q_5

where $Q_4 \equiv \exists$AIXM2:RWYhasLocation$^-$. ($\exists$ (AIXM2:RWYhasLength).{>13000}

$Q_5 \equiv \exists$AIRFRAMES:minRWYlen$^-$. ($\exists$ (AIRFRAMES: capacity).{>300})

6. Apply distributive property to restrictions in e_1^3 in order to partition into sub-conditions so that are possible to check its consistency (based on the application ontology constraints).

$Q_1' \equiv$ DAFIF:Airport$\sqcap e_1^4$

where $e_1^4 = \exists$ DAFIF:hasPart.($\exists$DAFIF:RWYhasLength.{>13000} $\sqcap$

$\exists$ DAFIF:hasPart.($\exists$ (DAFIF:supports.($\exists$ DAFIF:type. Q_3)

where $Q_3 \equiv \exists$(AIRFRAMES:type$^-$$\circ$ AIRFRAMES:capacity).{>300}))

7. Remove from Q' the sub-queries that are not consistent (a query is not consistent if its result is empty for any database state).

By reasoning tasks, we can show that the query Q_1'is not consistent (from constraint "DAFIF:Airport$\sqsubseteq \exists$DAFIF.hasPart.($\exists$DAFIF.RWYhasLength.{<5000})"). So, we have that:

$Q' \equiv Q_2'$

Fig. 8. (*continued*)

In our approach, a query has the form: **Q** $\equiv$ **A** $\sqcap$ **e**, where **A** represents an atomic concept and **e** represents a restriction over **A**. For example, consider a query **Q** that selects airports that have a runway whose length is greater than 13,000 meters, and support aircrafts with capacity greater than 300 passengers. This query in the DL $\mathcal{ALCQI}$ syntax is shown below:

Q $\equiv$ Airport $\sqcap$ **e**

where **e** = $\exists$hasPart . ($\exists$RWYhasLength.{>13000} $\sqcap$ $\exists$supports . ($\exists$capacity . {>300}))

Figure 8 illustrates the steps necessary for decomposing the query Q in sub-queries expressed in terms of the application ontologies.

In step 7, our approach takes advantage of DL reasoning to discard the sub-query Q_1' that is not consistent. After step 7, each subquery Q_i' will be converted into a WFS query and submitted over one data source. Data resulting from all sub-queries will be combined and encoded in XML format to produce the query's result. Finally, the resulting XML data will be used to populate the domain ontology. In the literature, the translation from XML to RDF/OWL is often called "lifting", and some solutions are already provided [9]. Due to space limitation, this research topic will be not discussed in this paper.

4 Conclusion

In this paper, we have presented an ontology-based framework for integration of geographic data. This framework takes a query on domain ontology and rewrites it into sub-queries submitted over multiples data sources. The query's result is obtained by the proper combination of data resulting from these sub-queries. We have illustrated, through an example, how our framework allows the combination of data from different sources, thus overcoming some limitations of other ontology-based approaches.

We showed in Section 4 how to decompose a query over the domain ontology in sub-queries expressed in terms of the application ontologies using the mediated mappings. Our approach takes advantage of DL reasoning to discard sub-queries that are not consistent.

Although our present work deals with some spatial aspects (e.g. FTS, spatial locations), we are aware that this approach can be applied to other domains. As a future work, we intend to investigate how to generate application ontologies and mappings. Besides, we want to implement and evaluate our query processing algorithm. In addition, we plan to study how to optimize query processing by incorporating progressive reasoning evaluation. Last but not least, we intend to investigate, in the near future, how to express spatial operations as built-in properties within the domain ontology.

References

1. Essid, M., Boucelma, O., Colonna, F., Lassoued, Y.: Query processing in a Geographic Mediation System. In: Proceedings of GIS, pp. 101–108 (2004)
2. Klien, E., Fitzner, D.I., Maué, P.: Baseline for Registering and Annotating Geodata in a Semantic Web Service Framework. In: Proceedings of the 10th Conference on Geographic Information Science, Aalborg, Denmark (2007)
3. Lutz, M.: Ontology-based Discovery and Composition of Geographic Information Services. Phd Thesis, Institut für Geoinformatik (2005)
4. Lutz, M., Kolas, D.: Rule-based Discovery in Spatial Data Infrastructures. Transactions in GIS 11(3), 317–336 (2007)
5. Wache, H., Vögele, T., Visser, U., Stuckenschmidt, H., Schuster, G., Neumann, H., Hübner, S.: Ontology-based Integration of Information - A Survey of Existing Approaches. In: Proceedings of the IJCAI 2001 Workshop: Ontologies and Information Sharing, pp. 108–117 (2001)
6. Xavier, E.M.A.: Serviços Geográficos baseados em Mediadores e Padrões Abertos para Monitoramento Participativo na Amazônia, Master's Thesis, INPE, Brasil (2008)
7. Calvanese, D., Lenzerini, M., Nardi, D.: Description Logics for Conceptual Data Modeling. In: Chomicki, J., Saake, G. (eds.) Logics for Databases and Information Systems. Kluwer Academic Publishers, Dordrecht (1998)
8. Casanova, M.A., Lauschner, T., Paes Leme, L.A., Breitman, K.K., Furtado, A.L.: A Strategy to Revise the Constraints of the Mediated Schema. Technical Report MCC34/09, Department of Informatics, PUC-Rio (2009)
9. Akhtar, W., Kopecky, J., Krennwallner, T., Polleres, A.: XSPARQL: Traveling between the XML and RDF worlds – and avoiding the XSLT pilgrimage. In: Bechhofer, S., Hauswirth, M., Hoffmann, J., Koubarakis, M. (eds.) ESWC 2008. LNCS, vol. 5021, pp. 432–447. Springer, Heidelberg (2008)

A Semantic Approach for the Modeling of Trajectories in Space and Time

Donia Zheni[1], Ali Frihida[1], Henda Ben Ghezala[2], and Christophe Claramunt[3]

[1] LTSIRS, Ecole Nationale d'Ingénieurs de Tunis, University Tunis El Manar, Tunisia
donia.zheni@isi.rnu.tn, ali.frihida@enit.rnu.tn
[2] RIADI, Ecole Nationale des Sciences de l'Informatique, University of Manouba, Tunisia
henda.BG@cck.rnu.tn
[3] Naval Academy Research Institute, France
christophe.claramunt@ecole-navale.fr

Abstract. The modeling and analysis of trajectories in space and time have been long a domain of social science studies since early developments of Time Geography. Early works have been mainly conceptual, but things are changing with recent advances in telecommunications and ubiquitous computing that allow representation of moving points and trajectories within spatial database systems. These have generated a large amount of research in formal and qualitative modeling of moving points, providing many opportunities to enrich emerging geometrical-based data structures with semantic approaches. This is the objective of the research presented in this paper that introduces a semantic-based model and manipulation language of trajectories. It is based on an algebraic Spatio-Temporal Trajectory data type (STT) endowed with a set of operations designed as a way to cover the syntax and semantics of a trajectory. The approach is formally presented and illustrated by a case study.

Keywords: Space-time modelling, trajectories, time-geography.

1 Introduction

Geographical spaces are complex systems that generate many interactions between humans and the environment, those being the scope of natural, social and behavioral sciences. Amongst many domains of study, the role of time in human activities, and individuals displacements have been conceptually considered in social sciences since early progress of time geography [1], [2]. In particular, the concept of time-path has been suggested by early principles of time geography to model human trajectories considered as a combination of dynamic (e.g., *trips*) and static properties (e.g., *activities*) (cf. Fig. 1). Time geography principles have been long applied to the measurement and modelling of dynamic patterns in space and time, e.g., [3]. Meanwhile, continuous advances in telecommunications and ubiquitous computing have generated growing interest in the representation of moving points within spatial database systems. A series of papers has contributed to the representation and manipulation of moving points and moving regions using Abstract Data Types (ADT) and algebraic approaches [4], [5], [6], [7], [8], [9]. Primitive operations include basic

C.A. Heuser and G. Pernul (Eds.): ER 2009 Workshops, LNCS 5833, pp. 347–356, 2009.

manipulations on the spatial semantics exhibited by moving objects (e.g., speed, acceleration and orientation), simulation of future trajectories [10], network-constrained displacements [11], and relative motions [12]. The semantics of moving points and trajectories have been also studied at a higher level of abstraction, and where the objective is to study and analyze emerging properties in space and time [13],[14],[15],[16]. Human trajectories generate complex spatio-temporal patterns that are of crucial interest for analyzing aggregate-disaggregate transportation applications [17], [18], [19].

The research presented in this paper proposes an integration of the semantic dimension, inspired from the concept of time-path, within a formal representation of space-time trajectories. We introduce an algebraic model that explicitly represents a *spatio-temporal trajectory* (STT) as an ADT, where a series of trajectory states is potentially observed and measured. At a higher level of abstraction, the STT type encapsulates the dynamic and semantic dimensions. The interest of the ADT representation is that it combines a formal definition with manipulation operations, and offers a primary and persistent identity to the STT. Henceforth, the STT is a natively defined data structure. Once integrated in a database management system, it acquires the same status than built-in and conventional data structures. This identity is supported by a set of operations covering the spatial, temporal, spatio-temporal and semantic properties. These manipulations should give insight on the way a moving entity is likely to evolve and possibly the reason behind such an evolution. The reminder of the paper is organised as follows. Section 2 defines the STT abstract data type, its syntax and semantics. Section 3 illustrates the STT type using query examples. Finally Section 4 concludes the paper and draws some perspectives.

2 Trajectory ADT

We introduce the STT ADT and operations, or algebra, suitable for representing and querying semantic-based trajectories. The modeling approach consists of two steps. The first step introduces the trajectory data type. The semantics of the proposed data type is given by a carrier set. The second step describes a collection of operations over the proposed data type. For each operation, its signature and semantics are given by defining a function on the carrier sets of the argument types.

2.1 STT ADT

The STT data type requires different sorts (i.e., types). We assume the following usual atomic, complex, spatial and temporal data types: *Integer*, *Real*, *Boolean, String, Enumeration constants* (*Enum*), *Alist*, *Point*, *Polyline*, *Polygon, Time* and *Interval.* An *Activity* is formally given as a quadruple $a = (l, t_s, t_e, purpose)$ where $l \in Point$ represents its location, t_s and $t_e \in Time$ represent, respectively, its starting and ending times, and $purpose \in Enum$ is its activity description (e.g., shopping, working). We formally define a *trip* as $d = (l_s, l_e, t_s, t_e, mode, path)$ where $l_s, l_e \in Point$ represent, respectively, its start and end locations; t_s and $t_e \in Time$ represent, respectively, its starting and ending times; $mode \in Enum$ is the mean used to make the trip. The attribute *path* represents the geometrical semantics of the trajectory, it can be directly considered as stepwise, for example using a *Polyline* data type, or in order to approximate a continuous trip as a spatio-temporal sub-trajectory of type *moving point* as suggested in [5]. At the semantic level, a valid activity must start before it ends. A similar constraint applies to trips. The set of possible activity values is then denoted by $D_a = \{a \mid a.t_s < a.t_e\}$. The set of possible trip values is denoted by $D_d = \{d \mid d.t_s < d.t_e\}$. Within a given STT, we assume that successive activities never temporally overlap. We denote the activity set of an STT as the temporally ordered subset A, as $A = \{a_i \mid \forall 1 \leq i < n : a_i \in D_a \wedge a_i.t_e < a_{i+1}.t_s\}$. Similarly, the trip set of an STT is the temporally ordered subset D with $D = \{d_i \mid \forall 1 \leq i < n : d_i \in D_d \wedge d_i.t_e < d_{i+1}.t_s\}$. Consequently, a value of type STT is a pair (A,D) of temporally ordered sets. The first state of an STT is a trip d_1 followed by an activity a_1. The last state of the STT is the activity a_n. The domain of the proposed type is then given as follows:

$$D_{STT} = \{(A,D) \mid (1,2)\ \forall 1 < i \leq n :\ d_i.l_s = a_{i-1}.l \wedge d_i.t_s = a_{i-1}.t_e$$
$$(3,4)\ \forall 1 \leq i \leq n :\ d_i.l_e = a_i.l \wedge d_i.t_e = a_i.t_s\}\ .$$

Constraints (1) and (3) model the spatial relations between a trip and its previous and following activities, respectively. Constraints (2) and (4) model the temporal relations between successive states. The combination of these constraints represents the chaining of an STT. Trips and activities are modeled at a given level of granularity which are applications dependent, e.g., a lifeline, a person diary or a series of shopping activities in a shopping mall.

2.2 Data Manipulation Operations

We introduce a manipulation language materialized by a set of operations on the STT data type, that is, semantic, spatial, temporal and set-based operations.

2.2.1 Semantic Operations

The following operations allow a user to formulate queries on the semantics of the STT data type. These include basic manipulation operations allowing an access to STT internal data, such as a specified activity location. Data retrieval operations return states (i.e., activities or trips) on the basis of their purpose, mode, and temporal precedence with other states or according to their position in the STT. They correspond to trajectory-based operations previously defined for the analysis of space-time paths [20]. For example, the operation *Activity_Before_Activity* returns the activity that precedes a given activity. Predicates and numeric operations are also included in the language. The predicate *Include_Activity* tests whether a given activity belongs to the STT; the operation *Activity_Count* returns the activity number of an STT. In fact, the principles of these operations can be applied to either activities or trips. The specification of a subset of operations is given in Table 1. Semantically, the behavior of these predicates can be reduced to the corresponding set-theoretic ones, and are defined as follows. Let *a* be an *Activity*, *stt* an *STT* value, then

$$f_{\text{Include\_Activity}}(stt, a) := (a \in A)\ .$$

The semantics of some of these operations is illustrated below (other operations such as *Activity_Before_Trip* are defined according to similar principles). Let *stt* be an *STT* value, *d* be a *Trip* value, *i* be an *Integer* value, and *p* be an *Enum* value. Then

$$f_{\text{Nth\_Activity}}(stt, i) := a_i \in A \text{ if } 1 \leq i \leq n .$$

$$f_{\text{Activity\_Before\_Trip}}(stt, d) := a_i \in A \text{ if } \exists 1 \leq i < n : d = d_{i+1} .$$

$$f_{\text{Activity\_With\_Purpose}}(stt, p) := a_i \in A \text{ if } \exists 1 \leq i \leq n : a_i.purpose = p .$$

Table 1. Semantic operations

Operation	Signature	
Activity_STime, Activity_ETime	*STT* × *Activity*	→ *Time*
Activity_Mode	*STT* × *Activity*	→ *Enum*
First_Activity, Last_Activity	*STT*	→ *Activity*
Activity_With_Purpose, Activity_With_Mode	*STT* × *Enum*	→ *Activity*
Activity_After_Activity, Activity_Before_Activity	*STT* × *Activity*	→ *Activity*
Include_Activity	*STT* × *Activity*	→ *Boolean*
Activity_Count, Trip_Count	*STT*	→ *Integer*
Activity_With_Mode_Nbr	*STT* × *Enum*	→ *Integer*
Activity_With_Purpose_Nbr	*STT* × *Enum*	→ *Integer*

2.2.2 Temporal Operations

These operations concern essentially those expressing binary relationships in time as introduced in [21]. Temporal operations are defined between STTs and temporal entities, and between two STTs. This gives operations between pairs of STT/Time, Time/STT, STT/Interval, Interval/STT and STT/STT. Table 2 illustrates some of the operations proposed (other operations such as *Interval_During_STT* are defined under similar principles).

Table 2. Temporal operations

Operation	Signature	
STT_EndedBy_Time, STT_Before_Time	*STT* × *Time*	→ *Boolean*
Time_Begins_STT, Time_During_STT	*Time* × *STT*	→ *Boolean*
STT_Overlaps_STT, STT_Contains_STT	*STT* × *STT*	→ *Boolean*

In order to formulate the semantics of these operations, the STT is projected to the temporal domain. At the temporal level, an STT value is equivalent to the time interval during which the trajectory is defined. This interval is denoted by

$I_{STT} = [d_1.t_s, a_n.t_e]$. For example, let stt_1 and stt_2 be two STT values, as shown in Fig 1. Let t be a *Time* value, then

$$f_{\text{STT\_Contains\_STT}}(stt_1, stt_2) := I_{stt1} \text{ Contains } I_{stt2} .$$

$$f_{\text{Time\_During\_STT}}(t, stt_1) := t \text{ During } I_{stt1} .$$

The predicates expressed above are true (i.e., hold) in the schema illustrated in Fig1.

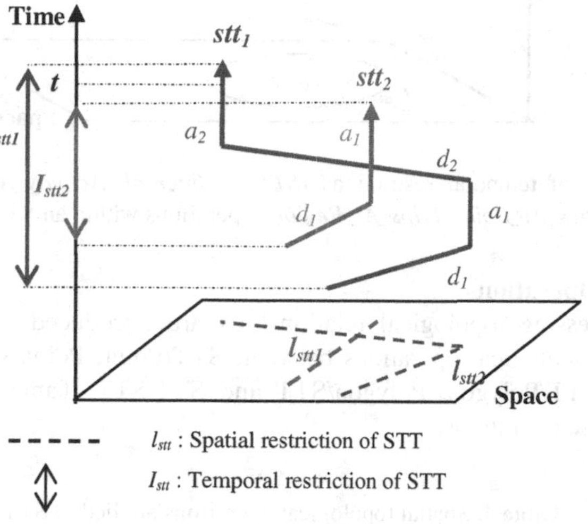

Fig. 1. Spatial and temporal projections of an STT

In the temporal domain, several operations apply a temporal restriction of the STT at a specific time instant or interval, e.g., operations *STT_At_Interval* and *Activity_At_Time* (see Table 3). The former returns a "*part*" of the STT (i.e., ordered sequence of trips and activities) occurring at a specific interval. Fig. 2 denotes an example of the operation *STT_At_Interval* applied to an interval I and a trajectory stt, and that returns the list (a_1, d_2, a_2, d_3). The operation *Activity_At_Time* returns the activity of the STT that occurred at the time t. As illustrated in Fig. 2, this operation returns the activity a_3 for the time instant t.

Table 3. Temporal operations applied to trips, activities and STTs

Operation	Signature
STT_At_Interval	*STT* × *Interval* → *Alist*
Activity_At_Time	*STT* × *Time* → *Activity*
Activities_At_Interval	*STT* × *Interval* → *Alist*
Trip_At_Time	*STT* × *Time* → *Trip*
Trips_At_Interval	*STT* × *Interval* → *Alist*

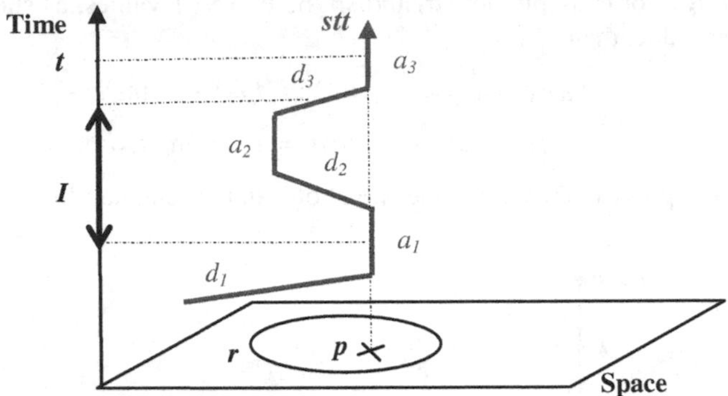

Fig. 2. Examples of temporal restrictions (*STT_At_Interval, Activity_At_Time*) and spatial restriction (*Activities_At_Point, Trips_At_Region*) operations within an STT

2.2.3 Spatial Operations

Predicates expressing topological relationships are introduced as defined in [22]. These include topological operations between STT/Point, Point/STT, STT/Polyline, Polyline/STT, STT/Polygon, Polygon/STT and STT/STT. Table 4 illustrates some examples of these operations.

Table 4. Spatial topological operations applied to STTs

Operation	Signature
STT_EndsBy_Point, STT_Contains_Point	*STT* × *Point* → *Boolean*
STT_Equal_Polyline, STT_Cross_Polyline	*STT* × *Polyline* → *Boolean*
STT_Equal_STT, STT_Cross_STT	*STT* × *STT* → *Boolean*

In the spatial domain, an STT value can be valued as a *Polyline* denoted by l_{STT}. This *Polyline* represents the path (i.e., itinerary) followed by a moving entity during its different trips. This path is generated by the concatenation of the polyline set $d_i.path$, for $1 \leq i \leq n$. In the case of a continuous change, a path attribute can be defined as a moving point. A single trip path can be generated using the *trajectory* : *mpoint* → *Polyline* operation. Then, l_{STT} is the concatenation of the polyline set *trajectory*($d_i.path$), for $1 \leq i \leq n$. Let stt_1 and stt_2 be two STT values, as shown in Fig. 1. Let l be a *Polyline* value, then

$$f_{\mathrm{STT\_Cross\_STT}}(stt_1,stt_2) := l_{stt1} \text{ Cross } l_{stt2} .$$

$$f_{\mathrm{Polyline\_Disjoint\_STT}}(l,stt_1) := l \text{ Disjoint } l_{stt1} .$$

The predicates expressed above are true (i.e., hold) in the schema illustrated in Fig.1. Elementary spatial retrieval operations are introduced, e.g., retrieving locations where moving objects either stop or complete partially or totally a given itinerary. Stop locations represent points of interest (i.e., activity locations) in the STT. Table 5 illustrates some examples of spatial manipulation, e.g., *Points*, *Itinerary*, *Path_From_To*, as well as spatial computations on an STT.

Table 5. Spatial manipulation operations applied to STTs

Operation	Signature
Points	*STT* → *Alist*
Itinerary	*STT* → *Polyline*
Path_From_To	*STT* × *Activity* × *Activity* → *Polyline*
Spatial_Length	*STT* → *Real*
Length_Path_From_To	*STT* ×*Activity* ×*Activity* → *Real*

Let *stt* be an STT value, then

$$f_{\text{Points}}(stt) := \{a_i.l \; \forall \; 1 \leq i \leq n\} \, .$$

Spatial restrictions to specific geometrical entities can be applied to STTs. The example of operation *Activities_At_Point* introduced in Table 6 returns a list of the activities of an STT taking place at a given point (e.g., activity (a_1, a_3) at the point p in Fig. 2). The operation *Trips_At_Region* returns a list of the trips of an STT taking place inside a given region (e.g., trip (d_2, d_3) at region r in Fig. 2).

Table 6. Spatial restriction operations applied to activities

Operation	Signature
Activities_At_Point, Trips_At_Point	*STT* × *Point* → *Alist*
Activities_At_Polyline, Trips_At_Polyline	*STT* × *Polyline* → *Alist*
Activities_At_Region,Trips_At_Region	*STT* × *Polygon* → *Alist*

2.2.4 Set Operations

Set operations are specified by analogy to set-theoretic operations:

Table 7. Set operations applied to STTs

Operation	Signature
Equal, Include	*STT* × *STT* → *Boolean*
Union, Intersect, Difference	*STT* × *STT* → *STT*

For example, the semantics of the operation equality can be informally defined as follows: two trajectories STT_1 and STT_2 are *equal* if these trajectories show a same spatio-temporal and semantic behavior over a given period of time. Let stt_1 and stt_2 be two STT values, then

$$f_{\text{Equal}}(stt_1, stt_2) := \text{true if } \text{card}(stt_1.A) = \text{card}(stt_2.A) \wedge (\forall\, 1 \leq i \leq n : stt_1.a_i = stt_2.a_i \wedge stt_1.d_i = stt_2.d_i)\,.$$

$$f_{\text{Union}}(stt_1, stt_2) := \{(stt_1.A \cup stt_1.A\,,\, stt_1.D \cup stt_2.D\,)\} \text{ if } (stt_2.d_1.l_s = stt_1.a_n.l\,) \wedge (stt_2.d_1.t_s = stt_1.a_n.t_e\,)\,.$$

3 Case Study

In order to illustrate the capabilities of the query language, its principles are applied to an illustrative case study. Let us formulate several illustrative queries on an application example that models daily time-paths of person activities. Every person generates several activity time-paths corresponding to given periods of time. After a sequence of activity data collection, spatial, temporal and semantic information on trajectory time-path are stored as STTs. Let us consider the following self-explanatory tuples:

person(id: *String*, name: *String*)
TimePath(id: *String*, idPer: *String*, traj: *STT*, day: *Date*)
city(name: *String*, zone: *Region*)
place(name : *String*, location: *Point*)
mode(name: *String*)

In particular the tuple *TimePath* models human activities and trips expressed by the STT attribute *traj* of a given person *idPer* over a day *d*. Let us formulate some query examples formulated without loss of generality using an algebraic based semantics:

Query 1. What is the transport mode that a person of identity *p* used just before its ith activity during the day *d*?

Trip_Mode(TP.traj, *Trip_Before_Activity*(TP.traj, *Nth_Activity*(TP.traj, *i*))) | TP ∈ TimePath ∧ TP.idPers = *p* ∧ TP.day = *d*

This query returns the transport mode of the trip occurring before the ith activity.

Query 2. Which activities, of the person *p* during the day *d*, takes place at a given location of the place *x* ?

Activities_At_Point(TP.traj, PL.x) | TP ∈ TimePath ∧ PL ∈ place ∧ TP.idPers = *p* ∧ TP.day = *d*

Query 3. Calculate the length of path the person of identity *p* followed to get from the first activity to the ith activity, and during the day *d*.

Length_Path_From_To(TP.traj, *First_Activity*(TP.traj), *Nth_Activity*(TP.traj, *i*)) | TP ∈ TimePath ∧ TP.idPers = *p* ∧ TP.day = *d*

Query 4. Which persons traversed the city *c* during the day *d?*

PR.name | PR ∈ person ∧ TP ∈ TimePath ∧ CT ∈ city ∧ PR.id = TP.idPers ∧ TP.day = *d* ∧ *STT_Cross_Polygon* (TP.traj, CT.zone)

The above query applies the topological predicate *STT_Cross_Polygon* to test whether person trajectories crossed the region describing the given city.

Query 5. Which persons have the same spatio-temporal behavior as the person of id *id* during the day *d?*

PR1.name | PR1 ∈ person ∧ PR2 ∈ person ∧ TP1 ∈ TimePath ∧ TP2 ∈ TimePath ∧ PR1.id = TP1.idPers ∧ PR2.id = TP2.idPers ∧ PR2.id = *id* ∧ TP1.day = *d* ∧ TP2.day = *d* ∧ *Equal*(TP1.traj, TP2.traj)

These examples illustrate the potential of the language with respect to the semantic, spatial, temporal and spatio-temporal aspects of a person displacement. One of the advantages of the model and language identified is that they integrate usual properties of spatial, temporal and set-based operators within a unified modeling approach. The specific semantics of space-time trajectories is reflected by an integration of the concepts of trips and activities. The language is flexible enough as it can be modulated to allow alternative trip and activity representations (e.g., for example by allowing multiple activities within a given displacement). Overall, the modeling approach favors study of emerging behaviors exhibited by a given series of trips and activities, and cross-comparison of the patterns resulting from several trajectories. It is the combination of the different categories of operations that provide the resulting semantics and interest of the language.

4 Conclusion

The database representation of spatio-temporal trajectories still requires integration of semantic-based approaches necessary for a successful application to the many scientific domains oriented to the study of human activities in space and time. The research presented in this paper introduces an algebraic data type for a semantic-based representation of spatio-temporal trajectories that integrate the thematic, spatial, temporal and spatio-temporal dimensions at the data representation and manipulation levels. The modelling approach is formally introduced; it represents the concept of space-time trajectories by a series of connected trips and activities that are the usual primitives used in the conceptual apprehension of space-time trajectories. A data manipulation level language is introduced, it integrates several categories of set-based, thematic, spatial and temporal operations that together offer several query capabilities. The aim of the modelling approach is to provide a support to reveal additional knowledge on the dynamics, changes and evolution of the represented trajectories. Further work concerns representation of continuous changes, extension of the language to data mining functions.

References

1. Hägerstrand, T.: What about people in regional science? Papers of the Regional Science Association 24, 6–21 (1970)
2. Lenntorp, B.: Paths in space-time environments: A time-geographic study of the movement possibilities of individuals. Lund Studies in Geography, Series B 44 (1976)

3. Miller, H.J.: A measurement theory for time geography. Geographical Analysis 37, 17–45 (2005)
4. Erwig, M., Güting, R.H., Schneider, M., Vazirgiannis, M.: Spatio-temporal data types: An approach to modelling and querying moving objects in databases. GeoInformatica 3, 269–296 (1999)
5. Güting, R.H., Erwig, M.C., Jensen, S., Lorentzos, N.A., Schneider, M., Vazirgiannis, M., Böhlen, H.: A foundation for representing and querying moving objects. ACM Trans. Database Syst. 25, 1–42 (2000)
6. Forlizzi, L., Güting, R.H., Nardelli, E., Schneider, M.: A data model and data structures for moving objects databases. In: ACM SIGMOD International Conference on Management of Data, pp. 319–330. Dallas, Texas (2000)
7. Cotelo Lema, J.A., Forlizzi, L., Güting, R.H., Nardelli, E., Schneider, M.: Algorithms for moving object databases. The Computer Journal 46, 680–712 (2003)
8. Güting, R.H., Behr, T., De Almeida, V.T., Ding, Z., Hoffmann, F., Spiekermann, M.: SECONDO: An extensible DBMS architecture and prototype. Informatik-Report 313, Fernuniversität Hagen (2004)
9. Pelekis, N., Theodoridis, Y., Vosinakis, S., Panayiotopoulos, T.: Hermes - A framework for location-based data management. In: Ioannidis, Y., Scholl, M.H., Schmidt, J.W., Matthes, F., Hatzopoulos, M., Böhm, K., Kemper, A., Grust, T., Böhm, C. (eds.) EDBT 2006. LNCS, vol. 3896, pp. 1130–1134. Springer, Heidelberg (2006)
10. Praing, R., Schneider, M.: A universal abstract model for future movements of moving objects. In: 10th AGILE International Conference on Geographic Information Science, Aalborg, Denmark (2007)
11. Güting, R.H., Almeida, V.T., Ding, Z.: Modelling and querying moving objects in networks. VLDB Journal 15, 165–190 (2006)
12. Noyon, V., Claramunt, C., Devogele, T.: A relative representation of trajectories in geographical spaces. GeoInformatica 4, 479–496 (2007)
13. Mouza, C., Rigaux, P.: Mobility patterns. Geoinformatica 9, 297–319 (2005)
14. Brakatsoulas, S., Pfoser, D., Tryfona, N.: Modeling, storing, and mining moving object databases. In: International Database Engineering and Applications Symposium, Washington, pp. 68–77 (2004)
15. Spaccapietra, S., Parent, C., Damiani, M.L., De Macedo, J.A.F., Porto, F., Vangenot, C.: A conceptual view on trajectories. Data and Knowledge Engineering 65, 126–146 (2008)
16. Raubal, M.: Representing concepts in time. In: Freksa, C., Newcombe, N.S., Gärdenfors, P., Wölfl, S. (eds.) Spatial Cognition VI. LNCS (LNAI), vol. 5248, pp. 328–343. Springer, Heidelberg (2008)
17. Frihida, A., Marceau, D.J., Thériault, M.: Spatio-temporal object-oriented data model for disaggregated travel behaviour. Transactions in GIS 6, 277–294 (2002)
18. Donggen, W., Cheng, T.: A spatio-temporal data model for activity-based transport demand modelling. Int. J. Geog. Inf. Sci. 15, 561–585 (2001)
19. Yu, H., Shaw, S.L.: Exploring potential human activities in physical and virtual spaces: a spatio-temporal GIS approach. Int. J. Geog. Inf. Sci. 22, 409–430 (2008)
20. Frihida, A., Marceau, D.J., Thériault, M.: Development of a temporal extension to query travel behavior time paths using an object-oriented GIS. GeoInformatica 8, 211–235 (2004)
21. Allen, J.F.: Towards A general theory of action and time. Artificial Intelligence 23, 123–154 (1984)
22. Egenhofer, M., Golledge, R.G.: Spatial and Temporal Reasoning in Geographic Information Systems. Oxford University Press, Oxford (1998)

Author Index